# 无机化学实验

## Inorganic Chemistry Experiment

## （英汉双语教材）

主　编　伍晓春　姚淑心

副主编　梁晓琴　宁光辉　郑荞佶　高道江

科学出版社

北京

# 内 容 简 介

　　本书精选了无机化学中最为重要的 35 个实验，内容包括了无机化学实验的基本操作和常用仪器的使用、无机化学实验基础知识、无机化学中常用的一些平衡常数的测定、元素及其化合物的性质、无机化合物的制备与提纯。同时，教材中适当加入了一些反映无机化学学科发展前沿的实验，体现出无机化学学科的系统性和科学性，有利于学生在掌握实验基本技术的同时，对无机化学学科的新进展、新技术有所了解，激发学习无机化学知识及相关学科的兴趣。

　　本书可作为普通高等学校化学、化工及相关专业的无机化学实验教材，也可供有关化学专业的工作人员及研究人员参考。

**图书在版编目（CIP）数据**

　无机化学实验　＝ Inorganic Chemistry Experiment
：英汉对照／伍晓春，姚淑心主编.—北京：科学出版社，
2010.7（2018.7 重印）
　ISBN　978-7-03-028322-1

　Ⅰ．①无…　Ⅱ．①伍…　②姚…　Ⅲ．①无机化学-化
学实验-英、汉　Ⅳ．①O61-33

中国版本图书馆 CIP 数据核字（2010）第 137678 号

责任编辑：韩卫军　于　楠　　封面设计：陈思思

*科学出版社* 出版

北京东黄城根北街 16 号
邮政编码：100717
http://www.sciencep.com

四川煤田地质制图印刷厂印刷
科学出版社发行　各地新华书店经销

\*

2010 年 7 月第 一 版　　开本：787×1092　1/16
2018 年 7 月第七次印刷　　印张：18 1/2
字数：440 千字

定价：36.00 元

# 前　言

跨入 21 世纪，人类进入了一个以科技革命推动生产力发展、以信息技术和高新技术为核心、以知识经济为主导的新世纪。而掌握先进的科学技术和生产力将意味着一个国家的超前发展。语言是知识和信息的载体，英语是目前世界上使用得最广泛的语言。人类积累的科学知识各国都有权分享，但分享多少还得依赖于一个国家的综合国力的发展，依靠全民素质的提高，其中包括国民的外语水平。

据德国出版的《语言学及语言交际工具问题手册》的统计，全世界的科技文献资料中约有 85% 是用英语出版的。当前，我国对外科技学术交流日益频繁，对科技英语的要求也越来越高。掌握专业英语的水平已经成为衡量科技人才素质的重要方面。在这样的形势下，我国教育部要求大学推广双语教学，其中 5%～10% 的专业课采用双语教学，并将高等院校开展双语教学作为教学评估的重要内容。

化学就其起源和本质来说是一门实验学科。无机化学是高等学校开设的一门基础学科，掌握无机化学的基本技能，对大学生来说不仅是为了以后更进一步的学习，对他们工作能力的提高来说也是大有好处的。

无机化学实验课程的目的是通过实验教学，加深对无机化学中基本理论、无机化合物性质和反应性能的理解，熟悉无机化合物的一般分离和制备方法，掌握基础无机化学的基本实验方法和操作技能，培养学生严谨的科学态度、分析问题和解决问题的能力。

紧扣无机化学实验的教学大纲，我们在本教材中加强了基本操作训练、基础理论、无机制备方面的内容。在部分制备实验中对产物进行了限量分析或定量分析，以建立学生"质"和"量"的观念。另外，在教材中，我们对无机实验中常见仪器，如水泵、酸度计等进行了一一介绍，要求学生掌握其正确的使用方法。对于元素及其化合物的验证实验，则删繁就简，力求突出它们的主要化学性质。

大学教育不同于基础教育的特点之一在于它的研究性。没有研究性，大学教学从本质上失去其特征。而在本实验教材的编写过程中，加入了一些目前科研领域内较为活跃的实验题目，如水热法合成纳米 $SnO_2$ 微粉。各学校可根据本校的实际情况，选择相应的实验项目。

参加本书编写的所有人员均是长期从事无机化学实验课教学的有经验的教师，该教材选用的实验内容均在四川师范大学化学与材料科学学院进行了长达十年的反复验证，从而保证了该书的质量。无机化学实验（英汉双语教材）编写组成员为：伍晓春、姚淑心、梁晓琴、宁光辉、郑莽佶、高道江、罗凤秀。特别感谢四川师范大学 2006 级硕士研究生王聪聪为本书中部分插图所做的努力。现在这本书能呈现在读者面前，是与编写组成员的辛勤劳动和无私奉献分不开的。

由于时间紧迫和水平有限，书中的缺点和错误在所难免，欢迎各院校师生批评指正，以便再版时能予以修订，使之能更好地为培养高素质的化学专业人才做贡献。

《无机化学实验（英汉双语教材）》编写组

2010 年 4 月

# Preface

Chemistry is a science of experiment in terms of its origin and nature. Inorganic chemistry is a basic course in university. Skills in inorganic chemistry are important for students not only in further studying of other courses but also acknowledging work ability. This book is suitable for students of Chinese universities who are majoring in chemistry, applied chemistry or other related disciplines. It provides a good link between the theory and practice of inorganic chemistry. These experiments in this book have been performed in inorganic chemistry laboratory course and improved continuously during the past decades. This bilingual textbook is complied on the basis of prior Chinese version.

Experiment teaching of inorganic chemistry plays an important role in training students' thinking scientifically and methodically as well as in improving their sense and ability of blazing new trails. The objectives of training in experiment are to cultivate the students a good habit of working carefully, methodically, practically and improving constantly. By observation of phenomena in experiments, students will be able to improve their abilities in examining, analyzing and solving problems.

Based on the summary of the reform in experiment teaching of inorganic chemistry as well as bilingual teaching of this course in both Chinese and English in recent years, by using for reference the experiences of the reform in experiment teaching of inorganic chemistry in other colleges and universities, we have complied this book, which focuses on the foundation and briefly describes the basic operations and principles of experiments in inorganic chemistry.

In order to improve students' language skill through practice and meet the demands of bilingual teaching in both Chinese and English, we have translated the 35 experiments into English.

The chief complier is Xiaochun Wu. Following are the compliers: Shuxin Yao, Xiaoqin Liang, Guanghui Ning, Qiaoji Zheng, Daojiang Gao, Fengxiu Luo. Espeically, Congcong Wang is appreciated for her work.

In compiling this textbook, we have tried our best to select suitable materials and provide the users with a fine English version of the 35 experiments. Since English is not our native tongue, there might still exist something improper or even erroneous due to our academic limitations. We would be most appreciative if anyone could give us further constructive comments or suggestions on improving this textbook.

# 目　录

# Content

# 实验一　仪器的认领和洗涤

## 实验目的

1. 认识化学实验常用仪器的名称，规格与用途，了解使用注意事项。
2. 学习并练习常用玻璃仪器的洗涤和干燥方法，掌握酒精灯的使用。
3. 学习常用仪器以及仪器装置图的绘画方法。

## 化学实验常用仪器介绍

### 1. 普通常用仪器（图 1-1）

试管与离心试管：它们均为玻璃质料，分硬质和软质，有普通试管和离心试管。普通试管又有翻口、平口、有刻度、无刻度、有支管、无支管、有塞、无塞等几种。离心试管也有有刻度和无刻度的。

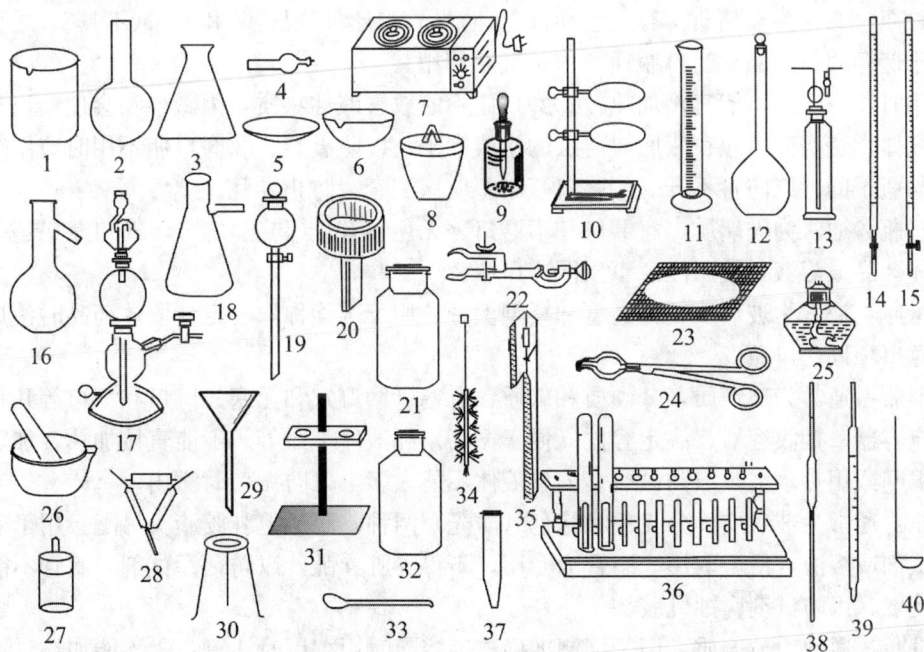

1. 烧杯 2. 烧瓶 3. 锥形瓶 4. 干燥管 5. 表面皿 6. 蒸发皿 7. 水浴锅 8. 坩埚 9. 滴瓶 10. 铁架和铁圈 11. 量筒 12. 容量瓶 13. 洗气瓶 14. 碱式滴定管 15. 酸式滴定管 16. 蒸馏烧瓶 17. 启普发生器 18. 吸滤瓶 19. 分液漏斗 20. 布氏漏斗 21. 广口瓶 22. 铁夹 23. 石棉铁丝网 24. 坩埚钳 25. 酒精灯 26. 研钵 27. 称量瓶 28. 泥三角 29. 漏斗 30. 三角架 31. 漏斗架 32. 细口瓶 33. 药匙 34. 试管刷 35. 试管夹 36. 试管架和试管 37. 离心试管 38. 移液管 39. 吸量管 40. 燃烧匙

图 1-1　普通常用仪器

一般情况下试管多用于常温或加热条件下用作少量试剂的反应容器，便于操作和观察；也用于收集少量气体用。支管试管可用于检验气体产物，也可接到装置中用。离心试管用于沉淀分离。

使用时应注意：①反应液体不超过试管容积 1/2，加热时不超过 1/3。以便防止液体振荡时溅出，或受热溢出。②加热前试管外面要擦干，加热时要用试管夹。防止由于有水滴附着受热不匀使试管破裂或烫手。③加热液体时，管口不要对人，并将试管倾斜与桌面成 45°，同时不断振荡，火焰上端不能超过管里液面。防止液体溅出伤人。扩大加热面可防止暴沸，防止因受热不均匀使试管破裂。④加热固体时，管口应略向下倾斜，避免管口冷凝水流回到灼热的管底而引起破裂。⑤离心试管不可直接加热，防止破裂。

烧杯：通常为玻璃质，分硬质和软质，有一般型和高型，有刻度和无刻度的几种。

一般情况下，烧杯多用于在常温或加热条件下作大量物质反应容器，反应物易混合均匀；也用于配制溶液或代替水槽。

使用时应注意：①反应液体不得超过烧杯容量的 2/3，防止搅动时或沸腾时液体溢出。②加热前要将烧杯外壁擦干，烧杯底要垫石棉网，防止玻璃受热不均匀而破裂。

烧瓶：通常为玻璃质，分硬质和软质，有平底、圆底、长颈、短颈、细口和广口几种。

圆底烧瓶通常用于化学反应，平底烧瓶通常用于作洗瓶或代替圆底烧瓶用于化学反应，它是平底能放置平稳。

使用时为防止受热破裂或喷溅，一般要求盛放液体量为烧瓶容量的 1/3 ～ 2/3，加热前要固定在铁架台上，不能直接加热，应当下垫石棉网等软性物。

广口瓶：通常为玻璃质，有无色和棕色(防光)，有磨口(具塞)和光口(不具塞)之分。磨口瓶用于储存固体药品，光口瓶通常作集气瓶使用。

使用时注意：①不能直接加热；②磨口瓶不能放置碱性物质，因碱性物会使瓶口和塞粘黏，作气体燃烧实验时应在瓶底放薄层的水或砂子，以防破裂。③磨口瓶不用时应用纸条垫在瓶塞与瓶子间，以防打不开。④磨口瓶与塞应该配套，防止弄乱。

细口瓶：通常为玻璃质，有磨口和不磨口、无色和有色(防光)之分，磨口瓶(具塞)用于盛放液体药品或溶液，使用注意事项同广口瓶。

称量瓶：通常为玻璃质，分高型和矮型两种，用于准确称取一定量固体药品时使用。使用时注意事项同广口瓶。

锥形瓶：通常为玻璃质，分硬质和软质、有塞（磨口）和无塞、广口和细口等几种。可用作反应容器、接收容器、滴定容器（便于振荡）和液体干燥等。不能直接加热，加热时应下垫石棉网或用热水浴，以防破裂。内盛液体不能太多，以防振荡时溅出。

滴瓶：通常为玻璃质，分无色和棕色（防光）两种。滴瓶上乳胶滴头另配。用于盛放少量液体试剂或溶液，便于取用。滴管为专用，不得弄脏弄乱，以防玷污试剂，滴管不能吸得太满或倒置，以防试剂腐蚀乳胶头。

容量瓶：通常为玻璃质，用于配制准确浓度溶液时用，用时注意：①不能加热，不能代替试剂瓶用来存储溶液，以避免影响容量瓶容积的准确性。②为使配制准确，溶质应先在烧杯内溶解后移入容量瓶。

洗气瓶：通常为玻璃质，用于洗涤净化气体。反接可作安全瓶使用。用于洗气时应将进气管通入洗涤液中。瓶中洗涤液量一般为容器高度的 1/3～1/2，太高易被气体冲出。

吸滤瓶：又称抽滤瓶，玻璃质，用于减压过滤。使用中应注意：①不能直接加热；②和布氏漏斗配套使用，并用橡皮管连接，确保密封性良好。

量筒：通常为玻璃质，用于量取一定体积的液体。使用时不可加热，不可量热的液体或溶液；不可作实验容器，以防影响容器的准确。为使读数准确，应将液面与视线置同一水平上并读取与弯月面相切的刻度。

漏斗：多为玻璃质，分短颈与长颈两种。用于过滤或倾注液体。不可直接加热。过滤时漏斗颈尖端应紧靠承接滤液的容器壁。用长颈漏斗往气体发生器加液时颈端应插至液面以下，以防气体泄露。

分液漏斗：玻璃质，有球形、梨形、筒形之分。用于加液或多相溶液分离。上口玻璃及下端旋塞均为磨口，不可掉换。用时旋塞可涂上凡士林，不用时磨口处应垫纸片。

研钵：瓷质，也有玻璃、玛瑙、石头或铁制品，通常用于研碎固体，或固－固、固－液的研磨。使用时应注意：①放入物体量不宜超过容积的 1/3，以免研磨时，物质溅出；②只能研，不能舂，以防击碎研钵或研杵，避免固体飞溅；③易爆物只能轻轻压碎，不能研磨以防爆炸。

坩埚：瓷质，也有石英、石墨、氧化锆、铁、镍、银或铂制品。用于强热、灼烧固体。使用时放在泥三角上或马弗炉中强热。加热后应用坩埚钳取下（出），以防烫伤。热坩埚取下（出）后应放在石棉网上，防止骤冷破裂或烫坏桌面。

蒸发皿：瓷质，也有玻璃、石英、铂制品，有平底和圆底之分。用于蒸发液体、浓缩。一般放在石棉网上加热使受热均匀。注意防止骤冷骤热导致其破裂。

表面皿：通常为玻璃质，多用于盖在烧杯上，防止杯内液体加热时迸溅或液体挥发污染空气。使用时不能直接接触热源。

干燥管：玻璃质，用于干燥气体。用时两端应用棉花或玻璃纤维填塞，中间装干燥剂。干燥剂受潮后应及时更换清洗。

滴定管：玻璃质，分碱式和酸式两种。用于滴定分析或量取较准确体积的液体。酸式滴定管还可以用作柱色谱分析中的色谱柱。使用时注意酸碱式不能调换使用，以免碱液腐蚀酸式滴定管中的磨口旋塞，造成旋塞粘连损坏。

移液管：又叫吸量管。通常为玻璃质，分刻度管型和单刻度大肚型两类，还有自动移液管。用于精确移取一定体积的液体时用。

坩埚钳：铁或铜制，用于夹持坩埚。

试管夹：有木制、竹制、钢制等，形状各不相同，用于夹持试管以免烫伤。

铁夹：铁制，夹内衬布或毡，用于夹持烧瓶等容器。

试管架：一般为木质或铝质，有不同形状与大小，用于放试管，加热后的试管应用试管夹夹住悬放在试管架上，不要直接放入试管架，以免因骤冷炸裂。

漏斗架：通常为木制，过滤时承接漏斗用。放置漏斗架时不要倒放，以免损坏。

三脚架：铁制，用于放置较大或较重的加热容器。放置容器（除水浴锅）时应先放石棉网，使受热均匀，并可避免铁器与玻璃容器碰撞。

铁架台：铁制品，用于固定反应容器。其上铁圈可代替漏斗架用，使用时应注意平稳和牢固，以防倾倒、松脱。

泥三角：由铁丝弯成，并套有瓷管。用于灼烧时放置坩埚。使用前应检查铁丝是否断裂。

石棉网：在铁丝网上涂石棉，容器不能直接加热时用，并使受热均匀。不可卷折，以防石棉脱落，不能与水接触，以免石棉脱落和铁丝锈蚀。

水浴锅：铜或铝制。用于间接加热或粗略控温实验。使用时注意防止水烧干，以免把锅烧坏，用完应把水倒净擦干，防止锈蚀。

燃烧匙：铜制，用于检验某些固体的可燃性。用完应立即洗净并干燥，以防腐蚀。

药匙：瓷质或用塑料、牛角制成，用于取用固体药品。用时只能取一种药品，不能混用。用后应立即洗净、干燥。

毛刷：分试管刷、烧瓶刷、滴定管刷等多种，用于洗刷仪器。使用时注意用力均匀适度，以免捅破仪器。掉毛（尤其竖毛）的刷子不能用。

酒精灯：多为玻璃质，灯芯套管为瓷质，盖子有塑料制或玻璃制之分。用于一般加热。

启普发生器：玻璃质，用于产生气体。

**2. 微型仪器**

微型仪器用于微型化学实验。它具有试剂用量少、操作简单、减小污染、节约经费等优点。因此越来越受到重视。在化学实验中运用微型实验已成为绿色化学的一个内容。

常用微型玻璃仪器（图1-2）中1、2、3、4、5、6均为微型实验台；折叠试管架包括2、3；烧杯及试管插孔装置包括5、6、7、8；简易气体发生器9、14、26；可控气体制备及性质系列反应装置14、19、28、29、30；升华装置20；微型滴定操作装置9、10、11、15；等等。

常用微型玻璃仪器（图1-3）包括1. 牛角管；2. 直玻管；3. 长柄V形管；4. V形管；5. 消膜V形管；6. W管；7. 消膜W管；8. 微型U形管；9. 消膜U形管；10. T形三通管；11. Y形三通管；12. 尾气吸收管；13. 加料滴管；14. 容量V形管；15. ψ形管；16. 微型漏斗；17. 大加料管；18. 缓冲燃烧尖嘴；19. 燃烧凝结管；20. 异径对接接口。

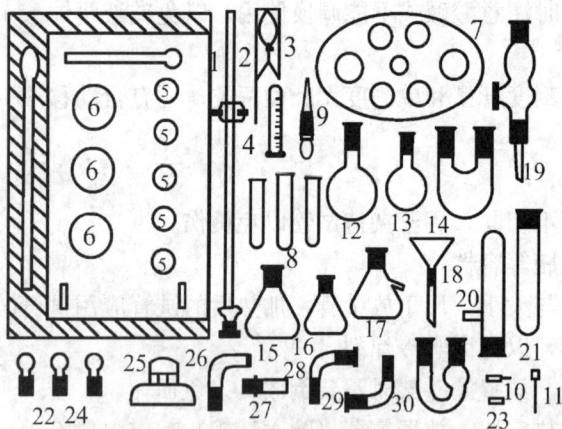

图 1-2　微型玻璃仪器　　　　　　　图 1-3　常用微型玻璃仪器

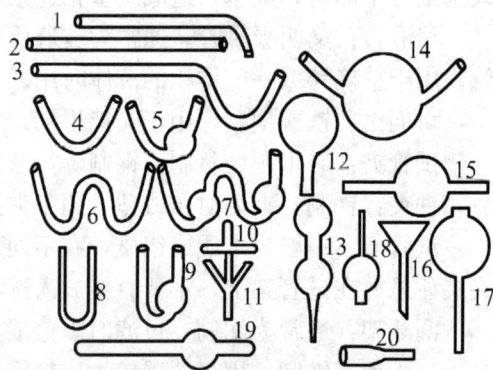

# 化学实验常用玻璃仪器的洗涤和干燥

为了使实验得到正确的结果，实验所用的玻璃仪器必须是洁净的，有些实验还要求是干燥的，所以需对玻璃仪器进行洗涤和干燥。要根据实验要求、污染物性质和污染的程度选用适宜的洗涤方法。玻璃仪器的一般洗涤方法有冲洗、刷洗及药剂洗涤等。对一般沾附的灰尘及可溶性污物可用水冲洗。洗涤时先往容器内注入约容积1/3的水，稍用力振荡后把水倒掉，如此反复冲洗数次。

当仪器内壁附有不易冲洗掉的污物时，可用毛刷刷洗，通过毛刷对器壁的摩擦去掉污物。刷洗时需要选用合适的毛刷。毛刷可按洗涤的仪器类型、规格（口径）大小来选择。洗涤试管和烧瓶时，端头无直立竖毛的秃头毛刷不可使用。（为什么？）刷洗后再用水连续振荡数次。冲洗或刷洗后，必要时还应用蒸馏水淋洗三次。对于上述方法都洗不去的污物，则需要洗涤剂或药剂来洗涤。对油污或一些有机污物，可用毛刷蘸取肥皂液或洗涤剂或去污粉来刷洗。对更难洗去的污物或因口径较小或管细长不便刷洗的仪器可用铬酸洗液或王水洗涤，也可针对污物的化学性质选用其他适当的药剂洗涤（例如可用 6 mol·L$^{-1}$ HCl 溶解碱、碱性氧化物、碳酸盐等）。用铬酸洗液或王水洗涤时，先往仪器内注入少量洗液，使仪器倾斜并慢慢转动，让仪器内壁全部被洗液湿润。再转动仪器，使洗液在内壁流动，经流动几圈后，把洗液倒回原瓶（不可倒入水池或废液桶，铬酸洗液变暗绿色失效后可另外回收再生使用）。对污染严重的仪器可用洗液浸泡一段时间，或者用热洗液洗涤。用洗液洗涤时，决不允许将毛刷放入洗液中！（为什么？）倾出洗液后，再用水冲洗或刷洗，必要时还应用蒸馏水淋洗。铬酸洗液的配制技术下文介绍。

仪器是否洗净可通过器壁是否挂水珠来检验。将洗净后的仪器倒置，如果器壁透明，不挂水珠，则说明已洗净；如器壁有不透明处或附着水珠或有油斑，则未洗净应予重洗。

洗净后的仪器，不可用布或纸擦拭，而应用晾干或烘烤的方法来使之干燥。晾干是让残留在仪器内壁的水分自然挥发而使仪器干燥。倒置后稳定性比较好的仪器可将之倒置在仪器柜内或放置在干净的搪瓷盘中，隔离灰尘放置干燥；倒置不稳的仪器可倒插在格栅板中或干燥板上干燥［图1-4(1)］。晾干的缺点是耗时较长。

（1）晾干

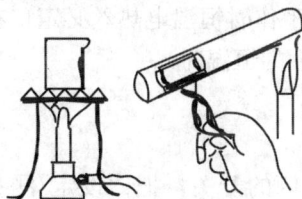
（2）烤干

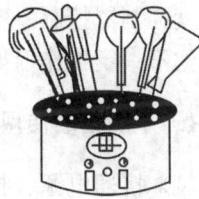
（3）气流烘干

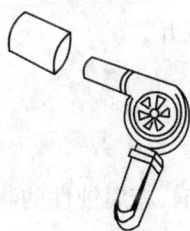
（4）吹干

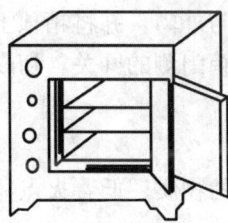
（5）干燥箱烘干

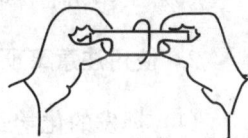
（6）快干

图 1-4　仪器的干燥

如将仪器倒插在气流烘干器上或用电吹风吹进热空气则可加速干燥［图1-4(3)、(4)］。急用时，可用有机溶剂助干，即往仪器内注入少量无水乙醇或丙酮等能与水互溶且挥发性较好的有机溶剂，然后转动仪器使溶剂在内壁流动，润湿全部内壁后倒出全部溶剂（有机溶剂应回收），操作方法见图 1-4(6)，再用电吹风吹干残留在内壁的有机溶剂，达到快干的目的。

对可加热或耐高温的仪器，如试管、烧杯、烧瓶等还可利用加热的方法使水分迅速蒸发而干燥。加热前先将仪器外壁擦干，然后用小火烤干，烤时注意不时转动以使受热均匀，烤干操作方法见图 1-4(2)。

如需干燥较多仪器，可使用电热干燥箱烘干，如图 1-4(5)。将洗净的仪器倒置沥出水滴后，放入干燥箱的隔板上，关好门，控制箱内温度在 105℃左右（以干燥箱温度示值为准），恒温烘干。

仪器干燥时需注意带有刻度的计量仪器不能用加热的方法进行干燥，以免影响仪器的精度；对于厚壁瓷质仪器不能用烤干，但可烘干。刚烤烘完毕的热仪器不能直接放在冷的、特别是潮湿的桌面上，以免因局部骤冷而破裂。

## 铬酸洗液的配制技术

称取 10 g 工业级重铬酸钾固体放入烧杯中，加入 30 mL 热水溶解，冷却后在不断搅拌下慢慢加入 170 mL 浓 $H_2SO_4$，即得暗红色铬酸洗液。将之储存于细口玻璃瓶中备用。取用后，要立即盖紧。

## 电热恒温干燥箱的使用技术

电热恒温干燥箱是利用电热丝隔层加热使物体干燥的设备。它适用于比室温高 5℃至 200℃范围的恒温烘焙、干燥、热处理等，灵敏度通常为 ±1℃。电热恒温干燥箱一般由箱体、电热系统和自动恒温控制系统三个部分组成。其电热系统一般由两组电热丝构成。一组为辅助电热丝，用于短时间内急升温和 120℃以上恒温时辅助加热。另一组为恒温电热丝，受温度控制器控制。辅助电热丝工作时恒温电热丝必定也在工作，而恒温电热丝工作时辅助电热丝不一定工作（如 120℃以下的恒温时）。

### 1. 干燥箱的使用操作步骤

（1）接通电源。打开烘箱门侧的开关，把"设定/测量"键按下，旋转温度调节旋钮，调节至所需温度，设置好干燥所需的温度。弹出"设定/测量"键测量当前温度。红色指示灯亮，表示加热，待红灯灭，绿灯亮，表示加热停止。待红、绿灯自动继熄，表示恒温。

（2）将物品放进箱内，关上玻璃门与外门，并将箱顶上的排气阀适当旋开。

（3）工作完毕，取出物品，关闭烘箱门侧的开关，切断主机电源。

### 2. 使用注意事项

（1）挥发的化学药品、低浓度爆炸的物质、低着火点物质等易燃易爆和具有腐蚀性的物质不能在电热干燥箱中使用；

（2）试剂和玻璃仪器要分开烘干，以免相互污染。干燥箱内物品之间应留有空间，不可过密；

（3）设置使用温度时应根据待烘干物品的特性选择温度。

（4）不允许将被烘物品放在烘箱底板上，因为底板受电热丝加热，温度超过干燥箱所控制的温度，且底板起传热作用，物品置于底板上会影响传热；

（5）鼓风装置的电热干燥箱，在使用过程中必须将鼓风机开启，否则影响工作室温度的均匀性，损坏加热元件；

（6）干燥箱使用时，顶部的排气阀应旋开一定间隙，以便于让水蒸气逸出，停止使用时应及时将排气阀关闭，以防潮气和灰尘进入；

(7) 当需要观察箱内物品情况时,可打开外门通过玻璃观察,但箱门应尽量少开,以免影响恒温,特别是工作温度超过200℃时,打开箱门有可能使玻璃门骤冷而破裂。

## 酒精灯的使用技术

### 1. 酒精灯及其灯焰的构造

酒精灯是实验室常用的加热工具,其加热温度为400~500℃,适用于温度不需要太高的实验。酒精灯由灯罩、灯芯(以及瓷质套管)和盛酒精的灯壶三个部分组成 [图 1-5(1)]。正常使用时酒精灯的火焰可分为焰心、内焰和外焰三个部分 [图 1-5(2)],外焰的温度最高,往内依序降低。故加热时应调节好受热器与灯焰的距离,用外焰来加热 [图 1-5(3)]。当有风或室内气流不太稳定时,酒精灯灯焰也不太平稳,为此可在酒精灯上加一个金属网罩 [图 1-5(4)]。

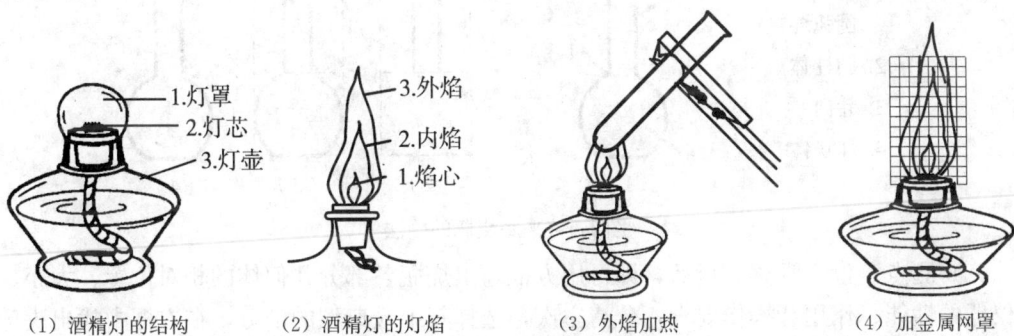

| (1) 酒精灯的结构 | (2) 酒精灯的灯焰 | (3) 外焰加热 | (4) 加金属网罩 |

图 1-5　酒精灯的构造及其使用

### 2. 酒精灯的使用注意事项

(1) 点燃酒精灯之前,先打开灯盖,并把灯头的瓷管向上提一下,使灯内的酒精蒸气逸出,这样才可避免点燃时酒精蒸气因燃烧受热膨胀而将瓷管连同灯芯一并弹出,从而引起燃烧事故。灯芯不齐或烧焦时,应用剪刀修整为平整。灯芯长度可控制在浸入酒精后再长4~5 cm。新换的灯芯应让酒精浸透后才能点燃,否则一点燃就会烧焦。

(2) 酒精灯应用火柴杆引燃,绝不能拿燃着的酒精灯去引燃另一盏酒精灯。因为这样做将使灯内的酒精从灯头流出,引起燃烧。

(3) 熄灭酒精灯时,把灯罩罩上,片刻后再把灯罩提起一下,然后再罩上,可避免灯帽揭不开(为什么?)。注意千万不能用口来吹熄。

(4) 添加酒精时应先熄灭灯焰,然后借助漏斗把酒精加入灯内。灯内酒精的储量以酒精灯容积的1/2~2/3为宜,不得超过。

## 实验仪器及装置图的画法

化学实验仪器及装置的图形是用来表示仪器的形象,装置组合形式和实验的操作方法,是化学交流的一种重要语言。因此对其总的要求是形状正确、线条分明、流程清楚、重点突出、整洁大方,还要符合科学性。

仪器及装置的主要绘制对象是容器,或有容量的管状体,夹持工具,热源工具等。绘画时我们一般采用线描式的纵断面图来表示。

**1. 化学实验仪器的画法**

观察分析实验仪器造型的特点，可以看出它们都由简单的几何形体所组成。如试管由空心圆柱体和空心半球体组成；圆底烧瓶由空心圆柱体和空心球体所组成；漏斗由空心圆柱体和空心圆锥体所组成等。因此，我们只要抓住构成这些几何体的主要点、线进行作图，就可以画出其图形。

仪器的绘画可分为三个步骤：观察分析、测点定位和起稿深描，以平底烧瓶为例。

（1）观察分析：先观察它的外形、高度、宽度，估计各部分的大小比例［图 1-6(1)］，接着分析它的几何形体的组合性质。从图 1-6(2) 可知：平底烧瓶是由高度基本相等的圆柱体和球缺两个部分连接而成，圆柱体的直径约为球缺体直径的 1/3。此外还要注意到它们的外形构造特征——瓶口具有"卷门"，肩部是直线-圆弧-圆弧平滑地连接，底为平底。

1. 直线-圆弧-
圆弧连接
2. 圆柱体
3. 卷门
4. 球缺体

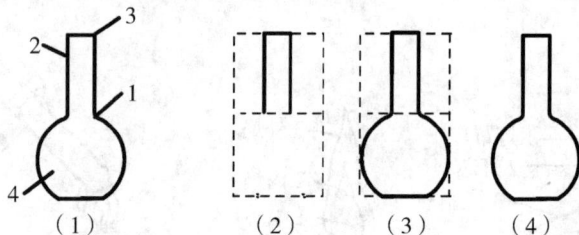

（1）　　　　（2）　　　（3）　　　（4）

图 1-6　平底烧瓶的画法

（2）测点定位：观察分析后，即可用方框勾出烧瓶各部分几何体的相对位置、大小、并按仪器的性能、作用和操作要求，分清主次，选择其中最典型的部分，在方框中找出决定其轮廓外形的各个主要点的位置，如图 1-6(2) 所示。

（3）起稿深描：用轻而细的线连接图 1-6(1) 中的各相应测点，勾出图形的大体轮廓，然后认真地勾深。描绘所画的对象的图形。并注意勾出它外形构造的特点，如图 1-6 中的 (3)、(4)。最后擦去不需要的线条，使图形整洁。

使用方格绘图纸（坐标纸）或计算机相关软件可方便地画出仪器图（图 1-7）。

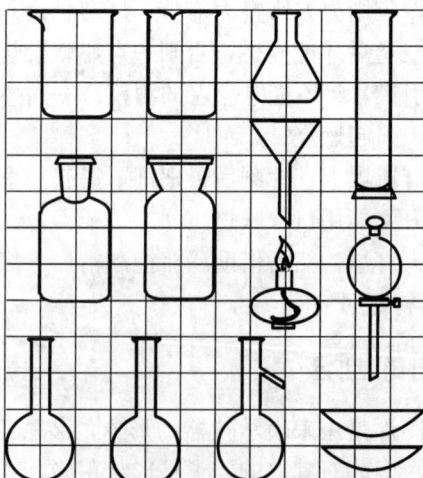

图 1-7　一些常用仪器的简易画法

**2. 装置图的画法**

装置图是实验仪器的组合图形，它必须表示出仪器的形状，所处的相对位置，相互连接的方式及作用。其基本的画法仍然按照上述三个步骤，不过在具体绘画时要注意整体安排，合理布局。一般先画主体图，后画配件图，最后画辅助设施，这样才能做到和谐统一。以排水法制取氧气的装置为例阐明其绘制步骤（图 1-8）。

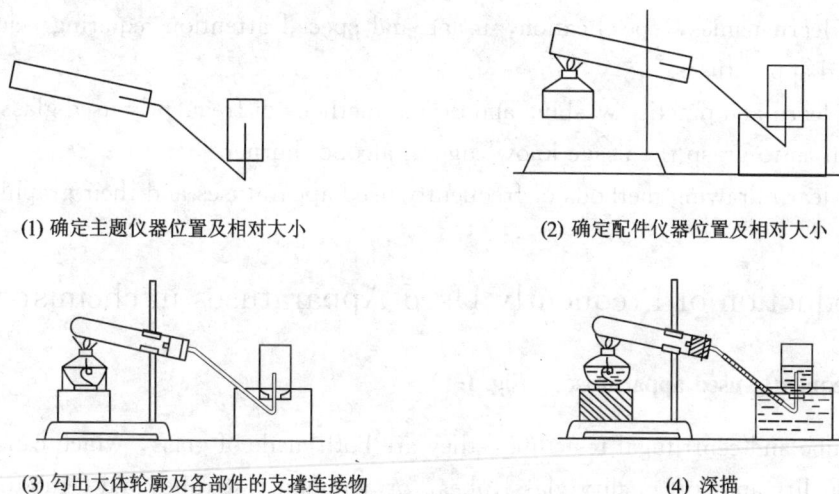

(1) 确定主题仪器位置及相对大小　　　　　　　　(2) 确定配件仪器位置及相对大小

(3) 勾出大体轮廓及各部件的支撑连接物　　　　　　(4) 深描

图 1-8　排水法制取氧气实验装置图的绘画

## 实验训练

**1. 认领仪器**

按照所发的仪器清单，结合本实验图 1-1 的有关内容，领取并逐个认识化学实验常用的仪器。若发现破损的仪器应立即向仪器室提出更换。

**2. 玻璃仪器的洗涤**

洗涤所领的仪器，其中至少用冲洗法和刷洗法洗涤两支试管；用去污粉洗涤一个 100 mL 的烧杯；用铬酸洗液洗涤一支试管。

请互相检查上述仪器是否洗净，洗净为止。

**3. 玻璃仪器的干燥**

对洗涤干净的仪器，至少用气流烘干器烘干一支试管及一个烧杯。将洗涤干净的玻璃仪器存放于实验柜内晾干。

## 思考题

正确画出下列仪器的简图，并按下表所示格式填表。

仪器：试管、烧杯、量筒、酒精灯、容量瓶、蒸馏烧瓶、分液漏斗。

| 仪器名称 | 仪器简图 | 规格 | 主要用途 | 注意事项 |
|---|---|---|---|---|
|  |  |  |  |  |

# Experiment 1　Adopting and Washing Apparatuses

## Objectives

1. To learn names, specification, usage and special attention-requiring issues of frequently used apparatuses.

2. To learn and practice washing and drying methods of frequently used glass apparatuses, as well as to grasp the usage knowledge of alcohol burner.

3. To learn drawing methods of frequently used apparatuses and their graphical setting pictures.

## Introduction of Frequently Used Apparatuses in chemistry

### 1. Frequently used apparatuses（Fig. 1-1）

Test tube and centrifugal test tube: they are both made of glass, which can be divided into hard quality and soft quality glass tubes, or common test tubes and centrifugal tubes. There are many kinds of common test tubes, such as, one with turning over mouth, leveling mouth one with graduation, without graduation; with branch, without branch; with stopper, without stopper. The centrifugal test tubes also have different kinds with or without graduation.

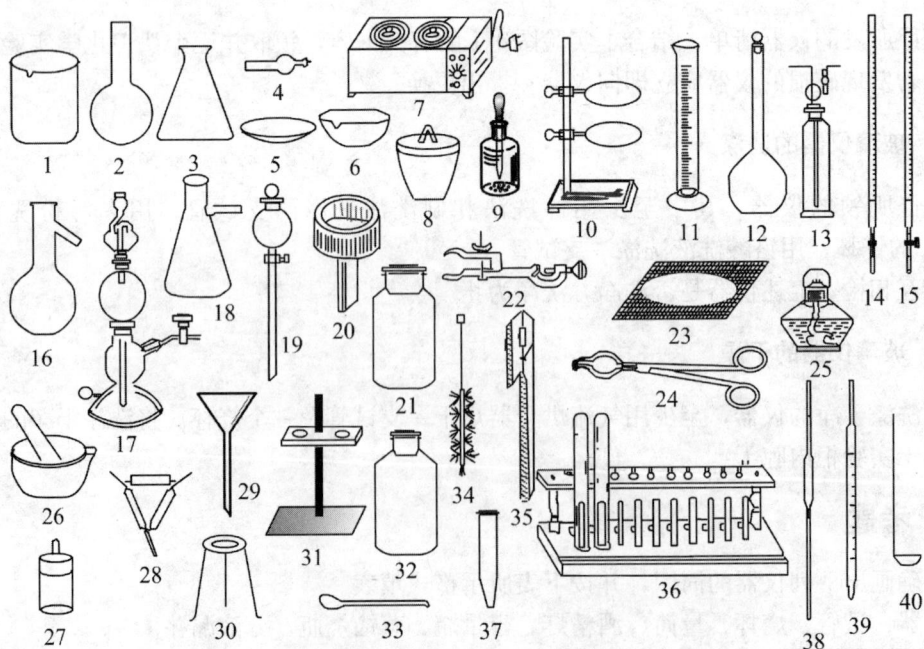

Fig. 1-1　Frequently used apparatuses

In Fig. 1-1, there are following apparatuses: 1. beaker 2. flask 3. conical flask 4. drying tube 5. watch-glass 6. evaporating dish 7. water bath 8. crucible 9. dropping bottle 10. hob and iron ring 11. measuring cylinder 12. volumetric flask 13. gas washing bottle 14. base burette 15. acid burette 16. distilled flask 17. kipp's apparatus 18. suction bottle 19. separatory funnel 20. büncher funnel 21. wide mouth bottle 22. iron clamp 23. asbestos gauge 24. crucible tongs 25. alcohol burner 26. mortar 27. weighing bottle 28. mud triangle 29. funnel 30. V-shaped rack 31. filter holder 32. narrow-necked bottle 33. medicine spoon 34. test tube brush 35. test tube clamp 36. test tube stand 37. centrifugal test tube 38. dropping pipet 39. measuring pipet 40. burning spoon.

Generally, test tube is used as reaction container for small quantity of reagents reacting at room temperature or being heated. It is easily to operate and observe, and it also can be used to collect small amount of gas. Test tube with branch can be used to identify gaseous product and used in setting up a device. Centrifugal test tube can be used to separate precipitate. When we use them, we should pay attention to: ① Reaction liquid should not exceed 1/2 volume of test tube, should not exceed 1/3 when heating so as to avoid splashing or overflowing. ② Before heating, the outside surface of test tube should be dried. When heating liquid, test tube clamp should be used to avoid water-drops on the tube's wall being heated unevenly, which would lead the tube to be overheated or broken. ③When liquid is heated, the mouth of test tube should not be directed at any person, and the test tube should keep an angle of 45° with the tabletop. At the same time, it should be vibrated continuously, while the top of flame shouldn't exceed the surface of liquid to avoid splashing liquid hurting people. Expanding the surface of heating can avoid liquid bumping and tube breaking. ④When solid is heated, the mouth of the test tube should be placed slightly downward to avoid condensed water regurgitating and breaking the test tube. ⑤Centrifugal test tube shouldn't be heated directly to avoid breaking.

Beaker: it is commonly made of glass. It is divided into hard quality one and soft quality one; general model and high model; with graduation and without graduation.

Beaker is usually used as reaction container for large amounts of reagents, which makes the reagent easier to be mixed evenly. It also can be used in preparing solution or as a replacement of gutter.

Special attention-requiring issues: ① Reaction liquid shouldn't exceed 2/3 of the beaker's volume to avoid overflowing of liquid when stirring or boiling. ②Dry the outside of the beaker; asbestos gauge should be laid under beaker when heating to avoid heating so unevenly that breaking the beaker.

Flask: it is commonly made of glass. which is divided into hard quality one and soft quality one, including leveling bottom , round bottom, dolichoderus, short neck, small mouth and wide mouth.

Round bottom flask is usually used in chemical reaction, while leveling bottom flask is usually used in solution preparation, used as washing bottle or taking the place of round bottom flask as chemical reaction container, as it has a level bottom and can be placed steady.

To avoid overheating, container breaking and splashing of liquid, the liquid's volume should be between $1/3 \sim 2/3$ that of the flask. Before heating, it should be fixed at a hob stand, which can not be heated directly, and soft material such as gauze asbestos should be laid under the flask.

Big mouth bottle: it is usually made of glass, including colorless one and brown one (light tight); one with grinded mouth (with stopper) and common mouth (without stopper). Grinded mouth bottle is used to store solid reagents, and common mouth bottle is used as gas gathering bottle.

Special attention-requiring issues: ① It can not be heated directly. ② Alkaline substances should not be stored in grinded mouth flasks, for alkaline substances would make the bottle and its stopper stick together. When we do gas combustion experiment, a thin layer of water or sand is necessary at the bottom of the bottle to avoid breaking. ③A small piece of paper should be placed between the bottle and its stopper to avoid sticking when not in use. ④ One grinded mouth bottle should be matched with its stopper, to avoid disarrangement.

Small mouth bottle: it is usually made of glass. It is divided into grinded mouth one and common one; colorless one and colorful one (light tight). Grinded mouth bottle (with stopper) is used to store liquid reagents or solution. The special attention-requiring issues of small mouth bottle are just the same as those of big mouth bottle.

Weighing bottle: it is usually made of glass. It is divided into high model and short model one. It is used to weigh solid reagents. The special attention-requiring issues of weighing bottle are just the same as those of big mouth bottle.

Conical flask: it is usually made of glass. It is divided into hard quality and soft quality one; with stopper (grinded mouth) and without stopper; big mouth and small mouth. It can be used as reaction container, receiving container, titration vessel (easy to vibrate) and liquid drying container etc. It can not be heated directly, and gauze asbestos should be laid under the flask or the flask should be heated in certain bath to avoid breaking. There should not be too much liquid in it to avoid splashing.

Dropping bottle: it is usually made of glass, it is divided into colorless and brown one. The rubber dripper of a dropping bottle should be purchased alone. It is used to store small amounts of liquid reagents or solution for convenience. The dropping bottle and its dripper should be matched, and the dripper should not be disarranged and stained to avoid polluting reagents. The dripper should not be used to absorb too much liquid or be placed up down to avoid reagents corroding the rubber dripper.

Measuring flask: it is made of glass. It is used for preparing solution with an accurate concentration. Something should be noted: ①It can not be heated nor used to store solution as a reagent bottle to avoid affecting the accuracy of its volume. ②To ensure the accuracy of solution's concentration, the solute should be dissolved before transferring into a measuring flask.

Gas washing bottle: it is usually made of glass. It is used for washing gas. when connected inversely, it also can be used as a safety flask. When it is used to wash gas, the inlet duct should be immersed in the eluant. The volume of eluant is usually about $1/3\sim1/2$ volume of the bottle. If there is too much eluant, gas may run the eluant out.

Filtering bottle: it is also called suction bottle, made of glass. It used for filtering under a reduced pressure. Special attention-requiring issues: ①It should not be heated directly. ②It should be matched with proper büchner funnel, and rubber tube should be used to connect funnel with bottle to ensure a good gas tightness.

Measuring cylinder: it is usually made of glass. It is used for measuring a certain volume of liquid. It can not be heated directly, nor used to measure hot solution or liquid. It can not be used as reaction container to avoid affecting its accuracy. To ensure an accurate measurement, the reader's eyes and the liquid surface should be on the same line and read a scale tangential with the surface of the liquid.

Funnel: it is usually made of glass. It is divided into short neck and long neck one. It is used for filtering or pouring liquid into a container. It should not be heated directly. And the peak of the funnel's neck should be next to the wall of the collecting container. When a long neck funnel is used for adding liquid to gasifier, its neck should be under the surface of liquid to avoid gas leaking.

Separating funnel: it is usually made of glass, including ball shape, pear shape, tube shape one. It is used for adding liquid or separating solutions with multi-phases. Its upper mouth glass stopper and the lower one are both grinded, which can not be changed, when it is used, vaseline should be put evenly on the lower stopper, while a piece of paper should be put around the upper one when not in use.

Mortar: it is made of porcelain. It can also be made of glass, agate, stone and iron. It is usually used to grind solid materials or make solid-solid, solid-liquid mixture. Special attention-requiring issues: ①The amount of substance should not exceed $1/3$ volume of mortar. ②It can only used for grinding. It should not be pounded to avoid it breaking or the substance splashing. ③Explosive substances can only be crushed to pieces. It should not be grinded to avoid explosion.

Crucible: it is made of porcelain, or made of quartz, graphite, zirconium oxide, iron, nickel, silver or platinum. It is used to burn or calor substances vigorously. When it is used, it should be placed on a mud triangle or a muffle furnace. After heating, it should be fetched by crucible tongs to avoid scalding. Hot crucible should be placed on gauze asbestos to avoid it being quenched or destroying the tabletop.

Evaporating dish: it is made of porcelain, can also be made of glass, quartz and platinum. It could be divided into flat bottom and round bottom one. It is used to evaporate and concentrate liquid. It should be used placed on gauze asbestos to be heated evenly. Do not cool or heat it rapidly or it will be broken.

Watch glass: it is usually made of glass. It is used to cover beaker to avoid the liquid in it splashing or being polluted. It can not be heated directly.

Drying tube: it is made of glass. It is used for gas drying. When it is used, a piece of cotton or fiberglass should be stuffed in its amphi-mouth, and the middle of the tube should be stuffed with dryers. When dryer is wet, it should be changed and the tube ought to be cleaned immediately.

Buret: it is made of glass. It is divided into base buret and acid buret. It is used in titrimetric analysis or used to measure accurate liquid volume. Acid buret is also can be used as chromatographic column in column chromatography analysis. Remember that the base and the acid buret can not be exchanged in use to avoid base eroding the grinded stopper and destroying of the acid buret.

Pipette: it is usually made of glass, which is also called transferring pipette or measuring pipette. It is divided into graduated tube pipette, mono-graduated pipette with a big bell, as well as automatic pipette. It is also named measuring pipette, which is used to measure certain volume of liquid accurately.

Crucible tongs: it is made of iron or copper, which is used to clasp crucible.

Tube clip: it is made of wood, bamboo, steel, *etc*. Its shape could be various. It is used to clasp test tube to avoid burning people.

Iron clip: it is made of iron. which is usually used to clasp reaction container, iron ring on it can also be used as filter holder, remember keeping it stable and secure to avoid tumbing and loosening.

Test tube holder: It is usually made of wood or aluminum. Its shape and size are various. It is used to support test tubes. The heated test tube should be suspended on the test tube holder with tube clip. Don't put it on the holder directly to avoid breaking with cold.

Filter holder: It is usually made of wood, used for holding funnel. When fixing, it should not be placed inversely to avoid breaking.

Tribranch: It is made of iron, used to fix some large or heavy heating containers. When it is used for fixing containers (except for water-bath pot), gauze asbestos should be placed first to make the container heated evenly, and to avoid iron crashing with glass container.

Iron stand: it is made of iron, used for fixing or placing reaction containers. Its iron circle can be used as a replacement of filter holder. When used, special attention should be given to its equability and fastness to avoid toppling and slaking.

Mud triangle: It is made of iron bars with porcelain tubes on them. It is used for placing crucible in ignition. Before use, check it to avoid broken iron bars.

Gauze asbestos: Asbestos is spread on a iron net. When containers can not be heated directly, the gauze asbestos should be used to make the container heated evenly. It can not be curved as asbestos may desquamate, nor be wetted to avoid asbestos desquamating and iron eroding.

Water bath: it is made of iron or aluminum. It is used for indirect heating or roughly controlling reaction temperature. When in use, water should not be dried out, after using, water should be poured and it should be dried to avoid eroding.

Combustion spoon: it is made of copper, used to identify the flammability of some solid materials. Wash and dry it after usage to avoid eroding.

Medicine spoon: it is made of porcelain, plastic or horns of cattle. It is used to fetch solid reagents. When used, it can only get one reagent at one time. It should not be messed up. Wash and dry it after using.

Brush: it is divided into test tube brush, flask brush and buret brush, used for washing apparatuses. When used, take care to avoid destroying apparatuses. Hair losing brushes (especially one with erected hairs) can not be used.

Alcohol burner: it is usually made of glass, while the tube on the lamp mount is made of porcelain, and its cover is usually made of plastic or glass. It is used for common heating.

Kipp's apparatus: it is made of glass, usually used to produce certain gas.

### 2. Mini-apparatuses

Mini-apparatus is applied in mini-chemistry experiments. Following advantages: only need small quantity of reagents and its operation is simple and economic with little pollution. So recently it gets more and more attention from chemistry researchers. Applying mini-apparatuses in chemistry experiment has become an important part of green chemistry.

Mini-glassware(Fig. 1-2) 1, 2, 3, 4, 5, 6 are all mini-benches; folding test tube holders includes 2, 3; beakers and test tube hub installations include 5, 6, 7, 8; simple gas generators are 9, 14, 26; controllable gas preparation and property testing reaction instrument include 14, 19, 28, 29, 30; while sublimation instrument is 20 and mini-titration operation instrument are 9, 10, 11, 15 *etc*.

Frequently mini-glassware(Fig. 1-3)1. pipe like a horn of cattle; 2. straight glass pipe; 3. V-tube with a long handler; 4. V-tube; 5. eliminating membrane V-tube; 6. W-tube; 7. eliminating membrane W-tube; 8. mini U-tube; 9. eliminating membrane U-tube; 10. T-tube; 11. Y-tube; 12. exhausted gas absorber; 13. reagent adding dropper; 14. volumetric V-tube; 15. $\psi$-tube; 16. mini-funnel; 17. big reagent adding tube; 18. bufferring combustion cute mouth; 19. combustion condenser; 20. different diameter joint.

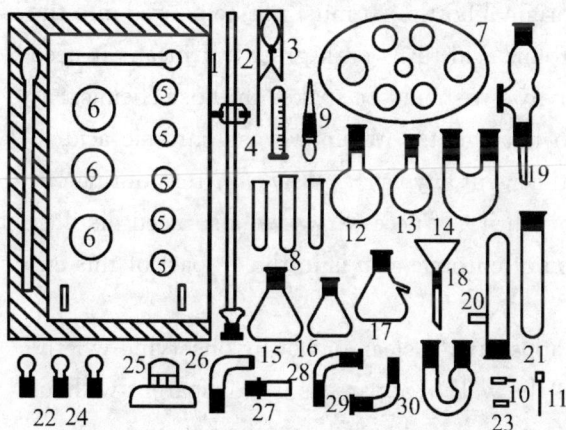

Fig. 1-2   mini-glassware

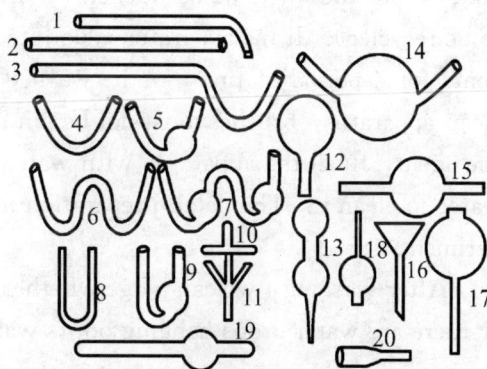

Fig. 1-3   frequently used mini-glassware

## Washing and Drying of Frequently Used Apparatuses in Chemistry Experiment

To guarantee correct experiment results, glassware used in experiment must be clean. For some experiments, the glassware also should be dry, so it is necessary to wash and dry glassware in right way. Choose proper washing methods according to the specific requirements of certain experiment, the nature of pollutants and the spotting extent. Common washing methods include bathing, scrubbing and washing with certain chemical agents *etc*. Flushing with water could wash common dusts and soluble pollutants. Add water to 1/3 of the container's volume, vibrate it, pour the dirty water and then repeat flushing it for several times.

If the pollutant can not be flushed out easily, we can use brush to scrub the apparatuses and clean it. Suitable brushes should be chosen in scrubbing, according to the type and specification (the caliber) of the apparatuses. When we wash test tubes and flasks, head bald brushes, which have no erected hairs, should not be used (Why?). After scrubbing, vibrate the apparatuses with water several times. After flushing or scrubbing, clean it with distilled water three times if necessary. If the pollutant can not be washed with the above two methods, special detergent or agent is needed. If there are grease stains or some organic pollutants, we can use brushes with soap solution or detergent to scrub the apparatus. In the following cases, such as the pollutant is too hard to clean, and apparatus with small caliber or long slim tube that can not be scrubbed with brushes, we can make use of chromic acid or aqua regia, or we can choose special chemical agents to clean it according the chemical properties of the pollutant. For example, bases, basic oxides, carbonates pollutants can be dissolved with $6 \, mol \cdot L^{-1}$ HCl. When we use chromic acid or aqua regia to clean apparatus, we need to add some of them in apparatus first, make the apparatus cline to one side and turn it around slowly to make the whole inside of the apparatus moisten by those chemical agents. Then turn the apparatus slowly to make the agent flowing alone its inside. After turning several circles, pour the agent into its original bottle. It must not be poured into the pool or into the waste liquid bucket. (When chromic acid turns dark green, it means it need to be recycled). If the apparatus is polluted heavily, it should be soaked in those chemical agents for a period of time, or be washed with hot agents. When we use chromic acid to clean apparatus, brushes must not be put into the agent anyway! (Why?) After pouring out the agent, flush the apparatus with water or brush it. If necessary, we also need distilled water to clean it. The detail preparation method of chromic acid is in the $4^{th}$ part of this experiment.

After washing, we can judge whether the apparatus is clean or not by observing whether there are water drops hanging on its wall. If the wall of apparatus is transparent without any water drop, which indicates it is clean, but if not, it should be washed again.

After washing, the apparatus should not be wiped by cloth or paper. Open-air drying or baking method should be used to dry it. Open-air drying could make the water inside the apparatus evaporating naturally. If the apparatus could keep stable when put upside down, we can place it upside down in the apparatus cabinet or on a clean enamel away from dust. But if the stability of the apparatus is not very good, we could put them up down on a grilling plate or drying plate [Fig. 1-4(a)]. But the shortcoming of open-air drying method is that it will waste a lot of time.

(a) drying naturally　　　(b) drying by baking　　　(c) drying by machine

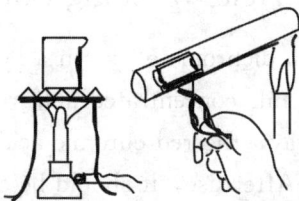

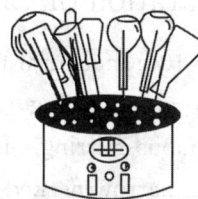

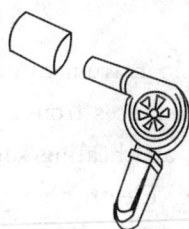

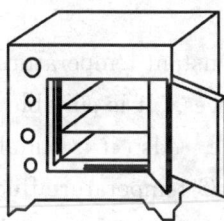

(d) drying by blowing　　　(e) drying by baking　　　(f) flash-dry (organic solvent method)

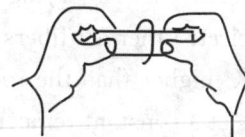

Fig. 1-4　Apparatus drying

If we put the apparatus upside down in an air current baker or blow hot air into it using electronic dryer, the apparatus could be dried faster [Fig. 1-4(c), Fig. 1-4(d)]. If in emergency, organic solvent could be used to help the apparatus dry faster, that is to say, adding a small amount of absolute alcohol, acetone which could be soluble with water and also could be evaporated fast, turning around the apparatus to make the organic solvent flowing along its inside wall, and then pouring out all the solvent after the inside wall all being moistened, the last blowing the remain solvent with a electronic dryer. Remember that the organic solvent should be recycled, and the detailed operation method see Fig. 1-4(f).

If the apparatus could be heated or be refractory, such as test tube, beaker, flask, they can be heated to accelerate water evaporation. Before heating, first wipe its outside dry, and then bake it with small fire. When it is baked, turn it around and around to heat it evenly. The baking method, see Fig. 1-4(b).

If there are a lot of apparatuses need to be dried, electro thermal drying cabinet can be used [Fig. 1-4(e)]. Put the washed apparatus upside down to leach the water, and then put it on the clapboard of a drying cabinet, close the door, control the temperature of the cabinet at about 105℃, taking the thermal measurement which is on the head of the cabinet as temperature controlling standard. Baking half an hour at a constant temperature could dry the apparatus.

When we drying a apparatus, we should know that metrology apparatus with graduation can not be dried by heating to avoid influencing their accuracy. If the apparatus has a thick porcelain wall, it can not be dried by oven, but can be dried by baking. If the apparatus is still very hot, it can not be placed on a cold, especially damp desk directly to avoid part quenching leading to its breaking.

## Preparation of Chromic Acid Washing Solution

Weigh 10 g industrial potassium dichromate (s) in a beaker, add 30 mL hot water to dissolve it, after cooling, add 170 mL concentrated sulfuric acid slowly into the solution with continuous stirring. In this way, the red chromic acid washing solution is prepared. Store it in a narrow-necked bottle. After use, it should be covered tightly.

## Use Technology of Electro Thermal Drying Cabinet with Constant Temperature

Electro thermal drying cabinet with constant temperature is an electronic instrument utilizing electro thermal fibers to dry substances. It is suitable for drying substances from 5℃ to 200℃ higher than the room temperature. Also it is suitable for baking and heating substances at a constant temperature with ±1℃ temperature fluctuating range. Electro thermal drying cabinet with constant temperature consists of a case body, a electro thermal system and an automatic thermostatic control system. Its electro thermal system is comprised of two groups of electro thermal fibers, one assistant electro thermal fiber group used for quickly rising temperature and assistant heating when the temperature is above 120℃, while another constant electro thermal fiber group controlled by the temperature controller. When the assistant electro thermal fiber group is working, the constant electro thermal fiber group should also be working at the same time, but when the latter group is working, the former one may not work with it, for example, in the case the temperature is keeping at a temperature below 120℃. The special attention-requiring issues of the drying cabinet goes as follows:

### 1. Usage procedure of cabinet drier

(1) Turn on the drier. Press the switch on the side of door, press 'set/measure' button, and rotate the adjusting temperature dial to a certain temperature. Pop the 'set/measure' button to measure current temperature. Red light is on, which indicates heating; while it is out and green light is on, which indicates stopping heating. After the red light and the green light put out one after another, the constant temperature is reached.

(2) Put the item into the cabinet, close the glass door and outside door, and open the valve on the top of cabinet properly.

(3) After all the items are dried thoroughly, take out of them, turn off the switch on the side of the cabinet door, and cut off the power.

### 2. Precautions for use

(1) Those reagents, such as flammable chemical reagent, explosive reagent, reagent with low fire point and caustic reagent shouldn't be dried in the cabinet drier.

(2) Reagent and glass apparatus should be dried separately to avoid polluting each other, the items in cabinet shouldn't be too crowded.

(3) The setting temperature should be decided by the particular properties of items.

(4) The item shouldn't be placed on floor of the cabinet, for the floor of the cabinet is heated by electroheat wire, whose temperaure exceeds the setting temperature, and the floor has the heat transfer function, the item on the floor will influence the heat transfer.

(5) For cabinet drier with blower unit, the unit should be turned on otherwise the homogeneity of the cabinet temperature will be influenced and the heating unit damaged.

(6) When the cabine drier is in use, the air escapevalve should be opened properly to let water vapor out, while it is not in use, close the valve to avoid moisture and dust entering.

(7) When we want to watch the items in cabinet, the outside door can be opened, watching can be performed through glass door. The cabinet door shouldn't be frequently to avoid influencing the constant temperature. The glass door may be broken for quenching when the temperature exceeds 200℃.

## Use technology of alcohol burner

### 1. Construction of an alcohol burner and its flame

Alcohol burner is a common tool for heating in lab. Its heating temperature is about 400 ~500℃, and suitable for experiments with moderate temperature. Alcohol burner is made up of a lamp shade, a lamp mount with a porcelain tube and a lamp kettle for alcohol storage [Fig. 1-5(a)]. Normally, when it is used, its flame can be divided into the center of flame, inside flame and outside flame [Fig. 1-5(b)]. The temperature of the outside flame is the highest, and declining from outside to inside of the flame. So the distance between the heated substance and flame is very important [Fig. 1-5(c)], we should always heat with outside flame. When there is some wind or unstable indoor air flows, the flame of alcohol burner could be also unstable, in this case, we can put a metal net on the alcohol burner [Fig. 1-5 (d)].

### 2. Special attention-requiring issues of the usage of an alcohol burner

(1) Take off the cap of the alcohol burner before kindling and lift the porcelain tube upward slightly to make the alcohol vapor emit, that can avoid the porcelain tube and lamp mount popping up for the alcohol vapor's heat expansion. If the lamp mount is uneven or burnt, you should cut it evenly with scissors. The length of the lamp mount should be 4~5 cm longer after immersing it in alcohol. Newly changed lamp mount should be immersed

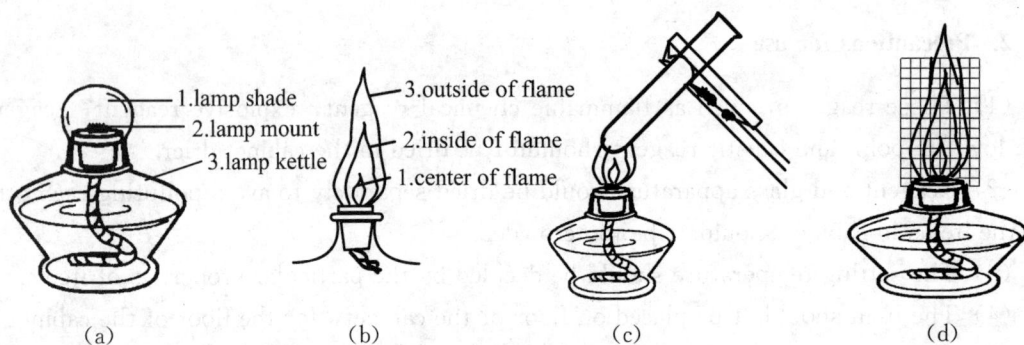

Fig. 1-5 Construction and usage of an alcohol burner

thoroughly by alcohol before kindling, otherwise it will be scorched.

(2) Alcohol burner should be ignited by match poles. Using a burning alcohol burner to ignite another one is banned, because this will make the alcohol in a burner flown out to cause fire.

(3) When we do not use the alcohol burner, we should cover it with its lamp shade. Lift the shade after several seconds, and then cover it again on the burner to avoid the situation that it can not be opened again. (Why?) Warning: it must not be blown out by mouth.

(4) When we add alcohol to the burner, we should firstly cover out the flame with its shade, and then add alcohol with a funnel. The volume of alcohol in the burner should be 1/2~2/3 of the total volume. Don't exceed.

## Draw Laboratory Apparatus and Installation Diagram

Figure of laboratory apparatus and installation diagram is presented the image of apparatus, the pattern of installation and the method of experimental operation, which is an important language of chemistry communication. So the overall requirement of it is shaped correctly, lined distinctly, preceded clearly, emphasized outstandingly, as well as drawn neatly and naturally. And at the same time, it should be scientific.

The main objects of drawing apparatus and installation are containers or graduated tubes, clamping tools, heating tools etc. we usually adopt profile diagram with line drawing style.

### 1. Pen craft of chemical experiment apparatus

We can find the character of experiment apparatus, that they are simply geometrically formed. For example, a test tube is made up of a hollow cylinder and a hollow hemispheroid; the round bottom flask is made up of a hollow cylinder and a hollow sphere; a funnel is made up of a hollow cylinder and a hollow cone, etc. So, if we can grasp the main points and lines of these geometry forms, we can easily draw those figures.

There are three steps in drawing apparatus: observing and analyzing, surveying station and deciding points, and then drawing carefully. We just take drawing a flat bottom flask as an example:

1. line-arc-arc
2. cylinder
3. roll door
4. nullisomic
    ball

(a)          (b)      (c)       (d)

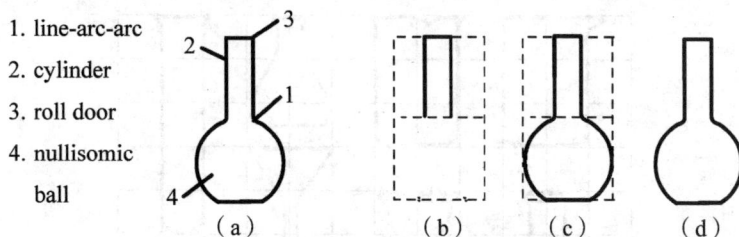

Fig. 1-6 Pen craft of flat bottom flask

(1) Observing and analyzing: First observing its shape, height and width, estimating the proportion of each partial form [Fig. 1-6(a)], and then analyzing its combination character. From Fig. 1-6(b), we can see that the flat bottom flask is made up of a cylinder and a sphere with equal heights, and the diameter of the cylinder is about 1/3 of the diameter of the nullisomic ball.

(2) In addition, we also must notice the character of their configuration—the mouth of the bottle is "a coiling mouth", and its shoulder is connected with a straight-line-an arc-a straight line, while the bottom of the flask is flat.

(3) Surveying station and deciding points: After observing and analyzing, we can draw out the relative position and size of its geometrical form, select out its most typical part according to its function, effect and operation requirement, distinguish major from secondary characters, and then decide the position of main points determining the shape of the apparatus. It is shown in Fig. 1-6(b).

(4) Drawing carefully: Connect the points with light and slim lines as shown in Fig. 1-6(a), line out the whole figure, and draw with thick lines. Depict the figure of drawn object. And keep in mind that you should draw out the character of its configuration, as Fig. 1-6(c), (d). At last, we should wipe out unneeded lines to make the figure clean and neat.

Apparatus figure can be drawn conveniently with check calculation paper or chemical software board (Fig. 1-7).

## 2. Pen craft of installation diagram

Installation diagram is the combination figure of reaction apparatus. It is used to express the shape of the apparatus, their relative positions, linking style and their functions. Its fundamental pen craft still follows the above-mentioned steps, but you must arrange those apparatuses suitably in the whole picture. Generally, you should draw main apparatuses first, and then draw fitments, at last draw ancillary devices. Just in this way, you can draw it in harmony and unity. Take the experiment of preparing oxygen with draining water as an example for illustrating the step of drawing (Fig. 1-8). In Fig. 1-8(a) determining the position and size of the main apparatuses (b) determining the position and size of the accessories (c) drawing the adumbration linkage on the whole and (d): completing drawing

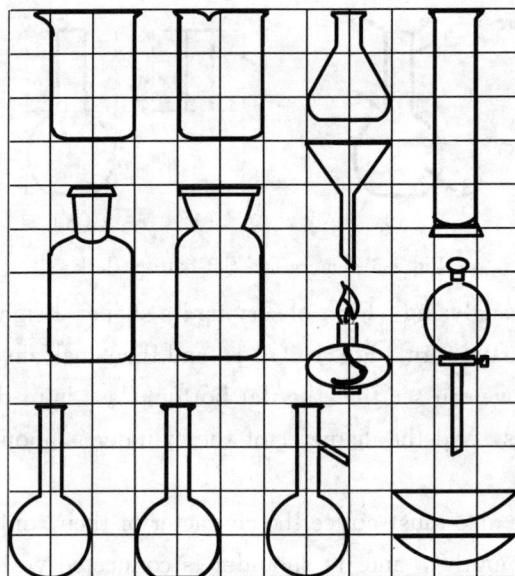

Fig. 1-7 Simple drawing of frequently used apparatus

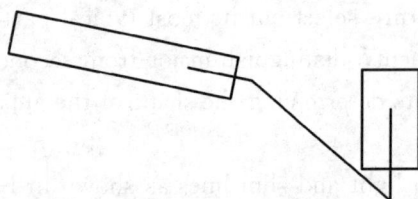

（a）determining the location and relative
size of the main instruments

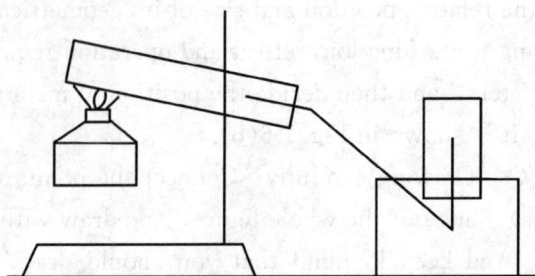

（b）determining the location and relative
size of the instruments parts

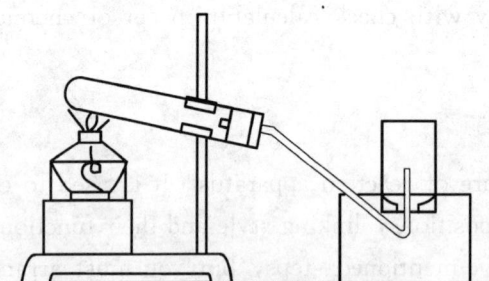

（c）outlining the support of the
components to connect objectives

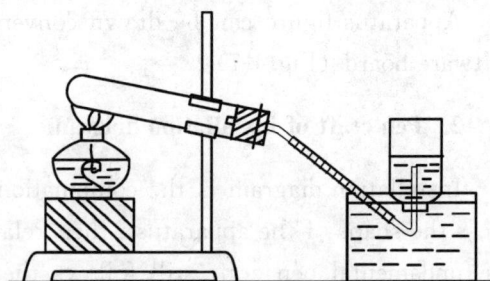

（d）thick description

Fig. 1-8 The method of drawing the instrument of $O_2$ generation by draining

# Experiment Training

## 1. Adopt apparatus

According to apparatus checklist, following the instruction in Fig. 1-1, adopt and recognize frequently used apparatus in chemistry experiment. If the apparatus is broken, you should ask to change at apparatus room immediately (Why?).

### 2. Washing glassware

Wash the adopted apparatuses. You must at least wash 2 test tubes with irrigating and scrubbing methods, wash one 100 mL beaker with cleanser, and wash a test tube with chromic acid solution.

Please mutually examine whether the apparatus are cleaned or not.

### 3. Drying of glassware

Among those cleaned apparatus, please dry a test tube by airflow baker and a beaker by the alcohol burner. And then store those cleaned apparatus open-air drying in the experiment cabinet or drawer.

## Experiment Exercise

Draw the diagram of the following apparatuses, and fill in the form with following pattern.

Apparatuses: test tube, beaker, cylinder, alcohol burner, measuring flask, distilled flask, separatory funnel.

| name | diagram of apparatus | specification | main use | special attention-requiring issues |
| --- | --- | --- | --- | --- |
|  |  |  |  |  |

# 实验二  由废铜屑制备硫酸铜

## 实验目的

1. 练习天平使用，蒸发浓缩，减压过滤，重结晶等基本操作。
2. 了解从金属制备它的某种盐的方法，弄清重结晶提纯物质的原理。

## 电热恒温水浴锅的使用技术

电热恒温水浴锅用于100℃以下的恒温加热，是常用的电热设备，有2、4、6、8孔等不同规格。

电热恒温水浴锅由电热恒温水浴槽和电器箱两部分构成。图2-1所示水浴锅左边为水浴槽，它为带有保温夹层的水槽，槽底搁板下装有电热管及感温管，提供热量和传感水温。槽面为有同心圆和温度计插孔的盖板。右边为电器箱，面板上装有工作指示灯（红灯表示加热，绿灯表示恒温）、温度调节开关、调温旋钮和电源开关。

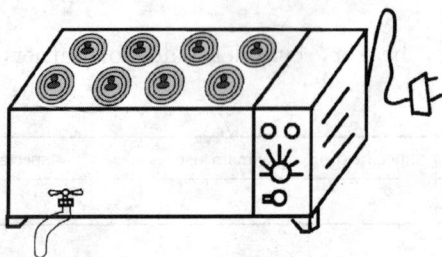

图 2-1  数显电热恒温水浴锅

使用时，先往电热恒温水浴锅内注入清洁的水至适当深度，然后接通电源，开启电源开关后，红灯亮表示电热管开始工作。按下温度调节开关，用"温度旋钮"将温度调至需要的控制值，按上温度调节开关，此时显示屏显示的温度是当前水浴的水温。待水温升到欲控制温度时绿灯切换变亮，这时就表示恒温控制器发生作用，达到恒定的水温。

使用电热恒温水浴锅时要注意爱护，必须切记要先加水，后通电，水位不能低于电热管。电器箱不能受潮，以防漏电损坏。使用时盐及酸、碱溶液不要撒入恒温槽内，如不小心撒入要立即停电，及时清洗，以免腐蚀。较长时间不用水浴锅时也应倒去水槽内的水，用干净的布擦干后保存。当水槽出现渗漏时要及时维修。

## 固、液分离方法（Ⅰ）

### 1. 倾析法

沉淀或结晶的比重或颗粒较大，静置后能沉降至容器底部时，可用倾析法进行分离和洗涤。沉淀（或晶体）沉降至容器底部后，小心地把上层清液沿玻璃棒倾入另一容器中。沉淀需要洗涤时，可往盛有沉淀的容器内加入少量洗涤液（如蒸馏水），充分搅拌，静置、沉降，倾去洗涤液。如此重复操作三次以上，即可把沉淀洗净。

## 2. 过滤法（Ⅰ）：减压过滤

（1）减压过滤用的仪器装置（图 2-2）

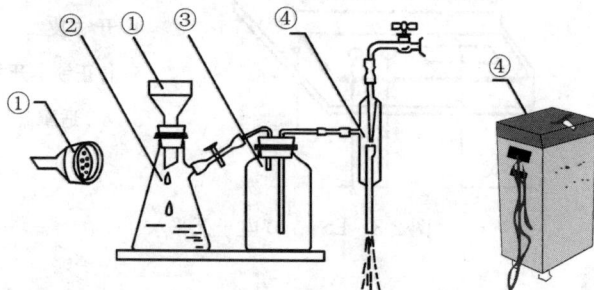

图 2-2　减压过滤装置图

①布氏漏斗：瓷质平底漏斗，平瓷底上面有很多小孔。漏斗下端颈部装有橡皮塞，借以和吸滤瓶相连。

②吸滤瓶：承受滤液用，有支管与抽气系统相连接。

③安全瓶：当减压过滤的操作完成而关闭真空泵时，都会由于吸滤瓶内的压强低于外界压强，使自来水压入吸滤瓶内，造成反吸，所以过滤时要在吸滤瓶和水泵之间装一个安全瓶，起缓冲作用。过滤完毕，应先打开连接在吸滤瓶和安全瓶间的开关或拔掉连接吸滤瓶的橡皮管，再关真空泵，以防反吸。

④水喷射泵（简称水泵）：泵内有一个逐渐收缩的喷嘴，水在此处高速喷出时形成低压，与水泵相连的系统内的气体由此吸入并和水一起排出，从而使系统内压强减小。实验中常用真空泵代替水喷射泵。开启的水泵或真空泵不断将吸滤瓶中的空气带走，从而使吸滤瓶内压强减小，在布氏漏斗内的液面与吸滤瓶内造成一个压强差，提高了过滤速度。

（2）减压过滤的操作方法

减压过滤时，所用的滤纸应比布氏漏斗的内径略小，但又能把瓷孔全部盖住。滤纸平铺在漏斗内，用少量蒸馏水润湿后将漏斗装到吸滤瓶上，使漏斗颈部的斜口对着吸滤瓶支管，以避免减压时，滤液被吸入吸滤瓶侧口；打开水龙头或开启真空泵，减压，让滤纸贴紧在瓷底上。在打开水龙头或开启真空泵的情况下，使溶液沿着玻棒流入漏斗中，注意加入溶液的量不要超过漏斗容积的 2/3。等溶液全部流完后，再向滤纸中间部位转移沉淀，不要把沉淀转移到滤纸的边缘上，否则沉淀易渗漏到滤液中，且使取下滤纸和沉淀的操作难以进行。继续减压抽滤到不再有滤液滴下为止。

过滤完毕后，先拔掉连接吸滤瓶的橡皮管，后关水龙头。用镊子轻轻揭起滤纸边，或取下布氏漏斗倒扣在表面皿上，轻轻拍打漏斗，以取下滤纸和沉淀。倒出滤液时，吸滤瓶的支管应朝上，支管只作连接减压装置用，不是滤液出口，应避免滤液由此流出。

# ES－100 电子天平使用方法

## 1. ES－100 电子天平的外观及说明

ES－100 电子天平的外观示意图及操作键简介如图 2-3 所示。

1. 秤盘
2. 显示器窗口
3. 关/模式
4. 开/去皮
5. 校正锁定开关
6. AC 适配器开关

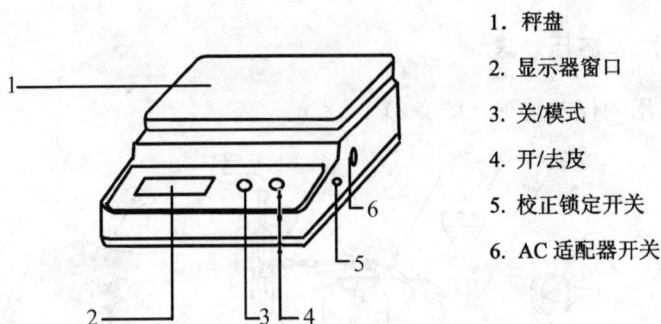

图 2-3　ES－100 电子天平

### 2. 操作方法

（1）将 AC 适配器的圆型插头插入天平右边的插孔内，AC 适配器插入 220 V 的两孔交流插座，接通电源。

（2）在秤盘上没有物品的情况下，按一次"开/去皮"键，天平显示"0"后，将干净且干燥的表面皿放在秤盘上，天平显示出质量后再按一次"开/去皮"，天平重新显示"0"后，将需要称取的试剂从试剂瓶中取出放在表面皿上，天平显示的质量即为所称试剂的净质量。记录所称质量。

（3）取下表面皿，再按一下"开/去皮"，让天平回到"0"。

（4）按住"关/模式"不动，显示器出现"OFF"，然后松开，天平关闭。

### 3. 天平维护

为使天平能正常工作，必须保持机壳和秤盘的干净，应避免异体物质的侵蚀。必要时可以用一块蘸取温和清洗剂的布进行清洗。不用时拔去 AC 插头。

## 由废铜屑制备硫酸铜实验

### 1. 基本原理

纯铜属不活泼金属，不能溶于非氧化性的酸中，但其氧化物在稀酸中却能溶解。因此在工业上制备胆矾（硫酸铜）时，先把铜烧成氧化铜，然后与适当浓度的硫酸反应而生成硫酸铜。本实验则采用浓硝酸作氧化剂，由废铜屑与硫酸、浓硝酸反应来制备硫酸铜。反应式为：

$$Cu+2HNO_3+H_2SO_4 =\!=\!= CuSO_4+2NO_2+2H_2O$$

产物中除硫酸铜外，还含有一定量的硝酸铜和其他一些可溶性或不溶性杂质。不溶性杂质可过滤除去，而硝酸铜则利用它和硫酸铜在水中溶解度的不同，通过结晶方法将其除去（留在母液中）。

表 2-1　硫酸铜和硝酸铜在水中的溶解度（克/100 克水）

| | 0℃ | 20℃ | 40℃ | 60℃ | 80℃ |
|---|---|---|---|---|---|
| $CuSO_4 \cdot 5H_2O$ | 23.3 | 32.3 | 46.2 | 61.1 | 83.8 |
| $Cu(NO_3)_2 \cdot 6H_2O$ | 81.8 | 125.1 | | | |
| $Cu(NO_3)_2 \cdot 3H_2O$ | | | ~160 | ~178.5 | ~208 |

由表 2-1 中的数据可知，硝酸铜在水中的溶解度不论在高温或低温下都比硫酸铜大得多，在本实验所得的产物中它的量又很小，因此，当热的溶液冷却到一定温度时，硫酸铜首先达到过饱和而硝酸铜却远远没有达到饱和。随着温度的继续下降，硫酸铜不断从溶液中析出，硝酸铜则绝大部分仍留在溶液中。有少量伴随硫酸铜结晶出来的硝酸铜可以和其他一些可溶性杂质一起，通过重结晶的方法除去，最后制得纯硫酸铜。

### 2. 仪器和药品

蒸发皿，烧杯(100 mL)，布氏漏斗，吸滤瓶，量筒(100 mL,10 mL)，$H_2SO_4$(3 mol·L$^{-1}$)，浓 $HNO_3$，废铜屑。

### 3. 实验步骤

(1) 称取 2.0 克铜屑，将它置于干燥的蒸发皿中，用酒精灯强烈灼烧至不再产生白烟为止（目的是除去附着在铜屑上的油污）。放冷。

(2) 往上述盛有铜屑的蒸发皿中加入 10 mL 3 mol·L$^{-1}$ $H_2SO_4$，然后缓慢分次加入 4 毫升浓硝酸（反应过程产生大量有毒的二氧化氮气体，操作应在通风橱内进行）待反应缓和后，盖上表面皿，放在水浴上加热。加热过程需要补加适量的 3 mol·L$^{-1}$ $H_2SO_4$ 和浓硝酸（由于反应情况不同，补加的酸量要根据具体反应情况而定，在保持反应继续顺利进行的情况下，尽量少加硝酸）。待铜屑全部溶解，趁热用倾析法将溶液转至一个小烧杯中。如果仍有一些不溶性残渣。可用少量 3 mol·L$^{-1}$ $H_2SO_4$ 洗涤后弃去，洗涤液合并于小烧杯中。将小烧杯放在热水浴中加热蒸发浓缩至有晶膜出现为止。取出，自然冷却至室温。用减压抽滤法分离出晶体。称量后计算产率。

(3) 将粗产品以每克加 1.2 毫升水的比例，溶于蒸馏水中。加热使其全部溶解，并趁热过滤。滤液收集在一个小烧杯中，让其慢慢冷却，即有晶体析出（如无晶体析出，可在水浴上加热蒸发，稍微浓缩）。冷却后减压过滤除去母液，取出晶体并称量，计算收率。母液回收。

## 思考题

1. 在天平上称量时必须注意哪几点？
2. 什么情况下可用倾析法，什么情况下用常压过滤或者减压过滤？
3. 在减压过滤操作中如果未开真空泵之前先把沉淀转入布氏漏斗内会产生何种影响？

# Exp 2  Preparation of Copper Sulfate from Waste Bronze Pieces

## Objectives

1. To exercise the use of balance and the basic operation of evaporation, vacuum filtration as well as recrystallization.

2. To know how to prepare salt from its related metal and understand the principle of purification by recrystallization.

## Use Technology of Electronic Thermostatic Water Bath

Electronic thermostatic water bath is used to evaporate and heat with constant temperature. It is a common constant temperature electroheat instrument, common ordinance including 2, 4, 6, 8 holes.

Electronic thermostatic water bath makes up of electro heat thermostatic water bath and electric appliance box. Fig. 2-1 shows: left side of waterbath is water thermostat-controlled waterbath, which is with heat preservation horse interbedded one. The shelf of bottom of waterbath equipped with electro heat tube and sensing temperature tube, it is applied to heat and sense temperature of water. The face of water bath is cover plate with concentric circles and thermometer hub. On the right of waterbath is electric appliance box. The face plate is equipped with active indicator (red light indicates heating, green light indicates constant temperature), turning button of temperature set and switch of power.

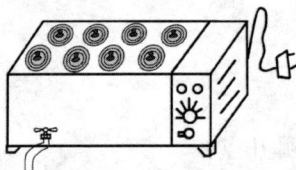

Fig. 2-1  Digital displaying electronic thermostatic water bath

For usage of water bath, water is added to the bath to a proper depth firstly, insert the power plug and turn on the power switch, as the red light is on, which indicates the heating element is working. Press down the temperautre adjusting switch, rotate the 'temperature' turn knob to the control value, then press up the temperature adjusting switch, the displayed temperature is just the water temperature. When the water temperature rises to the control temperature, green light is on, which indicates the thermostatic control unit works.

We must use electronic thermostatic water-bath carefully. Water must be added before power is on, water level must not be below pipe heater. Electrical box must be kept dry in order not to make damage of electrical leakage. Salts, acids or base solutions must not be sprinkled in thermostatic water-bath. As if they are sprinkled with imprudence, power will be cut-off immediately.  You should wash it in order to avoid being corroded.  If water-bath will not in use for long time, water in it should be dumpaged, wipe it with clean cloth.  If leakage occurs, it should be repaired immediately.

# Separation Method of Solid from Liquid （Ⅰ）

## 1.  Decantation

When the density of precipitate or crystal is heavier or with large particle and the solid can precipitate after standing still, such solid can be separated and washed by decantation. When the precipitate is arriving at the bottom of the container, pour the above layer into another container by glass rod carefully.  If the precipitate is needed to be washed, a small quantity of washing solution can be added to the precipitate with, stir thoroughly, stand still and precipitate, the washing solution can be decanted.  Repeat the operation more than 3 times to get the clean solid.

## 2.  Filter （I）: vacuum filtration

(1) Instrument of vacuum filtration (Fig. 2-2)

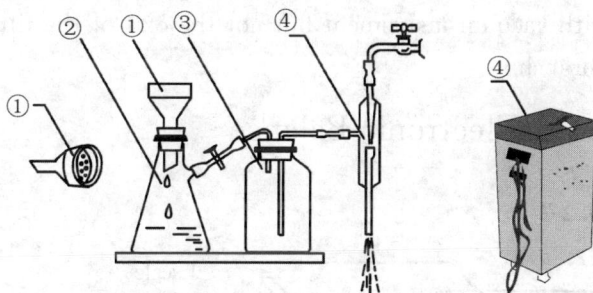

Fig. 2-2  Vacuum filtration instrument

① büncher funnel: flat bottom funnel made from porcelain, many holes on the bottom. The neck of the funnel below is surrounded by rubber stopple for linkage with suction bottle.

② suction bottle: for accepting filtrate, it has a branch for linkage with suction system.

③ safety flask: when the operation of vacuum filtration is over and the water pump is turned off, for the pressure in suction bottle is lower than atmospheric pressure, reverse suction will happen, the water will enter the suction bottle, so a safety flask should be equipped between the suction bottle and water pump, which has the buffer function.  When filtering is over, the rubber tube linking with suction bottle should be unplugged and the water pump be turned off finally.

④ water jet pump (abbreviated for water pump): there is a gradually contracted sprayer, water is sprayed at high velocity, so lower pressure is formed here.  The gas in the system is absorbed and pumped with water, thus lower pressure is formed.  Water jet pump is usually replaced by vacuum pump.  Water pump or vacuum pump at work absorb the air in the suction bottle consecutively, so the pressure in the suction bottle is degraded, pressure difference is produced, so the velocity is elevated.

(2) Operation of vacuum filtration

The inner diameter of the filter paper should be smaller than that of the funnel and can cover all the porcelain holes. Filter paper is placed in the filter, the paper firstly moistened by distilled water, let the sinking face to the branch of the bottle, avoid the filtrate to be sucked the side-mouth of the bottle; turn on the water tap or vacuum pump, after the filter paper is sticked to the funnel and ensure the vacuum is turning on, pouring the solution along the glass rod, pay attention to the amount of the solution less than 2/3 volume of the funnel. After all the solution going to the suction bottle, transfer the precipitate to the center of the filter paper but the edge of it, otherwise precipitate may leak into the filtrate, at the same time, the operation of taking out of the filter paper and precipitate becomes difficult. Go on vacuum filtering until no filtrate dropping.

After filtration, firstly unplug the rubber tube linking with the suction bottle, then turn off the vacuum pump. Unload the edge of the filter paper or put the funnel on the watch glass, tap the funnel slightly to take out of the filter paper and precipitate. When the precipitate is transferred, the branch of the suction bottle should be upward. As the branch only acts as linkage with vacuum instrument but not the exit of the filtrate, filtrate should not run out from the branch.

## Usage of ES-100 Electronic Balance

### 1. Structure (Fig. 2-3)

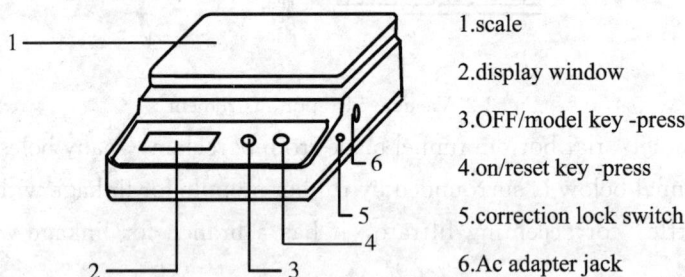

1. scale
2. display window
3. OFF/model key -press
4. on/reset key -press
5. correction lock switch
6. Ac adapter jack

Fig. 2-3 ES-100 electronic balance

### 2. Operation procedure

(1) Insert the round plug into the hole on the right of the balance, and adapter into the two-hole alternating current socket for linking power.

(2) When there is nothing on the pan, press the button 4, after balance displays, place the clean and dry watch glass on the pan, press the button 4 once more, and the balance displays the weight. When balance displaying '0', take out of the reagent from the bottle and place it on the watch glass, the weight that the balance displaying is just the weight of the reagent. Record the weight.

(3) Unload the watch glass, press button 4 once to make the balance display '0'.

(4) If the button 3 is pressed for several seconds, monitor will display 'OFF', the balance is power off.

### 3.　Maintenance of the balance

To keep the balance work well, the shell and the pan of the balance should be kept clean and erosion be avoided. If necessary, a piece of cloth moistened by mild abluent can be used to clean the balance. If the balance is not in use, the plug should not be inserted.

## Preparation of Copper Sulfate from Bronze Pieces

### 1.　Principles

Pure copper, which belongs to inactive metal, can not be dissolved in nonoxidizable acid. But its oxide could be dissolved in dilute acid. Therefore, while preparing $CuSO_4$ in industry, the copper is combusted into copper oxide first, and then react with sulfuric acid of proper concentration to produce $CuSO_4$. Nitric acid is selected as oxidizing agent, and copper is used to react with sulfuric acid and nitric acid to prepare copper sulfate. The chemical equation is as below:

$$Cu + 2HNO_3 + H_2SO_4 = CuSO_4 + 2NO_2 + 2H_2O$$

Besides copper sulfate, there is also small quantity of copper nitrate as well as some other soluble or insoluble impurities in product. The insoluble ones can be removed by filtration, and copper nitrate can be removed by recrystallization for its different solubility with copper sulfate in water (It will remain in water phase).

Table 2-1　Solubility of copper sulfate and copper nitrate in water (g /100 g water)

|  | 0℃ | 20℃ | 40℃ | 60℃ | 80℃ |
|---|---|---|---|---|---|
| $CuSO_4 \cdot 5H_2O$ | 23.3 | 32.3 | 46.2 | 61.1 | 83.8 |
| $Cu(NO_3)_2 \cdot 6H_2O$ | 81.8 | 125.1 |  |  |  |
| $Cu(NO_3)_2 \cdot 3H_2O$ |  |  | ~160 | ~178.5 | ~208 |

According to Table 2-1, solubility of copper nitrate in water, whether at high or low temperature, is much higher than copper sulfate. And in this experiment, the amount of it is smaller in the product. Therefore, when the hot solution is cooled to a certain temperature, copper sulfate will be over saturated, at the same time copper nitrate has not achieved saturation. While the temperature continues getting down, crystal of copper sulfate continues occurring and the copper nitrate can remain in the solution with majority of other soluble impurities. Minority of copper nitrate that gets out with other impurities can be removed by recrystallization.

### 2.　Instruments and Reagents

evaporating dish, beaker (100 mL), büchner funnel, suction bottle, graduated cylinder (100 mL,10 mL), $H_2SO_4$(3 mol·$L^{-1}$), concentrated $HNO_3$, copper.

### 3. Experimental Procedures

(1) 2.0 g copper is weighed and then placed in dry evaporating dish. And it is combusted vigorously in dry porcelain crucible by alcohol burner till no white smoke comes out (The aim is to remove the grease stain on the surface of copper), cool down.

(2) 10 mL 3 mol·$L^{-1}$ $H_2SO_4$ are added to the copper in the evaporating dish, and then 4 mL concentrated nitric acid are added. (operation should be done in the fume hood , for much $NO_2$ will be produced in this reaction). After the reaction is steady, cover the dish, and heat it in water bath. In the operation process of heating, proper amount of 3 mol·$L^{-1}$ $H_2SO_4$ and concentrated nitric acid are extra needed to add dropwise(due to the different situation among reaction, the quantity of acid needed is determined by the situation of reaction. When the copper reacts completely, nitric acid should be added as less as possible). After the copper being dissolved completely(within about 1 hour), hot solution is poured into a small beaker or directly transferred into another evaporating dish). If there are also some insoluble residues, it is possible to wash it with little 3 mol·$L^{-1}$ $H_2SO_4$ and abandon it, then the washed solution is combined with the previous solution in the small beaker. Heat the small beaker in the hot water bath until the crystal coating appears, take out of it and cool it to r. t. Separate the crystal by vacuum filtration, weigh and calculate the yield.

(3) Dissolve the crude product in distilled water in the portion of 1.2 mL water per 1 g product. Heat to make it dissolve completely, filter it while it is hot. The filtrate is collected in a small beaker, and cools down naturally and then crystal occurs (if there is no crystal, evaporate it in the water bath slightly). After it is cooled, remove the mother liquid by filtration, then weigh after the crystal is dry, calculate the yield. Recycle the mother liquid.

## Think and Answer the Following Questions

1. What should be noticed on weighing by balance?

2. In which situation the operation of decantation could be used and in which situation filtration or vacuum filtration could be selected?

3. What will happen if we transfer the precipitate to the büchner funnel before turning on the water pump?

# 实验三　硫酸亚铁铵的制备

## 实验目的

1. 了解复盐硫酸亚铁铵的制备原理。
2. 练习水浴加热、过滤（常压、减压）、蒸发、浓缩、结晶和干燥技术。
3. 学习用目测比色检验产品质量的技术。
4. 巩固电热恒温水浴锅的使用技术。

## 目视比色法技术简介

用眼睛观察比较溶液颜色深浅来确定物质含量的分析方法称为目视比色法。其原理是将标准溶液和被测溶液在同样条件下进行比较，当溶液液层厚度相同，颜色的深度一样时，两者的浓度相等。

应用目视比色法时先要配制系列标准溶液。以 $Fe^{3+}$ 含量测定为例，在测定之前，先要配制一组含 $Fe^{3+}$ 量不同的标准色阶。做法是：先确定显色剂（如选用 KSCN 等）和测量条件。准备好一组同样的比色管，然后在比色管中依次加入一系列不同量的 $Fe^{3+}$ 标准溶液，再分别加入等量的显色剂及其他辅助试剂（如 HCl），最后用蒸馏水稀释到同样体积，即配成一套颜色逐渐加深的标准色阶。

比色测定时取与配标准溶液色阶质地相同的比色管一支，称取一定量被测样品置于比色管中，加适量的蒸馏水溶解，然后按配标准溶液色阶相同的条件进行显色并稀释到相同的体积。

比色操作中，可在比色管下衬以白瓷板，然后从管口垂直向下观察，将被测样品管逐支与标准色阶对比，寻找确定颜色深浅程度相同者。若被测样品的颜色是介于相邻的两色阶之间，则两色阶含量的平均值就为样品中该物质含量的测定值。

## 固体的干燥技术

制备无机盐时，过滤所得的晶体总会含有一定的水分，需要干燥处理。常用的干燥方法有烘干、晾干和用吸水性物质吸干等。

对热较稳定的固体可放在表面皿上，在电热干燥箱中烘干，也可放在蒸发皿中，用水浴、沙浴或酒精灯加热烘干。

晶粒较大或受热易分解的固体，可用滤纸轻压吸去水分或置于表面皿中晾干；也可用有机溶剂如乙醇等洗涤后晾干，借助有机溶剂的挥发将吸附在晶体表面的水分带走。

有些易吸水潮解或需要长时间保持干燥的固体，可放在干燥器内保存。

## 固、液分离方法（Ⅱ）

**1. 过滤法（Ⅱ）：常压过滤**

（1）滤纸折叠（图 3-1）

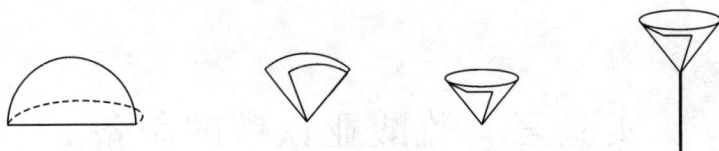

图 3-1　滤纸的折叠

　　折叠滤纸前将手洗净，把滤纸对折两次（为保证滤纸与漏斗密合、对折时不要折死），展开成圆钳形，放入漏斗（漏斗内壁应干净而且干燥）。如果滤纸与漏斗不十分密合，可以稍微改变滤纸的折叠角度，直到贴紧漏斗后把折边折死。滤纸锥体的一个半边为三层，另一个半边为一层。为了使滤纸和漏斗内壁贴紧且无气泡，常在三层滤纸的外层滤纸折角处撕下一小块，此小块滤纸保存在干净的表面皿上，用于擦拭烧杯内残留的沉淀。

　　折好的滤纸放入漏斗中，滤纸边缘应低于漏斗边缘 0.5～1 cm，三层的一边应放在漏斗出口短的一边，食指将滤纸按紧，用洗瓶吹入少量蒸馏水润湿滤纸，然后轻压滤纸边缘，赶去气泡，使滤纸的锥体与漏斗紧密贴合。用洗瓶加水至滤纸边缘，当漏斗中的水全部流尽后，漏斗颈内水柱应仍能保留并无气泡。由于液柱的重力可起到抽滤的作用，可以加快过滤速度。若不能形成水柱，可用手指堵住漏斗下口，稍掀起滤纸三层的一边，用洗瓶向滤纸和漏斗的空隙里加水，直到漏斗颈和锥体的大部分充满水，然后压紧滤纸边，慢慢松开堵住下口的手指，可形成水柱。

　　（2）沉淀的过滤、转移和洗涤

　　将准备好的漏斗放在漏斗架上，盖上表面皿，下面放一个承接滤液的洁净烧杯，斜盖上一块表面皿。漏斗颈口（长的一边）紧贴杯壁，使滤液能沿烧杯内壁流下。调节好漏斗架的高低，以漏斗颈下口不接触滤液为度。

　　为了避免沉淀堵塞滤纸上的空隙而影响过滤速度，待烧杯中的沉淀下沉以后，先将清液沿着玻璃棒引流入漏斗中。玻璃棒的下端对着三层滤纸处，并尽可能接近滤纸，但不要接触滤纸。一次倾入的溶液液面应低于滤纸边缘 1 cm，以免少量沉淀因毛细作用越过滤纸上缘而造成损失，且不便于洗涤。

　　暂停倾注溶液时，应将烧杯尖口沿玻璃棒向上提起，直到烧杯直立，以免使烧杯口上的液滴流失。

　　倾析完成后，把沉淀与剩余的少量溶液搅起混匀，将其一起通过玻璃棒转移至漏斗上。在烧杯中加入少量洗涤液，搅拌混匀再转移到漏斗内。如此重复几次后绝大部分沉淀转移到了滤纸上。残留的少量沉淀，按图 3-2 所示的沉淀的转移方法可将其全部转移干净。

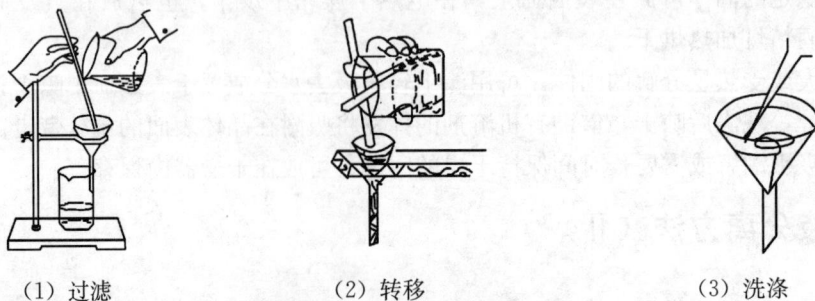

　　（1）过滤　　　　　　　（2）转移　　　　　　　（3）洗涤

图 3-2　沉淀的过滤、转移和洗涤

为了除去表面吸附的杂质和残留的母液，沉淀全部转移到滤纸上以后需要在滤纸上洗涤沉淀。洗涤时，应从滤纸的多重边缘稍下部位开始，按螺旋状向下移动。这样可使沉淀集中到滤纸锥体的下部。为提高洗涤效率，每次使用少量洗涤液，便于尽快沥干。沥干后，再次洗涤。如此反复多次，直至沉淀洗净为止。

洗涤剂的选择，应根据沉淀的性质而定。

**2. 过滤法（Ⅲ）：热过滤**

需要除去热、浓溶液中的不溶性杂质而又不能让溶质析出时，一般采用热过滤。

（1）减压热过滤：过滤前把布氏漏斗、吸滤瓶都放入烘箱（或在水浴）中加热，抽滤前用同一热溶剂润湿滤纸。抽滤方法与减压过滤相同。

（2）少量热溶液的过滤，可选一颈短而粗的玻璃漏斗放在烘箱中预热后放一折叠滤纸（折叠方法见图 3-3），用少量热溶剂润湿滤纸，以免干燥的滤纸吸附溶剂使溶液浓缩而析出晶体。然后迅速倒入溶液，用表面皿盖好漏斗，减少溶剂的挥发（图 3-4（1））。

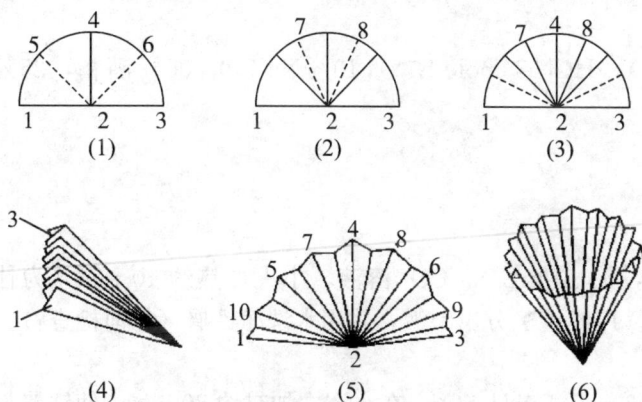

图 3-3　热过滤滤纸折叠方法　　　　　　　　图 3-4　热滤装置

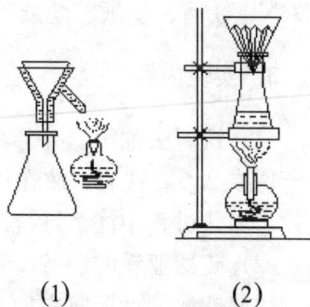

（3）过滤的溶液量较多，选择铜质的热滤漏斗。热滤漏斗是一种减少散热的夹套式漏斗，其夹套是金属套内安装一个玻璃漏斗而形成的。使用时将热水（通常是沸水）倒入夹套内，加热侧管。漏斗中放入折叠滤纸，用少量热溶剂润湿滤纸，再把热溶液分批倒入漏斗，未倒的溶液和保温漏斗用小火加热，保持微沸。热过滤时不用玻璃棒引流，以免加速降温。接受滤液的容器内壁不要贴紧漏斗颈，以免滤液迅速冷却而析出晶体（图 3-4（2））。

## 硫酸亚铁铵的制备实验

### 1. 实验原理

硫酸亚铁铵 $(NH_4)_2SO_4 \cdot FeSO_4 \cdot 6H_2O$ 又称摩尔盐，它是透明、浅蓝绿色单斜晶体，比一般的亚铁盐稳定，在空气中不易被氧化。

硫酸亚铁铵可由等物质的量的 $FeSO_4$ 和 $(NH_4)_2SO_4$ 反应制得，其反应如下：

$$FeSO_4 + (NH_4)_2SO_4 + 6H_2O = (NH_4)_2SO_4 \cdot FeSO_4 \cdot 6H_2O$$

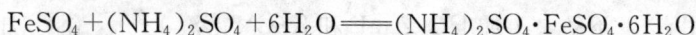

由于复盐在水中的溶解度比组成它的每一个组分的溶解度都要小，因此，只需要将 FeSO_4 与 $(NH_4)_2SO_4$ 的溶液混合后经浓缩结晶可以制得硫酸亚铁铵晶体。

本实验利用铁屑溶于稀 $H_2SO_4$，先制得 $FeSO_4$ 溶液。然后再在 $FeSO_4$ 溶液中加入 $(NH_4)_2SO_4$，使其全部溶解后，经浓缩、冷却、即得溶解度较小的硫酸亚铁铵晶体。

由于硫酸亚铁在中性溶液中能被溶于水中的少量氧气所氧化并进一步发生水解,甚至析出棕黄色的碱式硫酸铁(或氢氧化铁)沉淀,所以制备过程中溶液应保持足够的酸度。

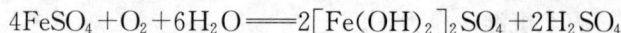

$$4FeSO_4 + O_2 + 6H_2O = 2[Fe(OH)_2]_2SO_4 + 2H_2SO_4$$

产品中 $Fe^{3+}$ 的含量可用比色法来测定。$Fe^{3+}$ 能与 $SCN^-$ 生成血红色的 $[Fe(SCN)_6]^{3-}$ 等。产品溶液加入 $SCN^-$ 后显较深的红色,则表明产品中含 $Fe^{3+}$ 较多;反之则表明产品含 $Fe^{3+}$ 较少。因而可将所制备的硫酸亚铁铵与 KSCN 在比色管中配成待测溶液。将它所呈现的红色与 $[Fe(SCN)_6]^{3-}$ 标准溶液色阶进行比较,找出与其红色深浅程度一致的那支标准溶液,则该支标准溶液所示 $Fe^{3+}$ 含量即为产品的杂质 $Fe^{3+}$ 含量。依此可确定出产品的等级(每克一、二、三级硫酸亚铁铵含 $Fe^{3+}$ 限量分别为 0.05 mg、0.10 mg 和 0.20 mg)。

### 2. 实验仪器、药品与材料

天平、锥形瓶、表面皿、玻棒、量筒、减压抽滤装置、电热恒温水浴锅、目视比色管 (25 mL)、蒸发皿

$(NH_4)_2SO_4$,$H_2SO_4$(3 mol·L$^{-1}$),HCl(3 mol·L$^{-1}$),10% $Na_2CO_3$,95%酒精,25% KSCN、铁屑、滤纸、滤纸碎片

### 3. 实验内容

(1) 铁屑去油污

称取 2.0 g 铁屑,放在锥形瓶中,加 10 mL $Na_2CO_3$ 溶液,小火加热约 10 min(为什么?),加热过程中注意维持溶液体积。用倾析法分出碱液,并用水洗净铁屑(如何检查?)。

(2) 硫酸亚铁的制备

往盛铁屑的锥形瓶中加入 15 mL 3 mol·L$^{-1}$ $H_2SO_4$,在水浴中加热约 30 min,使铁屑与 $H_2SO_4$ 反应。此反应过程应在通风橱或通风处进行。(为什么?)在加热过程中,应适当添加少量水,以补充失水。观察不再有气泡冒出时,加入 3 mol·L$^{-1}$ $H_2SO_4$ 1mL,(为什么?)趁热过滤。滤液转移至 100 mL 烧杯中,用少量热水洗涤锥形瓶及漏斗上的残渣。将残渣取出,并收集在一起,用滤纸碎片吸干后称量,算出已反应的铁屑的质量和理论上溶液中 $FeSO_4$ 的含量。

(3) 硫酸亚铁铵的制备

根据溶液中 $FeSO_4$ 的含量,按 $FeSO_4$:$(NH_4)_2SO_4$ = 1:1(以 mol 计),称取分析纯 $(NH_4)_2SO_4$ 加到 $FeSO_4$ 溶液中,然后在水浴上加热搅拌,使$(NH_4)_2SO_4$ 全部溶解。继续加热蒸发,浓缩直至溶液表面刚出现晶膜为止。静置让溶液自然冷却至室温,即有硫酸亚铁铵晶体析出。减压过滤后,用少量酒精洗涤晶体两次。(为什么?)将晶体取出;摊在干净表面皿上晾至晶体不再粘附玻璃棒为止。称量,(外观如何?)计算理论产量与产率。

(4) 产品检验

$Fe^{3+}$ 的限量分析:称取 1.0 g 样品置于 25 mL 比色管中,用 15 mL 不含氧蒸馏水(如何制取,为何要用它?)溶解,再加 HCl 2 mL 和 KSCN 溶液 1 mL,继续加入蒸馏水稀释至 25 mL 刻度,摇匀。与标准溶液进行目视比色,确定产品的等级。

$Fe^{3+}$ 标准溶液配制(实验室制备):先配制浓度为 0.01 mg·mL$^{-1}$ 的 $Fe^{3+}$ 标准溶液,然后用移液管取 $Fe^{3+}$ 标准溶液 5.0 mL 于比色管中,加 HCl 2 mL 和 KSCN 溶液 1 mL、再加入蒸馏水将溶液稀释到 25.0 mL,摇匀。这是一级试剂标准液(其中含 $Fe^{3+}$ 0.05 mg)。再

分别取 10.0 mL 和 20.0 mL $Fe^{3+}$ 标准溶液于比色管中，用同样的方法可配得二级和三级试剂的标准液，其中含 $Fe^{3+}$ 分别为 0.10 mg 和 0.20 mg。

## 思考题

1. 什么叫复盐？复盐与形成它的简单盐相比，有什么特点？

2. 硫酸亚铁铵的制备原理是什么？如何提高其产率与质量？

3. 在蒸发、浓缩过程中，若发现溶液变为黄色，是什么原因？应如何处理？

4. 硫酸亚铁铵的产率如何计算？计算时是以硫酸亚铁的量为准，还是以硫酸铵的量为准？为什么？

5. 目视比色法确定物质含量的要点是什么？

6. 固体的干燥方法有几种？硫酸亚铁能否用直接加热的方法干燥？

7. 试分析个别学生的产量会超出理论产量的原因。怎样防止此现象发生？

# Exp 3　Preparation of Ammonium Ferrous Sulfate

## Objectives

1.　To understand the preparation principle of $(NH_4)_2SO_4 \cdot FeSO_4 \cdot 6H_2O$.

2.　To practice heating in water bath, filter (normal pressure, reduced pressure), e-vaporation, concentration, recrystallization and drying.

3.　To learn the technology of product quality analysis by visual colorimetry.

4.　To learn how to use electroheat thermostatic waterbath.

## Introduction of Visual Colorimetry

The analytical method that compares the color shade of solution with eyes and determines the content of material is named visual colorimetry. Its principle is: comparing the standard solution with the test one under the same condition, if the thickness and the depth of color of the two solutions are the same, their concentration should be the same.

A series of standard solution should be prepared before visual colorimetry. Take analysis of $Fe^{3+}$ content as an example. Before determination, a series of standard solutions with different $Fe^{3+}$ concentration are needed to be prepared. The usual procedure is to determine color developing reagent (for example, KSCN) and measuring conditions. A series of color comparison tubes of the same size are prepared, and standard solutions with different $Fe^{3+}$ concentration are added, then the same amount of color developing reagent and other adju-vanticity reagent are added (for example, HCl). Finally distilled water is used to dilute the solution to the same volume. In this way, a series of standard color degree with gradual deepening color are gradually prepared.

As we do color selection experiment, we also need to prepare test solution with the same method. Take a color comparison tube with the same color as standard solution, weigh some test sample in the color comparison tube, dissolve it with distilled water, then dilute it to the same volume following the same condition as above method.

In color comparison operation, a white porcelain base plate is placed under the color comparison tube. Observe the solution down from canal orifice vertically, compare the test tube with every standard color degree, then look for and decide the tube with the same color shade extent. If the color is between two color degrees, the average concentration is the de-termined concentration.

## Dry Technology of Solid

Crystal from filtering in preparation of inorganic salt always contains some water. And it needs to be dried. Common dry method includes oven dry, open air drying and drying with hydroscopic material, etc.

Solid stable for heat can be placed on watch-glass, and then dried in electronic cabinet drier. It also can be placed on evaporating dish, then dried with water-bath, sand bath or alcohol burner.

Crystal with comparatively large size or readily to decompose can be dried with filter paper to absorb water or placed on watch glass to dry through open-air drying (As operation in the preparation of potassium nitrate). Or we can wash it with organic solvent then leave it in open-air drying as the evaporation of organic solvent can remove the water on surface.

Some solid readily to absorb water or deliquescence may be preserved in desiccator.

# Separation Method of Solid from Liquid ( II )

## 1. Filter ( II ): filter under normal pressure

(1) Folding filter paper (Fig. 3-1)

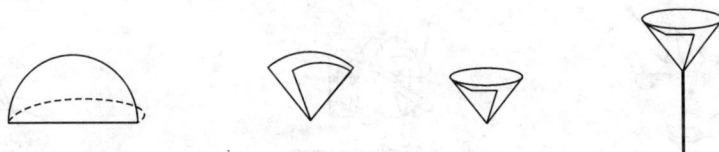

Fig. 3-1   Folding filter paper

Wash your hands before fold the filter paper, firstly, fold the paper along the center line twice (to ensure the filter paper close to the funnel, folding shouldn't be tight). Spread the paper, put it into the funnel (the wall should be clean and dry). If the paper is not close to the funnel, change the folding angle slightly until they are close to each other. One half of the pyramid of the filter paper is three-layer, another is one-layer. In order to keep the paper close to the funnel and no bubble, tear scrap from the outside layer at the threee-layer for cleaning the remaining precipitate in the beaker.

Put the filter paper in the funnel, the edge of the filter paper should be lower 0.5~1 cm than the edge of the funnel, press the filter paper tightly with forefinger, moisten the paper by water spraying from washing bottle, then press the edge of the paper slightly, drive off the bubble to keep the paper pyramid close to the the funnel. Add water to the funnel edge, when all of the water is running out, the water in the neck of the funnel should be remained and no bubble. Vacuum filtration may act for the gravity of the water column in the neck, and the velocity of filtering may be elevated. If the water column can not be formed, one finger can be used to block the below mouth of the funnel, lift the three-layer paper slightly, add water to the airspace between the paper and funnel until most of the space being filled with water, press the edge of paper tightly, loosen the forefinger, water column may be formed.

(2) Filter, transfer and wash

Place the funnel on the funnel holder, cover it by watch glass, a clean beaker is prepared for filtrate, a watch glass is leaned on it. The mouth of the funnel is close to the wall of the beaker to make filtrate run along the inner wall of beaker. Adjust the height of the holder to avoid the filtrate contacting with the mouth.

To avoid the precipitate blocking the holes of the paper and slow down the filtration ve-locity, add the limpid solution to the funnel along the glass rod when the precipitate de-scending. The inferior extremity of the glass rod faces to the three-layer and be close to the paper but contacting. The solution should lower 1 cm than the edge of the paper in case of precipitate exceeding the paper, if the precipitate is very near to the edge, washing the pre-cipitate is inconvient.

When pouring is paused, the sharp mouth of the beaker should be raised along the glass rod until the beaker is standing straight to avoid the solution drop running off.

After decantation, mix the precipitate and small quantity of solution, and transfer the mixture to the funnel. add a little washing solution, mix and transfer to the funnel. Repeat such operation several times until most of the precipitate is transferred to the filter paper. Remaining precipitate can be transferred thoroughly according to Fig. 3-2.

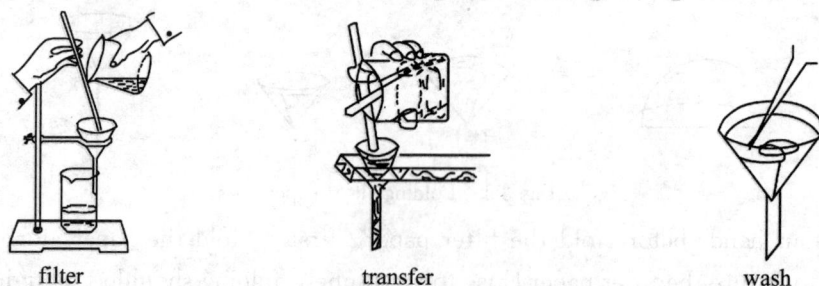

filter　　　　　　　　　　transfer　　　　　　　　　　wash

Fig. 3-2　Filter, transfer and wash

In order to remove the impurity and remaining mother liquid, precipitate should be washed after all the precipitate being transferred. Wash the precipitate below the edge of multi-layer, move downward helically. Thus, precipitate is concentrated on the bottom of the paper pyramid. To enhance the efficiency, small quantity of washing solution should be employed to guarantee the precipitate dry rapidly. Repeat above operation till the precipitate is washed thoroughly.

The property of the precipitate determines how to choose the washing solution.

## 2. Filtration (Ⅲ): hot filtering

When undissolvable impurity in hot and concentrated solution need to be removed and guarantee the solute not separate from the solution, hot filtering is employed.

(1) hot filtering in vacuum: heat the büncher funnel and suction bottle in the oven (or water bath); moisten the filter paper with same solvent, the method is same as that of vac-uum filtration.

(2) when hot solution is little, a short and thick glass funnel can be chosen to preheat in the oven . a folded filter paper may be employed (how to fold, see Fig. 3-3), very little hot solvent is used to moisten the filter paper to avoid solvent being absorbed by dry filter paper and crystal appearing. Pour the solution rapidly, cover the funnel by watch glass to reduce the evaporation of solvent[Fig. 3-4(a)].

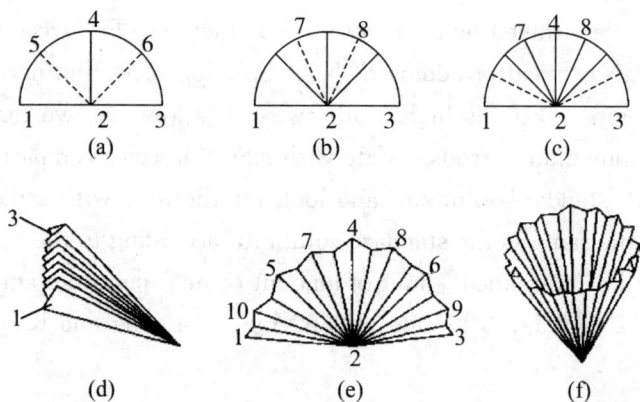

Fig. 3-3    folding procedure of filter paper for hot filtering          Fig. 3-4    instrument of hot filtering

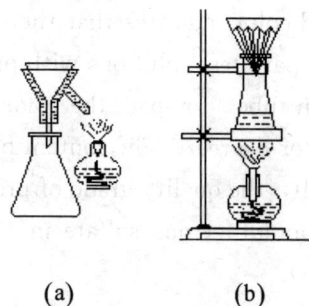

(3) If the solution has large volume, a copper hot filtering funnel may be chosen. Hot filtering funnel is a kind of jacket one which reduces heat loss, its jacket is a glass funnel fitting in the metal funnel. Pour the hot water (usually boiling water) into the jacket, and heat side-branch when in use. Folded paper was put in the funnel, small quantity hot solvent is used to moisten the filter paper, then pour the hot solution in portions, remaining solution and incubation funnel are heated by small fire to keep it boiling slightly. A glass rod is unnecessary to avoid temperature descending rapidly. The inner wall of container for accepting filtrate should be too close to the funnel neck to avoid crystal appearing for filtrate cooling rapidly [Fig. 3-4(b)].

## Preparation of Ammonium Ferrous Sulfate

### 1.  Principles

Ammonium ferrous sulfate is also named molar salt $(NH_4)_2SO_4 \cdot FeSO_4 \cdot 6H_2O$, which is a kind of transparent bluish-green monoclinic crystal, more stable than common ferrous salt, and not to be oxidized in air.

Ammonium ferrous sulfate is prepared with equivalent mole $FeSO_4$ and $(NH_4)_2SO_4$, reaction equation is as follow:

$$FeSO_4 + (NH_4)_2SO_4 + 6H_2O = (NH_4)_2SO_4 \cdot FeSO_4 \cdot 6H_2O$$

Because solubility of double salt is smaller than its component, so if concentrated $(NH_4)_2SO_4$ solution and concentrated $FeSO_4$ solution are mixed, crystal of ammonium ferrous sulfate can be obtained after concentration and crystallization.

In this experiment, scrap iron is dissolved in dilute $H_2SO_4$. Thus $FeSO_4$ solution is prepared. Then $(NH_4)_2SO_4$ is added to $FeSO_4$ solution. When it is dissolved completely, crystal of ammonium ferrous sulfate with small solubility forms after concentration and cooling.

Because ferrous sulfate can be oxidized by a small quantity of oxygen in water in neural medium and hydrolyze further, evenly buffy ferric subsulfate precipitate forms (or ferric hydroxide). So enough acidity should be kept in preparation procedures.

$$4FeSO_4 + O_2 + 6H_2O = 2[Fe(OH)_2]_2SO_4 + 2H_2SO_4$$

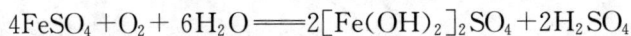

Content of $Fe^{3+}$ in product can be determined by color comparison method. $Fe^{3+}$ reacts with $SCN^-$ to form ensanguine $[Fe(SCN)_6]^{3-}$ after adding $SCN^-$. Solution appearing dark red color indicates that the concentration of $Fe^{3+}$ is high. Otherwise it is low. So we can prepare test solutions with prepared ammonium ferrous sulfate with KSCN in color comparison tube, compare the color of it with standard solution, and look for the tube with same color degree. The content of $Fe^{3+}$ is the same as the standard solution, according to the result, the quality grade of product can be determined (limit of amount of $Fe^{3+}$ per gram ammonium ferrous sulfate in $1^{st}$ grade, $2^{nd}$ grade, $3^{rd}$ grade is 0.05 mg, 0.10 mg and 0.20 mg).

## 2. Instruments, Reagents and Materials

electronic balance, conical flask, watch glass, glass rod, cylinder, instrument of reduced pressure filtering, electroheat thermostatic water-bath, visualing color comparison tubes (25 mL), evaporating dish.

$(NH_4)_2SO_4$, $H_2SO_4$ (3 mol·L$^{-1}$), HCl (3 mol·L$^{-1}$), 10% $Na_2CO_3$, 95% ethanol, 25% KSCN, scrap iron, filter paper, fragment of filter paper.

## 3. Experimental Procedures

(1) Wash the grease stain on scrap iron

Weigh 2.0 g scrap iron in conical flask, and 10 mL of $Na_2CO_3$ solution is added, then heat it with small fire for 10 min (why?), decant the alkalinous solution, lastly wash the scrap iron clean with water (how to detect it?).

(2) Preparation of ferrous sulfate

Add 15 mL $H_2SO_4$ to a conical flask with scrap iron, heat it in water-bath for about 30 min to make scrap iron react with it completely. The operation must be done in hood. (why?). When it is heating, small quantity of water must be added to replenish evaporated water. When no bubble emerges, another 1 mL $H_2SO_4$ is added, filter while it is hot. Transfer the filtrate to evaporating dish. And wash the residue on conical flask and funnel with small quantity of hot water. Take out of the residue, gather it together and weigh after using fragment of filter paper to absorb the residual water. In this way, the quality of reacting iron and the amount of $FeSO_4$ in solution can be calculated.

(3) Preparation of ammonium ferrous sulfate

According to the amount of $FeSO_4$ in solution, add $(NH_4)_2SO_4$ (A.R) to $FeSO_4$ solution following the ratio: $FeSO_4:(NH_4)_2SO_4 = 1:1$ (calculating by mol), heat and stir it in water bath to make all reagents dissolve completely. Continue to heat until solid appears on the surface of solution. Stand still, cool the solution to room temperature. Crystal of ammonium ferrous sulfate appears, filter under reduced pressure, and wash the crystal with small quantity of alcohol. (why?) Take out of the crystal, place it between two sheets of clean paper (or filter paper), squeeze it slightly to absorb mother liquid until crystal is no longer sticky to glass rod. Weigh, （what is its appearance?） calculate theoretical weight and yield.

(4) Analysis of product

Analysis dose limited of $Fe^{3+}$: weigh 1 g sample in a color comparison tube, dissolve it with 15 mL distilled water without oxygen, （how to prepare it and why must it be used?） another 2 mL of HCl and 1 mL of KSCN solution are added, continue to add distilled water to dilute it to the scale of 25 mL, shake it. Compare color with standard solution by visual colorimetry and determine the quality grade of product.

$Fe^{3+}$ standard solution may be prepared by the following method （it also can be prepared by teacher in lab）, first prepare standard solution with $0.01 \ mg \cdot mL^{-1} \ Fe^{3+}$ solution, then add 5 mL of the standard solution in color comparison tube, add 2 mL of HCl and 1mL of KSCN by pipet, lastly distilled water is used to dilute the mixture to 25 mL, shake it. This is the $1^{st}$ grade standard solution （0.05 mg of $Fe^{3+}$ in it）. Then add 10 mL and 20 mL $Fe^{3+}$ standard solution in color comparison tubes, following the same method to prepare $2^{nd}$ grade and $3^{rd}$ grade solution （in each tube, 0.10 mg and 0.20 mg $Fe^{3+}$）.

## Questions

1. What is double salt? Comparing it with simple salt forming it, what is the characteristic?

2. What is the principle of preparation of ammonium ferrous sulfate? How to elevate its yield and quality?

3. If the solution turns yellow in evaporation and concentration, what's the reason? How to deal with it?

4. How to calculate the yield of ammonium ferrous sulfate? Can we calculate it according to the amount of ferrous sulfate or ammonium sulfate? Why?

5. What are the main points in determining the concentration of substance by visual colorimetry?

6. How many methods are there in drying solid? Whether can ferrous sulfate be directly heated when drying?

7. Try to analyze the reason that the yield of product may exceed theoretical yield by some students. How to avoid it?

# 实验四　摩尔气体常数的测定

## 实验目的

1. 了解一种测定摩尔气体常数 $R$ 的原理与方法。
2. 熟练分压定律与气体状态方程式的计算。
3. 巩固电子天平和气压计的使用技术，学习气体体积的测量技术。

## 实验原理

气体状态方程可表示为：

$$PV=nRT=\frac{m}{M_{(G)}}RT \tag{1}$$

式中，$P$ 为气体的压强或分压（Pa）；$V$ 为气体的体积（m³）；$n$ 为气体的物质的量（mol）；$R$ 为摩尔气体常数（$Pa \cdot m^3 \cdot K^{-1} \cdot mol^{-1}$）；$T$ 为气体的温度（K）；$m$ 为气体的质量（g）；$M_{(G)}$ 为气体的摩尔质量（$g \cdot mol^{-1}$）。

从式（1）可知，只要测出一定温度下给定气体的体积 $V$、压强 $P$、物质的量 $n$ 或质量 $m$ 便可求得 $R$ 的数值。

本实验将已知质量的单质铝 Al 与过量的稀硫酸反应，在一定压强和温度下，用排水集气法收集反应所生成的氢气，测得其体积，便可求得氢气的物质的量，从而得到 $R$ 值。有关反应及计算公式如下：

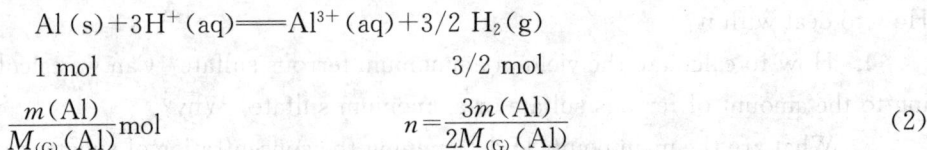

$$Al(s)+3H^+(aq)==Al^{3+}(aq)+3/2\ H_2(g)$$

$$1\ mol \qquad\qquad\qquad\qquad 3/2\ mol$$

$$\frac{m(Al)}{M_{(G)}(Al)}mol \qquad\qquad n=\frac{3m(Al)}{2M_{(G)}(Al)} \tag{2}$$

在实验条件下所收集的氢气是被水蒸气所饱和的。根据道尔顿分压定律，氢气的分压 $P_{H_2}$ 应是所收集的混合气体（即氢气与水蒸气的混合物）总压 $P$ 与水蒸气分压 $P_{H_2O}$ 之差：

$$P_{H_2}=P-P_{H_2O} \tag{3}$$

将（2）式与（3）式代入（1）式便得：

$$R=\frac{P_{H_2}V}{n_{H_2}T}=\frac{2M_{(G)}(Al)(P-P_{H_2O})V}{3m(Al)T}=\frac{17.987(P-P_{H_2O})}{m(Al)T} \tag{4}$$

式中，17.987 为 $2M_{(G)}(Al)/3$ 的值，$M_{(G)}(Al)$ 为 Al 的原子的摩尔质量。

（1）式中 $P$ 值取大气压值。实验中用量气管量气时要做到管内的气体与外界气体 $P_{实}=P_{大气}$。$P_{H_2O}$ 可由手册查取一定温度下的水的饱和蒸汽压值得到；$V$ 为量气管所收集到的 $H_2$ 的体积，由于 Al 与 $H^+$ 的反应为一放热反应，而气体的体积又与温度有关，故 $V$ 值的读取一定要等量气管冷却到室温；$m(Al)$ 为自己所称取的铝片的质量，称量时除了要刮净铝箔表面的氧化膜外，还要保证称准；$T$ 为温度计读数，可用实验室的室温代替。

由（4）式，$R$ 值的测定实际上通过测定 $P$，$V$，$m(Al)$ 和 $T$ 值来实现的，测准它们是做好本实验的关键。

## 实验仪器、药品与材料

分析天平、量气管（带水准瓶）、气压计、铁架台（带蝴蝶夹及铁圈等）、三角漏斗、剪刀、量筒（10 mL）、具单孔塞试管、铝箔、$H_2SO_4$（2 mol·$L^{-1}$）、砂纸、输血胶管、甘油。

## 实验内容

1. 用分析天平准确称取 2 至 3 份已擦去表面氧化膜的铝箔条，每份质量在 0.0240～0.030 g。

2. 按图 4-1 装配好仪器（本处用 50 mL 碱式滴定管代替量筒，用漏斗代替水准瓶）。打开试管的塞子，由漏斗往量气管内注水至略低于刻度"0.00"的位置，上下移动漏斗以赶尽附着在胶管和量气管内壁的气泡，然后把试管的塞子塞紧。

3. 检查装置是否漏气，把漏斗下移一段距离，并用铁圈固定在一定位置上，如果量气管中的液面只在开始时稍有下降，3～5 min 后即维持恒定，说明装置不漏气，如果液面继续下降，则表明装置漏气这时就要检查各接口处是否严密。经检查与调整后，再重复实验，直至不漏气为止，再把漏斗移至原来位置。

4. 取下试管，使量气管内液面保持在刻度"0.00"。然后用另一长颈三角漏斗将 5 mL $H_2SO_4$ 注入试管中（切勿使酸碰到试管壁上），稍倾斜试管、将铝箔用甘油蘸一下，贴在试管壁上部（图 4-2）确保铝箔不与稀 $H_2SO_4$ 接触。然后将试管的塞子塞紧，再检查一次装置是否漏气。

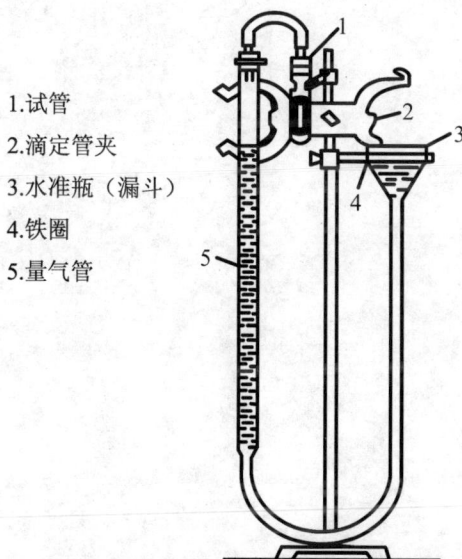

1.试管
2.滴定管夹
3.水准瓶（漏斗）
4.铁圈
5.量气管

1.铝箔
2.稀硫酸

图 4-1 摩尔气体常数的测定装置          图 4-2 铝箔贴在试管壁上半部

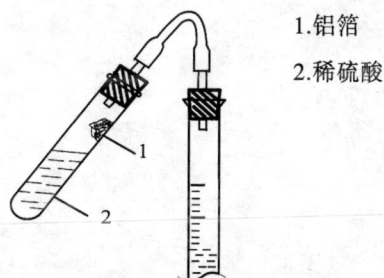

5. 把漏斗移至量气管的右侧，使两者的液面保持同一水平，记下量气管中的液面读数。然后把试管底部略微抬高（切勿使试管塞子松开），使铝箔与稀 $H_2SO_4$ 接触。这时由于反应产生的氢气进入量气管中，把管中的水压入漏斗内。为避免管内压力过大而造成漏气，在管内液面下降的同时，漏斗也应相应地向下移动，使管内液面和漏斗中液面大体上保持同一水平面。

6. 读数 铝箔反应完毕后，要等反应试管冷却至室温（约需 10 min），然后再移动漏斗,使漏斗与量气管中水的液面处于同一水平面,记下液面读数,稍等 2~3 min,再记录液面读数,如两次读数相等,说明管内气体温度已与室温一样。

记下室内温度和当时的大气压力。

用另一份铝箔重复实验一次。

查出该室温的饱和水蒸气压,计算摩尔气体常数 $R$ 及实验误差。

## 思考题

1. 为什么说测准 $P$,$V$,$m(\text{Al})$ 和 $T$ 值是做好本实验的关键?实验过程中应如何做才能测准它们?

2. 试分析下列情况对实验结果有何影响?

(1) 量气管及胶管内气泡未赶尽;

(2) 铝箔表面氧化膜未擦净,铝箔表面被手指上的汗垢所沾污但又未擦除就称量;

(3) 在开始收集氢气前不小心让贴在试管壁上的铝箔与稀硫酸有了接触;

(4) 反应过程中实验装置漏气了。

# Exp 4   Determination of Molar Gas Constant

## Objectives

1. To know the fundamental principles and method of determination of molar gas constant.

2. To be familiar with the calculation of Dalton's partial pressure law and ideal gas law.

3. To gain further insights into the technique of analytical balance and barometer; to learn measure technique of gas volume.

## Principles

Ideal gas law equation can be expressed as follow:

$$PV = nRT = \frac{m}{M_{(G)}}RT \tag{1}$$

Where $P$ is the pressure or partial pressure of gas (Pa); $V$ is the volume of gas ($m^3$); $n$ is the molar of gas (mol); $R$ is molar gas constant ($Pa \cdot m^3 \cdot K^{-1} \cdot mol^{-1}$); $T$ is the temperature of gas (K); $m$ is the mass of gas (g); $M_{(G)}$ is the molar mass of gas ($g \cdot mol^{-1}$).

According to equation (1), after measuring volume $V$, pressure $P$, molar $n$ or mass $m$ of the given gas at the given temperature, value of $R$ can be got.

Alumium of the given mass reacts with excessive dilute sulfuric acid at the given temperature and pressure in this experiment, hydrogen is collected by drain collection method, measure its volume, then the molar of hydrogen can be got. According to this, the value of $R$ can be got. The related reaction and calculation formula are as follows:

$$Al(s) + 3H^+(aq) = Al^{3+}(aq) + 3/2 \, H_2(g)$$

$$1 \text{ mol} \qquad\qquad\qquad\qquad 3/2 \text{ mol}$$

$$\frac{m(Al)}{M_{(G)}(Al)}\text{mol} \qquad\qquad n = \frac{3m(Al)}{2M_{(G)}(Al)} \tag{2}$$

Hydrogen collected is saturated by vapor. Partial pressure of hydrogen $P_{H_2}$ can be got by all pressure $P$ of mixed gas (the mixtures of hydrogen and vapor) subtracting the vapor pressure of water.

$$P_{H_2} = P - P_{H_2O} \tag{3}$$

(1) can be expressed as follow according to equation (2) and (3):

$$R = \frac{P_{H_2}V}{n_{H_2}T} = \frac{2M_{(G)}(Al)(P - P_{H_2O})V}{3m(Al)T} = \frac{17.987(P - P_{H_2O})}{m(Al)T} \tag{4}$$

Where $2M_{(G)}(Al)/3$ is 17.987, $M_{(G)}(Al)$ is atomic molar mass of Al.

P in equation (1) is atmosphere press. Pressure in tube shall be consistent with atmosphere press when measuring pressure using windpipe. $P_{H_2O}$ at certain temperature can be got in handbook. $V$ is the volume of hydrogen collected. It is noted that the value of $V$ shall be read when windpipe tube is cooled down to room temperature because the volume of the hydrogen is influenced by temperature and the reaction of Al and $H^+$ is an exothermic one. The mass of Al, which shall be weighed after oxidation film of Al surface scraped. The value of Al mass shall be proved to be accurate. $T$ is the value of thermometer, which may be replaced by room temperature of laboratory.

We can know from the equation (4), the determination of $R$ is actually through by determination of $P$, $V$, $m$ (Al) and $T$. So accurate measurement of them is the key of this experiment.

## Instruments, Reagents and Materials

analytical balance, eudiometer (with level ball), barometer, hob stand (attached butterfly nip and iron circle), triangle filter, scissors, graduate cylinder (10 mL), test tube with single bore stuff , aluminum foil, $H_2SO_4$ (2 mol·$L^{-1}$), sand paper, latex, glycerol.

## Experimental Procedures

1. Weigh 2~3 shares of aluminum foil with surface oxidation film scraped, whose mass varies at 0. 0240~0. 030 g.

2. Assemble the apparatus as Fig. 4-1 (graduated flask replaced by alkali buret, level bottle is replaced by filter). Open the plug of test tube, affusion from filter to eudiometer till to slightly under "0. 00" scale, move the filter to drive up air bladder attaching in glue tube and windpipe wall, and then fill plug of test tube strictly.

3. Check if instrument leaks air by moving filter down some distance and fix it up using iron ring at certain position. If liquid surface of eudiometer slightly falls in the beginning, then maintain invariableness after 3~5 min, which shows the instrument does not leak air. Otherwise you should check if the interface is rigor. Repeat the above trial to prevent leaking air after checking and adjusting. At last, move filter to original position.

4. Taking down the test tube to maintain liquid surface at "0. 00" scale. Then add 5 mL $H_2SO_4$ to the test tube using another filter (note: no touching the test tube wall), slightly incline the test tube, dipping aluminum foil in glycerol, paste it at the top of the test tube wall to prevent aluminum foil touching $H_2SO_4$ (see Fig 4-2). Then check whether air is leaked after filling plug strictly in the test tube.

5. Moving the filter to the right of eudiometer so that both of liquid surface are at the same level, note down the data of the liquid surface in eudiometer. Then drive up the bottom of test tube slightly (not to unclamp the plug of the test tube) to make aluminum foil touch $H_2SO_4$. Water in the test tube will be pressed into filter due to hydrogen produced of the entering eudiometer. To prevent leaking for excessive pressure, funnel shall move under correspond to liquid surface fall in the tube in order to make liquid surface of tube and funnel kept at the same level.

6. Reading data

After aluminum foil fully reacts, wait for the test tube cool down to room temperature (about 10 min), then move funnel so that liquid surface of the tube and filter are kept in the same level, record the data of liquid surface. Repeat reading data after 2~3 min. If these data are equal, it proves the temperature of gas in the tube to be consistent with the room temperature.

Note down room temperature and atmospheric pressure.

Repeat the experiment using another share of aluminum foil.

Look up the vapor pressure of water at this temperature; calculate molar gas constant $R$ and experiment errors.

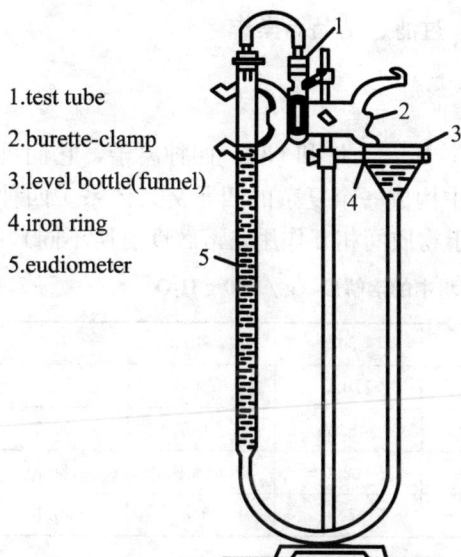

1.test tube
2.burette-clamp
3.level bottle(funnel)
4.iron ring
5.eudiometer

1.Aluminum foil
2.dilute $H_2SO_4$

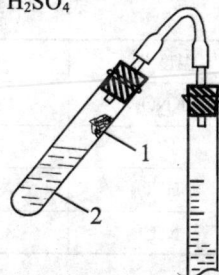

Fig. 4-1    Measuring instrument of molar
gas constant.

Fig. 4-2    Aluminum foil posted
at the top of the test tube wall

## Questions

1. Why the exact values of $P$, $V$, $m(Al)$ is the key to fulfill experiment successfully? How to measure them accurately in the experiment?

2. Try to analyze how those following factors will affect the experimental results.

(1) Air bladder doesn't be drove up completely in eudiometer and glue tube.

(2) Surface oxidation film of aluminum foil is not scraped completely; aluminum foil is weighted with surface sweat stained by the finger not being scraped completely.

(3) Aluminum foil touches $H_2SO_4$ at the top of the test tube wall before collecting hydrogen.

(4) Instrument leaks air in the process.

# 实验五　硝酸钾的制备、提纯及其溶解度的测定

## 实验目的

1. 学习利用各种易溶盐在不同温度时溶解度的差异来制备易溶盐的原理和方法。
2. 巩固重结晶法提纯物质的方法。
3. 学习测定易溶盐溶解度的方法；巩固溶解、过滤、结晶等操作。

## 实验原理

在 KCl 和 $NaNO_3$ 的混合溶液中同时存在 $Na^+$、$K^+$、$Cl^-$ 和 $NO_3^-$ 四种离子，它们可组成 $KNO_3$、KCl、$NaNO_3$ 和 NaCl 四种盐，在溶液中构成一个复杂的四元交叉体系。四种纯净盐在水中的溶解度列于表 5-1（在混合溶液中由于物质间相互作用，溶解度有所不同）。

表 5-1　纯 $KNO_3$、KCl、$NaNO_3$、NaCl 在水中的溶解度（g / 100 g $H_2O$）

| 温度/℃ | 0 | 20 | 40 | 60 | 80 | 100 |
| --- | --- | --- | --- | --- | --- | --- |
| $KNO_3$ | 13.3 | 31.6 | 63.9 | 110.0 | 169.0 | 246.0 |
| KCl | 27.6 | 34.0 | 40.0 | 45.5 | 51.5 | 56.7 |
| $NaNO_3$ | 73.0 | 87.6 | 102.0 | 122.0 | 148.0 | 180.0 |
| NaCl | 35.7 | 36.0 | 36.6 | 37.3 | 38.4 | 39.8 |

从表 5-1 的数据可以看出，在 20℃时，除 $NaNO_3$ 外其他三种盐的溶解度相差不大，因此不易使 $KNO_3$ 单独结晶出来。但是随着温度升高，NaCl 的溶解度几乎没有多大改变，而 $KNO_3$ 的溶解度却增大得很快。因此只要把 $NaNO_3$ 和 KCl 混合溶液加热蒸发，在较高温度下，NaCl 由于溶解度较小而首先析出，趁热把它滤去，然后将滤液冷却，利用 $KNO_3$ 的溶解度随温度下降而急剧下降的性质，使 $KNO_3$ 晶体析出。

易溶盐溶解度的测定方法主要有两种：分析法和定组成法。分析法的原理是用分析化学的手段，测定在一定温度下饱和溶液中易溶盐组分的含量，从而计算出该温度下该盐的溶解度。此法结果精确，但操作比较复杂、费时。定组成法的原理是观察已知含量易溶盐开始析出晶体的温度，从而计算出该温度下该盐的溶解度。根据定组成法，可用同一装置和一定量的溶质，通过连续补充定量水可以测得自高温至低温的一系列数据。

## 仪器、药品及材料

### 1. 仪器

烧杯（100，250 mL）、量筒（10 mL）、吸量管（1 mL，2 mL）、抽滤瓶（30 mL）、布氏漏斗（20 mm）、表面皿、温度计、电子天平、分析天平。

### 2. 药品和材料

$NaNO_3$（s），KCl（s），$AgNO_3$（0.1 $mol \cdot L^{-1}$），$HNO_3$（5 $mol \cdot L^{-1}$），滤纸。

## 实验步骤

### 1. 制备

在天平上称取固体 $NaNO_3$ 10.0 g，固体 KCl 8.8 g，放入 100 mL 烧杯中，加入 16.0 mL 蒸馏水，加热溶解。继续小火加热并不断搅拌，使溶液蒸发至原体积的 2/3，这时有晶体析出（是什么晶体？）。趁热用减压过滤（过滤前预先将布氏漏斗在蒸汽上或烘箱中预热）滤液转入 50 mL 洁净的小烧杯中（操作要迅速，若滤液在抽滤瓶中析出结晶，可水浴加热溶解之），自然冷却，随着温度下降，有结晶析出（是什么结晶？）。注意不要骤冷，以免结晶过于细小。用减压过滤分离母液，所得 $KNO_3$ 晶体置于表面皿上蒸汽浴烘干后称重，计算 $KNO_3$ 粗产品的产率。

### 2. 提纯

将粗产品（先称取 0.1 g 备纯度检验用）放入 50 mL 烧杯中，加入计算量的蒸馏水（按 $KNO_3$ 在 100℃时的溶解度计算）并搅拌之，用小火加热至沸，使结晶全部溶解，然后自然冷却至室温，待大量晶体析出后减压过滤，晶体置于表面皿中用蒸汽浴烘干后称重，计算提纯率。

### 3. 产品纯度的检验

定性检验，将提纯前取出的 0.1 g 粗产品和重结晶后得到的产品分别放入两支小试管中，各加入 2 mL 蒸馏水配成溶液。在溶液中分别滴入 1 滴 5 mol·$L^{-1}$ 硝酸酸化，再各滴入 0.1 mol·$L^{-1}$ 硝酸银溶液 2 滴，观察现象，进行对比。重结晶后的产品溶液应为澄清。

### 4. 溶解度测定

准确称取 $KNO_3$ 提纯品 1.00 g 放入一支干燥洁净的小试管中，用 1 mL 吸量管加入蒸馏水 1.00 mL，装上带有 100℃温度计的单孔软木塞，使温度计水银泡的位置处于液面下。将小试管置于水浴加热，使试管内的液面略低于水浴的液面，轻轻摇动试管内溶液至晶体全部溶解（注意不要使溶液长时间加热，以免管内水分蒸发）。将试管自水浴中取出，自然冷却并轻轻地水平摇动试管，在黑色背景下观察开始析出晶体时的温度。再次在水浴中加热溶解和冷却直到析出晶体的温度重复为止。

用吸量管向小试管中加入蒸馏水 0.50 mL，如前操作，测定另一饱和溶液的温度。再加水，每次 0.50 mL，如此反复直到已观察不到晶体析出为止。记下取得的溶液中开始析出晶体的温度数据。按表 5-2 整理，计算溶解度并作数据处理：

（1）以实验测得的 $KNO_3$ 的溶解度为纵坐标，温度为横坐标，绘制溶解度-温度曲线Ⅰ。
（2）根据表 5-2 $KNO_3$ 溶解度的文献值，绘制标准溶解度-温度曲线Ⅱ。
（3）比较曲线Ⅰ、Ⅱ，分析产生误差的原因。

表 5-2　纯 $KNO_3$ 在水中的溶解度 （g/100 g $H_2O$）

| 温度/℃ | 30 | 40 | 50 | 60 | 70 |
|--------|------|------|------|-------|-------|
| 溶解度 | 45.8 | 63.9 | 85.5 | 111.0 | 138.0 |

## 思考题

1. 用 KCl 和 NaNO$_3$ 来制备 KNO$_3$ 的原理是什么？

2. 根据溶解度数据，计算在本实验中应有多少 NaCl 和 KNO$_3$ 晶体析出？（不考虑其他盐存在时对溶解度的影响）。

3. 如所用 KCl 或 NaNO$_3$ 的量超过化学计算量，结果怎样？

4. KNO$_3$ 中混有 KCl 或 NaNO$_3$ 时，应如何提纯？

5. 本实验中为何要趁热过滤除去 NaCl 晶体？为何要小火加热？

6. 测定 KNO$_3$ 溶解度时，水的蒸发对本实验有何影响？应采取什么措施？溶解和结晶过程为什么要摇动？

# Exp 5   Preparation and Purification of KNO$_3$, Determination of Its Solubility

## Objectives

1. To learn the principle and method of drawing the different solubility of various soluble salts at different temperatures to prepare soluble salt.

2. To learn how to purify reagent by recrystallization.

3. To learn how to determine the solubility of soluble salt.

4. To further practice the operation of dissolving, filtering and recrystallization.

## Principles

$Na^+$, $K^+$, $Cl^-$ and $NO_3^-$ coexist in the mixture of KCl and $NaNO_3$, they can make up four kinds of salts: $KNO_3$, $KCl$, $NaNO_3$ and $NaCl$, which construct a complex four-element intersection system.

Solubility of four pure salts in water are listed in the following Table 5-1 (because interaction of substance in mixture, solubility will be somewhat different).

**Table 5-1   Solubility of pure KNO$_3$, KCl, NaNO$_3$, NaCl in H$_2$O   (g /100 g H$_2$O)**

| $T/℃$ | 0 | 20 | 40 | 60 | 80 | 100 |
|---|---|---|---|---|---|---|
| KNO$_3$ | 13. 3 | 31. 6 | 63. 9 | 110. 0 | 169. 0 | 246. 0 |
| KCl | 27. 6 | 34. 0 | 40. 0 | 45. 5 | 51. 5 | 56. 7 |
| NaNO$_3$ | 73. 0 | 87. 6 | 102. 0 | 122. 0 | 148. 0 | 180. 0 |
| NaCl | 35. 7 | 36. 0 | 36. 6 | 37. 3 | 38. 4 | 39. 8 |

From the table, we can see: four salts except $NaNO_3$ have similar solubilities at 20℃, so it is difficult for $KNO_3$ to crystallize from the solution alone. But with the increasing of temperature, solubility of NaCl is almost stable, but solubility of $KNO_3$ increases much. So if the mixture of $NaNO_3$ and KCl is evaporating, NaCl will crystallize first for its smaller solubility, filter while it is hot, cool down the filtrate, $KNO_3$ will be separated for its solubility descending with temperature descending.

Assay methods of soluble salts mainly include: analytic process and determining composition process. The principle of analytic process is determining the concentration of a constituent in soluble salt to calculate its solubility with analytical chemistry's means. The result is precise with this method but the operation is complex and time-wasting. The principle of determining composition process is to observe the temperature of starting crystallizing and calculate its solubility at this temperature. According to determining composition process, we can determine a series of data by adding quantitative $H_2O$ with the same instrument and fixed quantity of solvent.

## Instruments, Reagents and Materials

### 1. Instruments

beaker(25,250 mL), graduated cylinder(10 mL), pipet(1mL,2 mL), suction bottle (30 mL), büchner funnel(20 mm), watch glass, electronical balance, analytical balance.

### 2. Reagents and materials

$NaNO_3(s)$, $KCl(s)$, $AgNO_3(0.1\ mol \cdot L^{-1})$, $HNO_3(5\ mol \cdot L^{-1})$, filter paper.

## Experimental Procedures

### 1. Preparation

Weigh $NaNO_3(s)$ 10.0 g, $KCl(s)$ 8.8 g by electronical balance in a 50 mL beaker, add 8.0 mL distilled $H_2O$, heat to dissolve, continue to heat with small fire and stir, evaporate the solution until the volume is about 2/3 volume of original volume, at this time, crystal may appear (what is it?), filter under reduced pressure (pre-heating the büchner funnel by steam or in oven), transfer the solution to a 50 mL beaker (operate quickly, otherwise crystal will separate from the filtrate, if so, heat the solution with water bath), cool down naturally, crystal will appear as temperature descending (what is it?) Caution: don't cool down rapidly in order not to form tiny crystal. Separate the crystal under reduced pressure, dry the $KNO_3$ on watch-glass with steam, weigh, calculate the yield of crude $KNO_3$.

### 2. Purification

Add the crude product to a 50 mL beaker (weigh 0.1 g product to test its purity first), add calculated water (referring to solubility of $KNO_3$ at $100\,^{\circ}C$), stir and heat to boil with small fire, after the crystal dissolving, cool down naturally to room temperature, filter under reduced pressure while large quantity of crystal appeared, lay the crystal on watch-glass, dry it with steam, weigh, calculate the yield of purification.

### 3. Test the purity of product (qualitative analysis)

Take 0.1 g crude product and purified product after recrystallization one time in two small test tubes respectively, 2 mL water are added, one drop of 5 $mol \cdot L^{-1}$ nitric acid is added to acidify the solution, another two drops of 0.1 $mol \cdot L^{-1}$ silver nitrate is added, too. Observe the phenomena, compare them. The solution of recrystallized product should be clear.

### 4. Determination of solubility

Weigh recrystallized 1.00 g $KNO_3$ in a dry and clean small tube, add 1.00 mL with a 1 mL pipet, the test tube is fitted with a thermometer with a hole-cork, let the location of mercury bubble is close to the bottom of the tube. Heat the tube in water bath, keep the solution surface a little lower than water bath, shake the solution slightly until all the crystal dissolves (don't heat the solution for long-time in order to avoid evaporation of the water in the tube). Take out of the tube from the water bath, cool down naturally and shake the tube levelly, observe the temperature of crystal appearing. Heat the solution to make crystal be dissolved and cool to crystallize until the temperature that crystal appears repeated.

Add 0.5 mL distilled water to the small tube with pipet, determining the temperature of another solution as above operation process. Add water again, 0.5 mL each time, until no crystal is observed, record the temperatures that the crystal appears in the saturated solution.

Calculate solubility and process data:

(1) Take the solubility of as ordinate, temperature as abscissa, draw solubility-temperature curve I.

(2) According to the literature value of KNO₃ solubility in Table 5-2, draw solubility-temperature curve II.

(3) Compare curve I, II, analyze the reason of error.

**Table 5-2   Solubility of pure KNO₃ in water  (g/100 gH₂O)**

| $T/℃$ | 30 | 40 | 50 | 60 | 70 |
|---|---|---|---|---|---|
| Solubility | 45.8 | 63.9 | 85.5 | 111.0 | 138.0 |

## Questions

1. What is the principle of preparing KNO₃ from KCl and NaNO₃?

2. According to data of solubility, how much NaCl and KNO₃ crystal of will crystallize from the solution in this experiment? (Ignoring the influence with other salts coexisting).

3. If the amount of KCl or NaNO₃ exceeds stoichiometric amount, what will the result be? If KCl or NaNO₃ is mixed with KNO₃, how to purify KNO₃?

4. In this experiment, why we should filter the NaCl crystal while the mixture is hot? Why should we heat it with small fire?

5. In determining the solubility of KNO₃, how does the evaporation of water influence the result? What measures shall we take? Why should we shake in dissolving and crystallizing process?

# 实验六　　氯化钠的提纯

## 实验目的

1. 巩固减压过滤，蒸发浓缩等基本操作。
2. 了解沉淀溶解平衡原理的应用。
3. 学习在分离提纯物质的过程中，定性检验某种物质是否已除去的方法。

## 固、液分离方法（Ⅲ）：离心分离法

### 1. 离心机的使用

（1）打开离心机顶盖，把盛有沉淀和溶液的离心管放入离心机的套管内，与此套管相对位置上的套管内再放入一支同样大小的离心管，管内装与前混合物等体积的水以保持转动平衡。离心机内可以同时放入多支离心管，但应保持对称平衡。

（2）盖好离心机的顶盖，打开离心机开关，将变速旋钮由最慢速挡开始缓慢加速逐步过渡到快速挡。

（3）离心机的转动速率和时间由沉淀的性状而定，晶形沉淀用 1000 r/min 转动 1~2 min；对于非晶形沉淀，转速可提高到 2000 r/min 甚至 3000 r/min，转动 3~4 min。

（4）关机时，要逐档缓慢地把变速旋钮转至关闭。待离心机自然停下后打开顶盖取出离心管。

使用离心机时，发现离心机猛烈震动或发出异常声响，应立即停机检查。

### 2. 吸取清液

吸取离心管内的上层清液时，先在离心管外挤扁毛细滴管的橡皮胶头，排出其中的空气。然后把滴管伸入离心管内直至尖嘴恰好进入上清液液面为止，慢慢地减小对橡皮胶头的挤压力量使溶液慢慢进入滴管。随着离心管中液面的降低，滴管逐渐下移，至全部清液吸出为止。

### 3. 沉淀的洗涤

沉淀和溶液分离后，往离心管中加入适量的蒸馏水或其他洗涤液（大约是沉淀体积的 2~3 倍），用玻璃棒充分搅动，再进行离心分离，用滴管将上层清液吸出（图 6-1）。如此反复操作 2~3 次即可。

## 实验原理

氯化钠试剂或氯碱工业用的食盐，都是以粗盐为原料进行提纯的。粗盐中除了含有泥沙等不溶性杂质外，还含有 $K^+$，$Ca^{2+}$，$Mg^{2+}$ 和 $SO_4^{2-}$ 等可溶性杂质。不溶性杂质可用过滤法除去，可溶性杂质中的 $Ca^{2+}$，$Mg^{2+}$ 和 $SO_4^{2-}$ 则通过加入 $BaCl_2$，$NaOH$ 和 $Na_2CO_3$ 混合溶液，生成难溶的硫酸盐、碳酸盐或碱式碳酸盐而除去，也可加入 $BaCO_3$ 固体和 $NaOH$ 溶液进行如下反应除去：

电动离心机　　　　　　　吸取清液

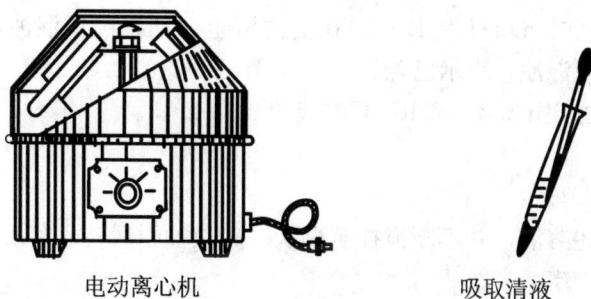

图 6-1　离心机与上清液吸取

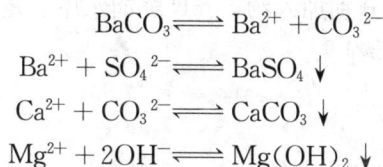

$$BaCO_3 \rightleftharpoons Ba^{2+} + CO_3^{2-}$$
$$Ba^{2+} + SO_4^{2-} \rightleftharpoons BaSO_4 \downarrow$$
$$Ca^{2+} + CO_3^{2-} \rightleftharpoons CaCO_3 \downarrow$$
$$Mg^{2+} + 2OH^- \rightleftharpoons Mg(OH)_2 \downarrow$$

## 仪器和药品

电子天平，温度计。

HCl（$6\ mol \cdot L^{-1}$），NaOH（$2\ mol \cdot L^{-1}$），$BaCl_2$（$1\ mol \cdot L^{-1}$），$(NH_4)_2C_2O_4$（饱和），粗盐，$BaCO_3(s)$，NaOH（$6\ mol \cdot L^{-1}$）和 $Na_2CO_3$（饱和）混合溶液（$50\%_{(v)}$），镁试剂。

## 实验步骤

### 1. 粗盐溶解

称取 10.0 g 粗盐于烧杯中，加入约 40 mL 蒸馏水，加热搅拌使其溶解。

### 2. 除 $Ca^{2+}$，$Mg^{2+}$ 和 $SO_4^{2-}$

（1）$BaCl_2$—NaOH，$Na_2CO_3$ 法

①除 $SO_4^{2-}$：加热溶液至沸，边搅拌边滴加 $1\ mol \cdot L^{-1}$ $BaCl_2$ 溶液至 $SO_4^{2-}$ 除尽为止。加热继续煮沸数分钟。冷却，过滤。

②除 $Ca^{2+}$，$Mg^{2+}$ 和过量的 $Ba^{2+}$：将滤液加热至沸，边搅拌边滴加 NaOH—$Na_2CO_3$ 混合液至溶液 pH 约等于 11，以除去溶液中的 $Ca^{2+}$，$Mg^{2+}$ 和过量的 $Ba^{2+}$。取清液检验 $Ba^{2+}$ 除尽后，继续加热煮沸数分钟。冷却，过滤。

③除剩余的 $CO_3^{2-}$：加热搅拌溶液，滴加 $6\ mol \cdot L^{-1}$ HCl 至溶液的 pH=2～3。

（2）$BaCO_3$—NaOH 法

①除 $Ca^{2+}$ 和 $SO_4^{2-}$：在粗食盐水溶液中，加入约 1.0 g $BaCO_3$（比 $SO_4^{2-}$ 和 $Ca^{2+}$ 的含量约过量 $10\%$(m)），在 363 K 左右搅拌溶液 20～30 min。取清液，用饱和的 $(NH_4)_2C_2O_4$ 检验 $Ca^{2+}$，如尚未除尽，需继续加热搅拌溶液，至除尽为止。

②除 $Mg^{2+}$：用 $6\ mol \cdot L^{-1}$ NaOH 调节上述溶液至 pH 为 11 左右。取清液，分别加入 2～3 滴 $6\ mol \cdot L^{-1}$ NaOH 和镁试剂，证实 $Mg^{2+}$ 除尽后，再加热数分钟，冷却，过滤。

检查 $SO_4^{2-}$，$Ca^{2+}$ 和 $Mg^{2+}$ 等是否除尽，可取少量上层溶液于离心试管中进行离心分离，加入几滴沉淀剂(如1 $mol \cdot L^{-1}$ $BaCl_2$)在上清液上。如果出现混浊，说明未除尽，需再加沉淀剂。如果不出现混浊，表示已除尽。

③除 $CO_3^{2-}$：滴加 6 $mol \cdot L^{-1}$ HCl 至溶液的 pH＝2～3。

### 3. 蒸发、结晶

加热蒸发浓缩上述溶液，并不断搅拌至稠状。趁热抽干后转入蒸发皿内用小火烘干。冷至室温，称重，计算产率。

### 4. 产品质量检验

取粗盐和产品 1 g 左右，分别加入约 5 mL 蒸馏水中。定性检验溶液中是否有 $SO_4^{2-}$，$Ca^{2+}$ 和 $Mg^{2+}$ 的存在，比较实验结果。

## 思考题

1. 能否用重结晶的办法提纯氯化钠？

2. 能否用氯化钙代替毒性大的氯化钡来除去食盐中的 $SO_4^{2-}$？

3. 试用沉淀溶解平衡原理，说明用碳酸钡除去食盐中的 $SO_4^{2-}$ 和 $Ca^{2+}$ 的根据和条件。

4. 在实验中，如果以 $Mg(OH)_2$ 沉淀形式除去粗盐中的 $Mg^{2+}$，则溶液的 pH 应为何值？

5. 在提纯粗盐溶液的过程中，$K^+$ 将在哪一步除去？

# Exp 6    Purification of Sodium Chloride

## Objectives

1. To consolidate the operation of vacuum filtration, evaporation and concentration.
2. To understand the application of precipitate-dissolving principle.
3. To learn the method of qualitative test about whether impurity has been removed.

## Separation Method of Solid and Liquid (Ⅲ): Centrifugal Separation

### 1.  Usage of centrifugal machine

(1) Open the top cover of the centrifugal machine, put the tube containing precipitate and solution into the double-tube, another tube is placed at the opposite position, and the last tube should contain same volume of water to ensure the machine rotate stable. A few tubes can be placed in the machine, but they should be placed symmetrically.

(2) Cover the machine properly, turn on it, rotate the ramp knob slowly from the most slow step to faster one gradually.

(3) The velocity of rotating and time are determined by the property of precipitate, for crystalline precipitate 1000 r/min, 1~2min; noncrystalline precipitate, 2000 r/min even 3000 r/min, 3~4 min.

(4) If the machine is turned off, rotate the ramp knob from the fastest step to the slowest step, when the machine stops naturally, open the top cover and fetch the centrifugal tube. If violent vibration or abnormal noise happens, the machine should never be used, stop using and examine.

### 2.  Absorbing clear solution

When above layer clear solution is absorbed, firstly press the rubber head of the capillary dropper to drive the air in it. Secondly, insert the dropper into the centrifugal tube until the sharp mouth entering the surface of above layer. Loosen the rubber head to make the solution enter the dropper. As the surface of the centrifugal tube descends, dropper should move downward slowly till all the clear solution is absorbed.

### 3.  Washing the precipitate

After the precipitate separates from the solution, add certain volume of distilled water or other washing solution (the volume is about 2~3 times that of the precipitate), stir the solution thoroughly, and repeat the centrifugal separation operation, absorb the above layer of the solution (Fig. 6-1), repeat the above operation 2~3 times.

Electric centrifugal machine　　　absorbing clear solution

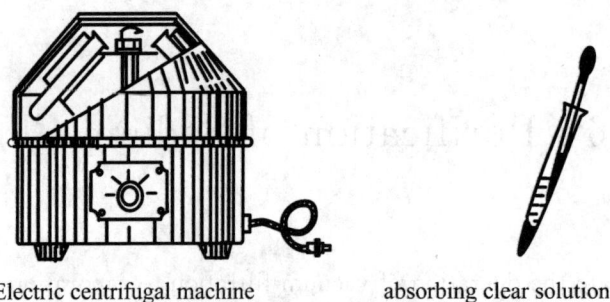

Fig. 6-1　Electric centrifugal machine and absorbing clear solution

## Principles

Sodium chloride, which is used in chlorine-base industry, is purified from crude salt. There is not only some indissoluble impurity in crude salt, such as sediment, but also some soluble impurity, such as $K^+$, $Ca^{2+}$, $Mg^{2+}$ and $SO_4^{2-}$ and so on. Indissoluble impurity can be filtered, soluble impurity such as $Ca^{2+}$, $Mg^{2+}$ and $SO_4^{2-}$ can be removed by adding solution of $BaCl_2$, $NaOH$ and $Na_2CO_3$ to precipitate indissoluble sulfate, carbonate or subcarbonate. $BaCO_3(s)$ and $NaOH(aq)$ can also be used to remove those soluble ions, the reactions are as follows:

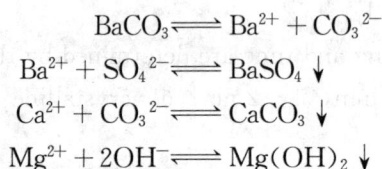

$$BaCO_3 \rightleftharpoons Ba^{2+} + CO_3^{2-}$$
$$Ba^{2+} + SO_4^{2-} \rightleftharpoons BaSO_4 \downarrow$$
$$Ca^{2+} + CO_3^{2-} \rightleftharpoons CaCO_3 \downarrow$$
$$Mg^{2+} + 2OH^- \rightleftharpoons Mg(OH)_2 \downarrow$$

## Instruments and Reagents

electronical balance, thermometer.

$HCl(6\ mol \cdot L^{-1})$, $NaOH$ $(2\ mol \cdot L^{-1})$, $BaCl_2(1mol \cdot L^{-1})$, $(NH_4)_2C_2O_4$ (saturated), crude salt, $BaCO_3(s)$, $NaOH$ $(6\ mol \cdot L^{-1})$ and $Na_2CO_3$ (saturated)$(50\%_{(V)})$, magnesium reagent.

## Experimental Procedures

### 1.　Dissolving of crude salt

Weigh 10 g crude salt in a beaker, add about 40 mL water, heat and stir to make it dissolve.

### 2.　Removing $Ca^{2+}$, $Mg^{2+}$ and $SO_4^{2-}$

(1) $BaCl_2$—$NaOH$, $Na_2CO_3$ method
①Removing $SO_4^{2-}$

Heat the solution to boil, add $1\ mol \cdot L^{-1}$ $BaCl_2$ until $SO_4^{2-}$ is removed completely with stirring. Heat and until it boils for a few minutes, cool, filter.

②Removing $Ca^{2+}$, $Mg^{2+}$ and excess of $Ba^{2+}$

Heat the solution to boil, add 1 mol·$L^{-1}$ NaOH—$Na_2CO_3$ until the pH value is about 11 with stirring. Take a sample to test whether $Ba^{2+}$ is removed completely. Heat until it boils for a few minutes, cool, filter.

③Removing residuary $CO_3^{2-}$

Heat and stir the solution, add 6 mol·$L^{-1}$ HCl until the pH value of the solution is 2~3.

(2) $BaCO_3$—NaOH method

①Removing $Ca^{2+}$ and $SO_4^{2-}$

Add about 1.0 g $BaCO_3$ in crude salt solution (exceed 10%(m) than content of $SO_4^{2-}$ and $Ca^{2+}$), stir 20~30 min at 363 K. Take a sample to test whether $Ca^{2+}$ is removed completely with saturated $(NH_4)_2C_2O_4$, if not, continue to heat and stir the solution.

②Removing $Mg^{2+}$

Adjust the pH value of the solution to about 11 with 6 mol·$L^{-1}$ NaOH. Take a sample to test whether $Mg^{2+}$ is removed completely with 2~3 drops of 6 mol·$L^{-1}$ NaOH and magnesium reagent, if not, continue to heat and stir the solution, filter.

If you want to determine whether $SO_4^{2-}$, $Ca^{2+}$ and $Mg^{2+}$ are removed thoroughly, above layer solution may be fetched to be centrifugalized, add several drops of precipitant to the above layer solution, if the solution becomes cloudy, which indicates ions are not removed thoroughly, more precipitant is needed, if not, the ions have been removed.

③Removing $CO_3^{2-}$

Adjust the pH value of the solution to pH=2~3 with 6 mol·$L^{-1}$ HCl.

### 3. Evaporation, crystallization

Heat and evaporate the above solution, stir continuously to dense, filter while it is warm, transfer it to evaporating dish, dry with small fire. Cool down to room temperature, weigh and calculate the yield.

### 4. Product quality analysis

Take crude salt and purified product each 1 g, add about 5 mL distilled water. Test whether $SO_4^{2-}$, $Ca^{2+}$ and $Mg^{2+}$ exist in solution by qualitative analysis, compare the experiment results.

## Questions

1. Can we use the method of recrystallization to purify sodium chloride?

2. Can we substitute calcium chloride with toxicant barium chloride to remove the $SO_4^{2-}$ in salt?

3. Make an attempt with precipitate-dissolving principle to illuminate the foundation and condition when using barium carbonate to remove $SO_4^{2-}$ and $Ca^{2+}$ in crude salt?

4. If $Mg^{2+}$ is removed in the form of $Mg(OH)_2$, what is the pH value of the solution?

5. In which step is $K^+$ removed in purification of crude salt solution?

# 实验七  化学反应速率、级数与活化能的测定

## 实验目的

1. 加深对化学反应速率、反应级数和活化能等概念的理解。

2. 了解过二硫酸钠氧化碘化钾的反应速率测定的原理；学习通过数据处理及作图求算反应级数和反应的活化能。

## 实验原理

在水溶液中过二硫酸钠和碘化钾反应的方程式为：

$$Na_2S_2O_8 + 3KI \Longrightarrow Na_2SO_4 + K_2SO_4 + KI_3 \text{ （慢）}$$

其离子方程式是：

$$S_2O_8{}^{2-} + 3I^- \Longrightarrow 2SO_4{}^{2-} + I_3^- \tag{1}$$

根据速率方程其反应速率可表示为：

$$v = k[S_2O_8{}^{2-}]^\alpha [I^-]^\beta \tag{2}$$

式中，$v$ 是在此条件下反应的瞬时速率，若 $[S_2O_8{}^{2-}]$ 和 $[I^-]$ 是起始浓度，则 $v$ 表示起始速率。$k$ 是速率常数，$\alpha$ 和 $\beta$ 之和是反应级数。

实验能测定的速率是在一段时间（$\Delta t$）内反应的平均速率 $\bar{v}$。如果在 $\Delta t$ 时间内 $S_2O_8{}^{2-}$ 浓度的改变为 $\Delta[S_2O_8{}^{2-}]$，则平均速率：

$$\bar{v} = \frac{-\Delta[S_2O_8{}^{2-}]}{\Delta t} \tag{3}$$

若控制 $\Delta t$ 很小，可以近似地用平均速率代替起始速率：

$$\lim_{\Delta t \to 0} \bar{v} = v = \frac{-\Delta[S_2O_8{}^{2-}]}{\Delta t} = k[S_2O_8{}^{2-}]^\alpha [I^-]^\beta \tag{4}$$

为了能够测出反应在 $\Delta t$ 时间内 $S_2O_8{}^{2-}$ 浓度的改变值，需要在混合 $Na_2S_2O_8$ 和 KI 溶液的同时，加入一定体积的已知浓度的 $Na_2S_2O_3$ 溶液和作为指示剂的淀粉溶液，这样在反应（1）进行的同时还进行下面的反应：

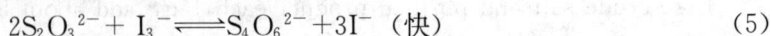

$$2S_2O_3{}^{2-} + I_3^- \Longrightarrow S_4O_6{}^{2-} + 3I^- \text{ （快）} \tag{5}$$

这个反应几乎瞬间完成。而反应（1）比反应（5）慢得多，因此由反应（1）生成的 $I_3^-$ 立即与 $S_2O_3{}^{2-}$ 反应，生成无色的 $S_4O_6{}^{2-}$ 和 $I^-$。所以在反应阶段看不到碘与淀粉反应而显示特有的蓝色。一旦 $Na_2S_2O_3$ 耗尽，反应（1）继续生成的微量 $I_3^-$ 就与淀粉反应而显现出特有的蓝色。

从反应（1）和（5）的关系可以看出，$S_2O_8{}^{2-}$ 减少的物质的量为 $S_2O_3{}^{2-}$ 减少的物质的量的 1/2。由于在 $\Delta t$ 时间内 $S_2O_3{}^{2-}$ 基本上全部耗尽，故有下列关系：

$$\Delta[S_2O_8{}^{2-}] = \frac{1}{2}[S_2O_3{}^{2-}] = \frac{1}{2}(0 - [S_2O_3{}^{2-}]_{始}) = -\frac{1}{2}[S_2O_3{}^{2-}]_{始}$$

记下反应开始到溶液显出蓝色所需的时间 $\Delta t$，就可以求出反应速率。

$$v = -\frac{\Delta[S_2O_8{}^{2-}]}{\Delta t} = \frac{[S_2O_3{}^{2-}]}{2\Delta t} \tag{6}$$

求算 $\alpha$、$\beta$ 值的方法如下：

保持 $[I^-]$ 不变，将式（2）两边取对数，则

$$\lg v = \alpha \lg [S_2O_8{}^{2-}]_{始} + 常数 \tag{7}$$

由实验测定 $Na_2S_2O_8$ 初始浓度不同时的 $v$，以 $\lg v$ 为纵坐标，$\lg[S_2O_8{}^{2-}]$ 为横坐标作图可得一直线，其斜率为 $\alpha$。

同样，保持 $[S_2O_8{}^{2-}]$ 不变，将式（2）两边取对数，则

$$\lg v = \beta \lg [I^-]_{始} + 常数 \tag{8}$$

由实验测定 KI 初始浓度不同时的 $v$，以 $\lg v$ 为纵坐标，$\lg[I^-]$ 为横坐标作图可得一直线，其斜率为 $\beta$。

$\alpha + \beta$ 即为过二硫酸根氧化碘离子反应（1）的反应级数。

已知反应级数，则通过式（2）可求出反应速率常数 $k$。

反应的活化能 $E_a$ 和反应速率常数及温度的关系由阿仑尼乌斯（Arrhenius）公式表示：

$$\lg k = -\frac{E_a}{2.303RT} + B \tag{9}$$

式中，$R$ 为摩尔气体常数，$B$ 是常数项，$E_a$ 的单位是 $J \cdot mol^{-1}$.

由实验测出不同温度下反应（1）的反应速率常数。以 $\lg k$ 为纵坐标，$1/T$ 为横坐标作图，可得一直线，其斜率即为 $-\dfrac{E_a}{2.303R}$。求出此斜率便可算得反应活化能。

## 仪器、药品及材料

### 1. 仪器

烧杯，大试管（规格 mm：$10 \times 100$），试管架，量筒，秒表，温度计，恒温水浴或水浴加热装置。

### 2. 药品

$Na_2S_2O_8$（$0.2 \ mol \cdot L^{-1}$），KI（$0.20 \ mol \cdot L^{-1}$），$KNO_3$（$0.2 \ mol \cdot L^{-1}$），$Na_2SO_4$（$0.20 \ mol \cdot L^{-1}$），$Na_2S_2O_3$（$0.010 \ mol \cdot L^{-1}$），$Cu(NO_3)_2$（$0.02 \ mol \cdot L^{-1}$），淀粉溶液（$0.2\%$）。

## 实验内容

### 1. 浓度对化学反应速率的影响

（1）在室温下，按表 7-1 所示剂量用移液管取一定量的 KI，$Na_2S_2O_3$ 溶液，用量筒量取定量的 $KNO_3$，$Na_2SO_4$ 和淀粉溶液加入已编号的 50 mL 干燥的烧杯中，混合均匀。

（2）用装有 $Na_2S_2O_8$ 溶液的加液器，（或用移液管）按表 7-1 所示剂量将一定量的 $Na_2S_2O_8$ 溶液迅速加到相应编号的烧杯中，同时启动秒表并不断搅拌。注意观察，当溶液刚出现蓝色时，立刻按停秒表，将时间记入表 7-1 中。

**表 7-1　浓度对反应速率的影响　室温：℃**

| 实验编号 | | I | II | III | IV | V |
|---|---|---|---|---|---|---|
| 试剂用量 (mL) | 0.20 mol·L$^{-1}$ Na$_2$S$_2$O$_8$ | 10.0 | 5.0 | 2.5 | 10.0 | 10.0 |
| | 0.20 mol·L$^{-1}$ KI | 10.0 | 10.0 | 10.0 | 5.0 | 2.5 |
| | 0.010 mol·L$^{-1}$ Na$_2$S$_2$O$_3$ | 4.0 | 4.0 | 4.0 | 4.0 | 4.0 |
| | 0.2%淀粉溶液 | 1.0 | 1.0 | 1.0 | 1.0 | 1.0 |
| | 0.20 mol·L$^{-1}$ KNO$_3$ | 0.0 | 0.0 | 0.0 | 5.0 | 7.5 |
| | 0.20 mol·L$^{-1}$ Na$_2$SO$_4$ | 0.0 | 5.0 | 7.5 | 0.0 | 0.0 |
| 混合液中反应物的起始浓度 | Na$_2$S$_2$O$_8$ | | | | | |
| | KI | | | | | |
| | Na$_2$S$_2$O$_3$ | | | | | |
| 反应时间 $\Delta t$（秒） | | | | | | |
| $v = [Na_2S_2O_3]/2\Delta t$ | | | | | | |
| $\lg[S_2O_8^{2-}]$ | | | | | | |
| $\lg[I^-]$ | | | | | | |
| $\alpha$ | | | | | | |
| $\beta$ | | | | | | |
| $k = v/[S_2O_8^{2-}]^\alpha[I^-]^\beta$ | | | | | | |

## 2. 温度对反应速度的影响

（1）按表 7-1 中 IV 号的剂量把 KI，Na$_2$S$_2$O$_3$，KNO$_3$ 和淀粉溶液加入一个 50 mL 干燥的烧杯中，混合均匀。

（2）将 10 mL 0.20 mol·L$^{-1}$ Na$_2$S$_2$O$_8$ 溶液加入一只试管中。

（3）将盛有试液的烧杯和试管一起置于恒温水浴中加热。调节水浴温度使之高出室温 10℃，待烧杯和试管中溶液达到此温度时，将 Na$_2$S$_2$O$_8$ 溶液加入混合溶液中，计时，搅拌，记下溶液刚出现蓝色所需的反应时间和温度于表 7-2 中（编号 VI）。

**表 7-2　温度对反应速率的影响**

| 实验编号 | 室温的平均反应速率常数 $k$ | VI | VII |
|---|---|---|---|
| 反应温度 $T$（K） | | | |
| 反应时间 $\Delta t$（s） | | | |
| 反应速率 $v$（mol·L$^{-1}$·s$^{-1}$） | | | |
| $\lg k$ | | | |
| $1/T$ | | | |
| 活化能 $E$（kJ·mol$^{-1}$） | | | |

（4）在高出室温 20℃ 的条件下重复上述实验。将盛有混合试液的烧杯和试管放入恒温水浴中升温，待试液温度高于室温 20℃ 时，将 Na$_2$S$_2$O$_8$ 溶液加入混合溶液中，计时，搅拌，记下溶液刚出现蓝色所需的反应时间和温度于表 7-2 中（编号 VII）。

**3. 催化剂对反应速率的影响**

按表 7-1 中 Ⅳ 号的剂量把 KI，$Na_2S_2O_3$，$KNO_3$ 和淀粉溶液加入一个 100 mL 的烧杯中，混合均匀。再滴加 2 滴 0.02 $mol \cdot L^{-1}$ $Cu(NO_3)_2$ 溶液，搅拌均匀后迅速将 $Na_2S_2O_8$ 溶液加入混合溶液中，计时，搅拌，记下溶液刚出现蓝色所需的反应时间。

## 数据处理与讨论

**1. 级数和速率常数**

由表 7-1 的数据，算出 $\lg v$，$\lg[S_2O_8^{2-}]$ 和 $\lg[I^-]$，按式（7）和（8）两关系式作图，求出 α 和 β。

将 α、β 和一定浓度时的 $v$ 代入（4）式求出 $k$。

$$\lim \bar{v} = v = \frac{-\Delta[S_2O_8^{2-}]}{\Delta t} = k[S_2O_8^{2-}]^{\alpha}[I^-]^{\beta}$$

**2. 求算反应的活化能**

由表 7-2 的数据，按（4）式求出各温度时的 $k$ 值，再按式（9）作 $\lg k \sim 1/T$ 图，求出活化能（文献值 51.8 $kJ \cdot mol^{-1}$）。列表示出上述计算结果并讨论之。

## 思考题

1. 根据化学方程式，是否能确定反应级数？用本实验的结果加以说明。

2. 若不用 $S_2O_8^{2-}$，而用 $I^-$ 或 $I_3^-$ 的浓度变化来表示反应速率，则反应速率常数 $k$ 是否一样？

3. 实验中为什么可以由反应出现蓝色的时间长短来计算反应速率，反应溶液出现蓝色后，反应是否就终止了？

4. 下列操作情况对实验结果有何影响？

（1）先加 $Na_2S_2O_8$ 溶液，最后加 KI 溶液。

（2）没有迅速连续加入 $Na_2S_2O_8$ 溶液。

（3）本实验 $Na_2S_2O_3$ 的用量过多或过少。

# Exp 7　Determination of Chemical Reaction, Reaction Order and Activation Energy

## Objectives

1. To have a better understanding of the concepts of the chemical reaction rate, reaction order and activation energy.

2. To understand the principle and method of determination of sodium persulfate $Na_2S_2O_8$ oxidizing potassium iodide; study the method of calculating reaction order and activation energy through data process and drawing graph.

## Principles

In aqueous solution the reaction equation between $Na_2S_2O_8$ and KI is:

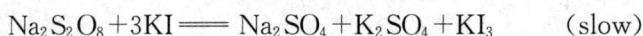

$$Na_2S_2O_8 + 3KI \Longrightarrow Na_2SO_4 + K_2SO_4 + KI_3 \qquad \text{(slow)}$$

The ionic equation is:

$$S_2O_8{}^{2-} + 3I^- \Longrightarrow 2SO_4{}^{2-} + I_3{}^- \tag{1}$$

The reaction rate law can be expressed as follows according to reaction equation

$$v = k[S_2O_8{}^{2-}]^\alpha [I^-]^\beta \tag{2}$$

Where $v$ is the instantaneous rate under this condition, if $[S_2O_8{}^{2-}]$, $[I^-]$ are initial concentrations, then $v$ means the initial velocity. $k$ is reaction rate constant. $\alpha$ and $\beta$ are reaction orders.

This experiment could determine the rate, which is the average rate $\bar{v}$ in a period of time ($\Delta t$). If the change of concentration of $S_2O_8{}^{2-}$ is $\Delta[S_2O_8{}^{2-}]$, then the average rate is

$$\bar{v} = \frac{-\Delta[S_2O_8^{2-}]}{\Delta t} \tag{3}$$

If we control $\Delta t$ to be very small, then the average rate can be take place of initial rate approximately.

$$\lim_{\Delta t \to 0} \bar{v} = v = \frac{-\Delta[S_2O_8^{2-}]}{\Delta t} = k[S_2O_8^{2-}]^\alpha [I^-]^\beta \tag{4}$$

In order to determine the value of $[S_2O_8{}^{2-}]$, mix the solution of $Na_2S_2O_8$ and KI, add definite of $Na_2S_2O_8$ with definite concentration value and starch solution for indicator, then the other reaction accompanying with the reaction also exists:

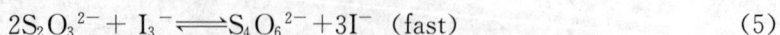

$$2S_2O_3{}^{2-} + I_3{}^- \Longrightarrow S_4O_6{}^{2-} + 3I^- \quad \text{(fast)} \tag{5}$$

The reaction completes almost in no time. The rate of reaction (5) is much more quicker than reaction (1), so the producing of $I_3{}^-$ in reaction (1) will react with $S_2O_3{}^{2-}$ immediately forming $S_4O_6{}^{2-}$ (colorless) and $I^-$. We can see the blue color during the reaction but as soon as the reactant $Na_2S_2O_3$ in reaction has been exhausted, the production of mini $I_3{}^-$ (in reaction then) will react with starch solution at once, then unique blue color appears.

From the reaction (1) and reaction (5), we can write:

$$\Delta[S_2O_8{}^{2-}] = \Delta[S_2O_3{}^{2-}]/2$$

Now that $S_2O_3{}^{2-}$ has exhausted during the time ($\Delta t$). We can conclude that:

$$\Delta[S_2O_8{}^{2-}] = \frac{1}{2}[S_2O_3^{2-}] = \frac{1}{2}(0 - [S_2O_3^{2-}]_{init}) = -\frac{1}{2}[S_2O_3^{2-}]_{init}$$

Record the time interval $\Delta t$ from the beginning of the reaction to the appearance of blue color emerging, then we can obtain reaction rate:

$$v = -\frac{\Delta[S_2O_8{}^{2-}]}{\Delta t} = \frac{[S_2O_3{}^{2-}]}{2\Delta t} \tag{6}$$

The method of obtaining $\alpha$ and $\beta$ values is as follows:

When $[I^-]$ is constant, taking the logarithm of equation (2) gives:

$$\lg v = \alpha \lg[S_2O_8{}^{2-}]_{init} + constant \tag{7}$$

Determine the $v$ in variety of $Na_2S_2O_8$ initial concentrations. Assign $\lg v$ to Y-coordinate, $\lg[S_2O_8{}^{2-}]$ to X-coordinate and we can get a straight line whose value of the slope is $\alpha$ through drawing. Also when $[S_2O_8{}^{2-}]$ is a constant, taking the logarithm of equation (2), then

$$\lg v = \beta \lg[I^-]_{init} + constant \tag{8}$$

Determine the $v$ in variety of KI initial concentrations. Draw the graph and we can get a straight line whose value of the slope is $\beta$. $\alpha + \beta$ is the reaction order of equation (1) that $Na_2S_2O_8$ oxides $I^-$. In this way, we can get reaction order. We can calculate reaction rate constant $k$ through equation (2). Arrhenius equation gives the relationship between $E_a$, $k$ and $T$:

$$\lg k = -\frac{E_a}{2.303RT} + B \tag{9}$$

$R$ is gas constant. $B$ is a constant. $E_a$ unit is $J \cdot mol^{-1}$.

Determine the reaction rate constant at different temperatures through the experiment. Assign $\lg k$ to Y-coordinate, $1/T$ to X-coordinate, we can draw a straight line whose slope is $-E_a/2.303R$, so the value of $E_a$ is obtained.

## Instruments, Reagents and Materials

### 1. Instruments

test tube (mm: $10 \times 100$), test tube rack, stopwatch, thermometer, electroheat thermostatic waterbath.

### 2. Reagents

$Na_2S_2O_8$(0.2 mol $\cdot L^{-1}$), KI(0.20 mol $\cdot L^{-1}$), $KNO_3$(0.2 mol $\cdot L^{-1}$), $Na_2SO_4$(0.20 mol $\cdot L^{-1}$), $Na_2S_2O_3$ (0.010 mol $\cdot L^{-1}$), $Cu(NO_3)_2$ (0.02 mol $\cdot L^{-1}$), starch solution (0.2%).

## Experimental Procedures

### 1. Concentration effects on the reaction rate

(1) At the room temperature, according to Tab 7-1, measure certain volume of KI solution, $Na_2S_2O_3$ solution with pipette and $KNO_3$ solution, $Na_2SO_4$ solution, starch solution with measuring cylinders. Then add all the solutions into a dry beaker (50 mL) which has been numbered and mix homogeneously.

(2) Add certain volume of $Na_2S_2O_8$ into corresponding numbered beakers quickly with measuring cylinders, turn on stopwatch at the same time and stir continuously. Pay attention to observe, as soon as blue color appears, please turn off the stopwatch, and record the passing time in the Tab 7-1 immediately.

**Table 7-1　Concentration effects on the reaction rate (room temperature: ℃)**

| | No. | I | II | III | IV | V |
|---|---|---|---|---|---|---|
| | $0.2$ mol·$L^{-1}$ $Na_2S_2O_8$ | 10.0 | 5.0 | 2.5 | 10.0 | 10.0 |
| | $0.20$ mol·$L^{-1}$KI | 10.0 | 10.0 | 10.0 | 5.0 | 2.5 |
| Reagent amount (mL) | $0.010$ mol·$L^{-1}$ $Na_2S_2O_3$ | 4.0 | 4.0 | 4.0 | 4.0 | 4.0 |
| | $0.2\%$ starch solution | 1.0 | 1.0 | 1.0 | 1.0 | 1.0 |
| | $0.20$ mol·$L^{-1}$ $KNO_3$ | 0.0 | 0.0 | 0.0 | 5.0 | 7.5 |
| | $0.20$ mol·$L^{-1}$ $Na_2SO_4$ | 0.0 | 5.0 | 7.5 | 0.0 | 0.0 |
| Initial concentration (mol·$L^{-1}$) | $Na_2S_2O_8$ | | | | | |
| | KI | | | | | |
| | $Na_2S_2O_3$ | | | | | |
| $\Delta t$(s) | | | | | | |
| $v=[Na_2S_2O_3]/2\Delta t$ | | | | | | |
| $\lg[S_2O_8^{2-}]$ | | | | | | |
| $\lg[I^-]$ | | | | | | |
| $\alpha$ | | | | | | |
| $\beta$ | | | | | | |
| $k=v/[S_2O_8^{2-}]^\alpha[I^-]^\beta$ | | | | | | |

### 2. Temperature effects on the reaction rate

(1) According to the given amount of number IV in the Tab 7-1, add KI solution, $Na_2S_2O_3$ solution, $KNO_3$ solution, $Na_2SO_4$ solution and starch solution into a beaker (50 mL) and mix homogeneously.

(2) Add 10.0 mL 0.2 mol·$L^{-1}$ $Na_2S_2O_8$ solution into a test tube.

(3) Place the beaker and the test tube into a water bath and heat, regulating water bath temperature, which makes its temperature, be 10℃ higher than room temperature. When

the temperature of the solution in the beaker and tube reach the same one, add $Na_2S_2O_8$ solution into the beaker and turn on the stopwatch, stir continuously. Record the time interval $\Delta t$ and temperature in the Table 7-2(No. VI) when the solution just turns blue.

(4) Repeat the above experiment at the temperature which is 20℃ higher than the room temperature in the same way. At last record the passing time interval $\Delta t$ and temperature in the Table 7-2(No. VII) when the solution just turns blue.

Table 7-2  Temperature effects on the reaction rate (room temperature: ℃)

| Experiment number | Reaction rate constant  $k$ | VI | VII |
| --- | --- | --- | --- |
| Reaction temperature (K) | | | |
| Reaction time $\Delta t$ (s) | | | |
| Reaction rate $v$ (mol·L$^{-1}$·s$^{-1}$) | | | |
| lg$k$ | | | |
| $1/T$ | | | |
| Activation energy (kJ·mol$^{-1}$) | | | |

### 3. Catalyst effects on the reaction rate

According to the amount of the reagents in Table 7-1 No. IV, add KI, $Na_2S_2O_3$, $KNO_3$ and starch solution to a 100 mL beaker, mix thoroughly. 2 drops of 0.02 mol·L$^{-1}$ Cu $(NO_3)_2$ are added to above solution, mix thoroughly, add $Na_2S_2O_8$ solution rapidly, stir, start to record the time, record the time that from last mixing to the blue color appearing.

## Data and comment

### 1. Order and rate constant

According to data in Table 7-1, calculate lg$v$, lg$[S_2O_8{}^{2-}]$, and lg$[I^-]$. According to the relationship between equation (7) and (8), draw the graph, so we can get the value of α and β. Then reaction rate constant $k$ can be obtained as following equation:

$$\lim \bar{v}=v=\frac{-\Delta[S_2O_8{}^{2-}]}{\Delta t}=k\ [S_2O_8^{2-}]^\alpha\ [I^-]^\beta$$

### 2. Calculate the reaction activation energy

According to the data in Table 7-2 we can get the value of $k$ at different temperatures by the relation of equation (4). Draw the graph (assign lg$k$ to Y-coordinate, $1/T$ to X-coordinate) and the value of $E_a$ can be obtained.

## Questions in Previewing

1. Can you get the reaction order according to chemical equation? Illustrate this experiment's result.

2. If we use $[I^-]$ or $[I_3{}^-]$, not $[S_2O_8{}^{2-}]$ to show the reaction rate, is the reaction

rate constant $k$ the same?

3. Why we should make use of the interval time $\Delta t$ from the reaction beginning to blue color appearing to calculate the reaction rate? Is the reaction terminated when the reaction solution takes on blue color?

4. How will the following operation influence the result of experiment?

(1) Add $Na_2S_2O_8$ solution first and KI solution finally.

(2) Don't add $Na_2S_2O_8$ solution quickly and continuously.

(3) The amount of $Na_2S_2O_3$ is too much or too little in this experiment.

# 实验八　$I_3^- \rightleftharpoons I_2 + I^-$ 体系平衡常数的测定

## 实验目的

1. 了解测量 $I_3^- \rightleftharpoons I_2 + I^-$ 体系平衡常数的原理和方法，加深对化学平衡和平衡常数的理解。

2. 巩固滴定操作。

## 实验原理

碘可以溶解于可溶性碘化物溶液而生成 $I_3^-$ 离子。该反应是可逆反应，在一定温度下建立平衡：

$$I_3^- \rightleftharpoons I_2 + I^-$$

其平衡常数的表达式如下：

$$K_a^\ominus = \frac{a(I_2) \cdot a(I^-)}{a(I_3^-)} = \frac{c(I_2) \cdot c(I^-)}{c(I_3^-)} \cdot \frac{\gamma(I_2) \cdot \gamma(I^-)}{\gamma(I_3^-)} \tag{1}$$

式中 $a$，$c$ 和 $\gamma$ 分别表示各物质的活度、物质的量的浓度和活度系数。$K_a^\ominus$ 越大，表示 $I_3^-$ 越不稳定，故 $K_a^\ominus$ 也称 $I_3^-$ 的不稳定常数。

在离子强度不大的溶液中，由于

$$\frac{\gamma(I_2) \cdot \gamma(I^-)}{\gamma(I_3^-)} \approx 1$$

故

$$K_a^\ominus \approx \frac{c(I_2) \cdot c(I^-)}{c(I_3^-)} = K_c^\ominus \tag{2}$$

因此，通过实验测定 $c(I_2)$、$c(I^-)$ 和 $c(I_3^-)$ 便可计算 $K_a^\ominus$。

将已知浓度 $c$ 的 KI 溶液与过量的固体碘一起摇荡，达到平衡后用标准 $Na_2S_2O_3$ 溶液滴定，可求得溶液中碘的总浓度 $c_{总}$ [即 $c_{平}(I_3^-) + c_{平}(I_2)$]。其中，$c_{平}(I_2)$ 可用 $I_2$ 在纯水中的饱和浓度代替，因此可以确定

$$c_{平}(I_3^-) = c_{总} - c_{平}(I_2)$$

由于每生成一个 $I_3^-$ 要消耗一个 $I^-$，因此平衡时溶液中 $I^-$ 的浓度为

$$c_{平}(I^-) = c - c_{平}(I_3^-)$$

将 $c_{平}(I_2)$，$c_{平}(I_3^-)$ 和 $c_{平}(I^-)$ 代入式（2），便可以求出该温度下的平衡常数 $K_a^\ominus$。

注意，为简洁起见，表达式中略去 $c^\ominus$。

## 仪器和药品

天平，振荡器，吸量管(5 mL)，移液管(1 mL，10 mL)，锥形瓶(25 mL)，碘量瓶(25 mL，50 mL)，量筒(25 mL)，酸式滴定管(5 mL)，洗耳球。

$I_2$，KI（0.3000 $mol \cdot L^{-1}$），$Na_2S_2O_3$ 溶液（0.05000 $mol \cdot L^{-1}$）（KI 和 $Na_2S_2O_3$ 溶液可由实验室预先标定），0.5%淀粉溶液。

## 实验步骤

1. 取三个 25 mL 和一个 50 mL 干燥的碘量瓶，按表 8-1 中所列的量配好溶液（1~3 号用吸量管，4 号可用量筒量取水）。

**表 8-1　试剂及加入量**

| 编号 | 1 | 2 | 3 | 4 |
|---|---|---|---|---|
| $c(KI)$ $(mol \cdot L^{-1})$ | 0.3000 | 0.3000 | 0.3000 | — |
| $V(KI)$ (mL) | 1.00 | 2.00 | 3.00 | — |
| $m(I_2)$ (g) | 0.2 | 0.2 | 0.2 | 0.2 |
| $V(H_2O)$ (mL) | 2.00 | 1.00 | - | 25 |

2. 将上述配好的溶液在室温下强烈振荡 25 min，静置，待过量的固体碘沉于瓶底后，取清液分析。

3. 用移液管分别在 1~3 号瓶中吸取上层清液 1.00 mL 于三只锥形瓶中，各加入约 30 mL蒸馏水，用标准 $Na_2S_2O_3$ 溶液滴定至淡黄色，然后加入 4 滴淀粉溶液，继续滴定至蓝色刚好消失。记下 $Na_2S_2O_3$ 溶液消耗的体积。

用移液管于 4 号瓶中量取 10.00 mL 清液，以标准 $Na_2S_2O_3$ 溶液滴定，记录消耗的体积。

由于碘容易挥发，取清液后应尽快滴定，不要放置太久，在滴定时不宜过于剧烈地摇动溶液。

## 数据记录及处理

1. 设计一表格记录实验数据及计算结果。

2. 根据实验原理由实验数据计算 1~3 号溶液的 $c_{总}$，$c_{平}(I_2)$，$c_{平}(I^-)$，$c_{平}(I_3^-)$，$K_a^\ominus$ 及 $K_a^\ominus$ 的平均值 $\overline{K_a^\ominus}$。

## 思考题

1. 在固体碘和 KI 溶液反应时，如果碘的量不够，将有何影响？碘的用量是否一定要准确称量？

2. 在实验过程中，如果有以下情况：（1）吸取清液进行滴定时不小心吸进一些碘微粒，（2）饱和碘水放置很久后才滴定，（3）振荡时间不够，对实验结果将会产生什么影响？

# Exp 8   Determination of Balance Constant of $I_3^- \rightleftharpoons I_2 + I^-$ System

## Objectives

1. To know the fundamental principles and methods of determination of balance constant of $I_3^- \rightleftharpoons I_2 + I^-$ system; to consolidate comprehension on chemical balance and balance constant.

2. To consolidate the operation of titration.

## Principles

Iodine can dissolve in soluble iodide solution to form $I_3^-$, which is reversible. At certain temperature the balance is as follow:

$$I_3^- \rightleftharpoons I_2 + I^-$$

Its balance constant can be expressed as (1)

$$K_a^\ominus = \frac{a(I_2) \cdot a(I^-)}{a(I_3^-)} = \frac{c(I_2) \cdot c(I^-)}{c(I_3^-)} \cdot \frac{\gamma(I_2) \cdot \gamma(I^-)}{\gamma(I_3^-)} \tag{1}$$

Where $a$, $c$ and $\gamma$ are the expression of live concentration, molar concentration and live concentration coefficient of all substances. The higher $K_a^\ominus$, of the more instability of $I_3^-$. So $K_a^\ominus$ is also expressed as the instability constant of $I_3^-$.

For the solution in which ion intensity isn't too great, owing to

$$\frac{\gamma(I_2) \cdot \gamma(I^-)}{\gamma(I_3^-)} \approx 1$$

Thus

$$K_a^\ominus \approx \frac{c(I_2) \cdot c(I^-)}{c(I_3^-)} = K_c^\ominus \tag{2}$$

So $K_a^\ominus$ can be calculated by $c(I_2)$, $c(I^-)$ and $c(I_3^-)$ value of experiment measure.

Shake the mixture of KI solution of certain concentration $c$ and excessive solid iodine, titrate the solution by $Na_2S_2O_3$ solution when the system reach balance. Thus general concentration $c_{tota}$[namely $c_{eq}(I_3^-) + c_{eq}(I_2)$] can be got. Where $c_{eq}(I_2)$ can be replaced by saturation concentration of $I_2$ in pure water. So $c_{eq}(I_3^-)$ may be expressed as follows:

$$C_{eq}(I_3^-) = c_{tota} - c_{eq}(I_3^-)$$

For forming one $I_3^-$ ion will consume one $I^-$ ion, $I^-$ concentration in the balance is

$$C_{eq}(I^-) = c - c_{eq}(I_3^-)$$

Equilibrium constant at this temperature can be obtained by equation (2), in which $c_{eq}(I_2)$, $c_{eq}(I_3^-)$ and $c_{eq}(I^-)$ value is replaced.

Note: for simplification, $c$ is omitted in the formula.

## Instruments and Reagents

electronic balance, shaker, pipet(5 mL), dropping piette(1 mL, 10 mL), conical flask(25 mL), iodine bottle(25 mL, 50 mL), cylinder(25 mL), acid buret(5 mL), aurilave.

$I_2$, KI (0. 3000 mol·$L^{-1}$), $Na_2S_2O_3$ solution (0. 05000 mol·$L^{-1}$) (KI and $Na_2S_2O_3$ solution can be demarcated in advance), 0. 5% starch solution.

## Experimental Procedures

1. Take three dry 25 mL-flasks of iodine measuring and one 50 mL one, formulating following solution for No. 1~3, pipet is employed to measure water, for No. 4, measuring cylinder is ok.

Table 8-1  The amount of reagents in determination of $K_a^\ominus$

| No. | 1 | 2 | 3 | 4 |
|---|---|---|---|---|
| $c$(KI) (mol·$L^{-1}$) | 0. 3000 | 0. 3000 | 0. 3000 | — |
| $V$(KI) (mL) | 1. 00 | 2. 00 | 3. 00 | — |
| $m$($I_2$) (g) | 0. 2 | 0. 2 | 0. 2 | 0. 2 |
| $V$($H_2O$) (mL) | 2. 00 | 1. 00 | - | 25 |

2. Stir the above solution vigorously for 25 min at r. t, stand, and take clear liquid for analysis after excessive solid iodine sinking down at the bottom of bottle.

3. Extract 1. 00 mL clear liquid of above layer from the bottle of 1~3 to three conical flasks, add 30 mL distilled water, respectively, titrate them till solution turning yellow with $Na_2S_2O_3$ solution respectively. Then add 4 drops of starch solution into it, titrate until the color of blue exactly vanishes. Note down the volumes of $Na_2S_2O_3$ solution consumed.

Take 10. 00 mL clear liquid from the fourth bottle by dropping pipettes, titrate it by standard $Na_2S_2O_3$ solution. Note down the volume of $Na_2S_2O_3$ solution consumed.

For iodine is volatile, the clear solution should be titrated, don't keep it too long and the solution shouldn't be vibrated violently as titration.

## Data Analysis

1. Design a table to record the experiment data and results of calculation.

2. Calculate $c_{tot}$, $c_{eq}(I_2)$, $c_{eq}(I^-)$, $c_{eq}(I_3^-)$, $K_a^\ominus$ and $\overline{K}_a^\ominus$ of the solution labeled 1~3 with experiment data according to the principle of experiment.

## Questions

1. When iodine (s) reacts with KI solution, what effect will bring about if iodine is not enough? Whether should the amount of iodine be needed being weighed accurately?

2. How will the following operations affect the experimental results in the procedures?

(1) Some iodine particles are extracted when extracting clear liquid for titration.

(2) Saturated iodine water is titrated after placement for a long time.

(3) The time of shaking is not enough.

# 实验九　电离平衡和沉淀反应

## 实验目的

1. 加深对电离平衡，同离子效应，盐类水解等理论的理解。
2. 学习缓冲溶液的配制并了解它的缓冲作用。
3. 了解沉淀的生成和溶解的条件。

## 基本原理

弱电解质（弱酸或弱碱）在水溶液中部分发生电离，电离出来的离子与未电离的分子间处于平衡状态，例如醋酸（HAc）：

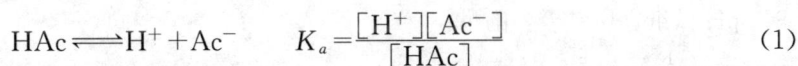

$$HAc \rightleftharpoons H^+ + Ac^- \qquad K_a = \frac{[H^+][Ac^-]}{[HAc]} \tag{1}$$

如果往溶液中增加更多的 $Ac^-$（比如加入 NaAc）或 $H^+$，都可以使平衡向左方向移动，增加 HAc 浓度，这种作用称为同离子效应。

在 $H^+$ 浓度（$mol \cdot L^{-1}$）小于 1 的溶液中，其酸度常用 pH 表示，其定义为：

$$pH = -lg[H^+]$$

在中性溶液或纯水中，$[H^+] = [OH^-] = 10^{-7} mol \cdot L^{-1}$，即 pH = pOH = 7，在碱性溶液中 pH = 14 - pOH > 7，在酸性溶液中 pH < 7。

如果溶液中同时存在着弱酸以及它的共轭碱，例如 HAc 和 NaAc，这时加入少量的酸可被 $Ac^-$ 结合为电离度很小的 HAc 分子，加入少量的碱则被 HAc 所中和，溶液的 pH 始终改变不大。这种溶液称为缓冲溶液。同理弱碱及其共轭酸也可组成缓冲溶液。缓冲溶液的 pH（以 HAc 和 NaAc 为例）为：

$$pH = pK_a - lg\frac{[酸]}{[盐]} = pK_a - lg\frac{[HAc]}{[Ac^-]}$$

弱酸和强碱，或弱碱和强酸以及弱酸和弱碱所生成的盐，在水溶液中都发生水解。例如

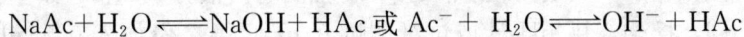

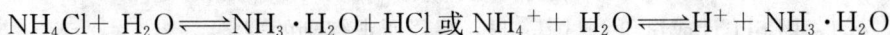

$$NaAc + H_2O \rightleftharpoons NaOH + HAc \text{ 或 } Ac^- + H_2O \rightleftharpoons OH^- + HAc$$

$$NH_4Cl + H_2O \rightleftharpoons NH_3 \cdot H_2O + HCl \text{ 或 } NH_4^+ + H_2O \rightleftharpoons H^+ + NH_3 \cdot H_2O$$

根据同离子效应，在溶液中加入 $H^+$ 或 $OH^-$ 就可以阻止它们（$NH_4^+$ 或 $Ac^-$）水解。另外，由于水解是吸热反应，所以加热则可促进盐的水解。

难溶强电解质在一定温度下与它的饱和溶液中的相应离子处于平衡状态。例如：

$$AgCl \rightleftharpoons Ag^+ + Cl^-$$

它的平衡常数就是饱和溶液中两种离子浓度的乘积，称为溶度积 $K_{sp}(AgCl)$。只要溶液中两种离子浓度乘积大于其溶度积，便有沉淀产生。反之如果降低饱和溶液中某种离子的浓度，使两种离子浓度乘积小于其溶度积，则沉淀便会溶解。所以在上述饱和溶液中加入 $NH_3 \cdot H_2O$，使 $Ag^+$ 转为 $[Ag(NH_3)_2]^+$，AgCl 沉淀便可溶解。根据类似的原理，往溶液中加入 $I^-$。它便与 $Ag^+$ 结合为溶解度更小的 AgI 沉淀。这时溶液中 $Ag^+$ 浓度减小了，对于 AgCl 来说已成为不饱和溶液，而对 AgI 来说，只要加入足够量的 $I^-$，便是过饱和溶液。结果，

一方面 AgCl 不断溶解,另一方面不断有 AgI 沉淀产生。最后 AgCl 沉淀可全部转化为 AgI 沉淀。

## 仪器和药品

pH 计,离心机,试管,烧杯(100 mL)。

NaAc(s),NH₄Cl(s),Fe(NO₃)₃·9H₂O(s),HCl(0.1 mol·L⁻¹,2 mol·L⁻¹,6 mol·L⁻¹),HAc(0.1 mol·L⁻¹,2 mol·L⁻¹),HNO₃(6 mol·L⁻¹),NaOH(0.1 mol·L⁻¹,2 mol·L⁻¹),NH₃·H₂O(0.1 mol·L⁻¹,2 mol·L⁻¹),FeCl₃(0.1 mol·L⁻¹),Pb(NO₃)₂(0.1 mol·L⁻¹),Na₂SO₄(0.1 mol·L⁻¹),K₂CrO₄(0.1 mol·L⁻¹),AgNO₃(0.1 mol·L⁻¹),NaAc(0.1 mol·L⁻¹),NaCl(0.1 mol·L⁻¹),NH₄Cl(0.1 mol·L⁻¹,saturated),Na₂CO₃(0.1 mol·L⁻¹),SbCl₃(0.1 mol·L⁻¹),NH₄Ac(0.1 mol·L⁻¹),(NH₄)₂C₂O₄(saturated),CaCl₂(0.1 mol·L⁻¹),MgCl₂(0.1 mol·L⁻¹),NaHCO₃(0.1 mol·L⁻¹),Al₂(SO₄)₃(0.1 mol·L⁻¹)。

pH 试纸,甲基橙,酚酞。

## 实验内容

### 1. 溶液的 pH

用 pH 试纸测试浓度各为 $0.1\ mol\cdot L^{-1}$ 的 HCl,HAc,NaOH 和 NH₃·H₂O 的 pH,并与计算值作一比较(HAc 和 NH₃·H₂O 的电离常数均为 $1.8\times10^{-5}$)。

### 2. 同离子效应和缓冲溶液

(1) 取约 $2\ mL\ 0.1\ mol\cdot L^{-1}$ HAc 溶液倒入小试管中,加 1 滴甲基橙,观察溶液的颜色,然后加入少量固体 NaAc,观察颜色有何变化?解释之。

(2) 取约 $2\ mL\ 0.1\ mol\cdot L^{-1}$ NH₃·H₂O 倒入小试管中,加 1 滴酚酞,观察溶液的颜色,再加入少量固体 NH₄Cl,观察颜色变化,并解释之。

(3) 在一小烧杯中加入 $15\ mL\ 0.1\ mol\cdot L^{-1}$ HAc 和 $15\ mL\ 0.1\ mol\cdot L^{-1}$ NaAc,搅拌均匀后,用 pH 计或精密 pH 试纸测试溶液的 pH。然后将溶液分成二份,第一份加入 $0.5\ mL$ $0.1\ mol\cdot L^{-1}$ HCl,摇匀,用 pH 计或试纸测试其 pH。第二份加入 $0.5\ mL\ 0.1\ mol\cdot L^{-1}$ NaOH,摇匀,用 pH 计或试纸测试其 pH。解释所观察到的现象。

(4) 在一小烧杯中加入 30 mL 蒸馏水,用 pH 试纸测其 pH。现将其分成二份,在一份中加 $0.5\ mL\ 0.1\ mol\cdot L^{-1}$ HCl 溶液,测其 pH。在另一份中加 $0.5\ mL\ 0.1\ mol\cdot L^{-1}$ NaOH 溶液,测其 pH。与上一实验作比较,得到什么结论?

### 3. 盐类水解和影响水解平衡的因素

(1) 用精密 pH 试纸测试浓度各为 $0.1\ mol\cdot L^{-1}$ 的 NaCl,NH₄Cl,Na₂CO₃ 和 NH₄Ac 的 pH。解释所观察到的现象。

(2) 取少量(两粒绿豆大小)固体 Fe(NO₃)₃·9H₂O,用 6 mL 水溶解后观察溶液的颜色,然后分成三份,第一份留作比较,第二份加几滴 $6\ mol\cdot L^{-1}$ HNO₃,第三份小火加热煮沸,观察现象。$Fe^{3+}$ 的水合离子为无色,由于水解生成了各种碱式盐而使溶液显棕黄色。

加入 $HNO_3$ 或加热对水解平衡各有何影响？加以说明。

（3）取约 0.5 mL $SbCl_3$ 溶液，加水稀释，观察有无沉淀生成？加入 6 mol·L$^{-1}$ HCl 沉淀是否溶解？再加水稀释，是否再有沉淀生成？加以解释。$SbCl_3$ 的水解过程总反应式为：

$$SbCl_3 + H_2O \Longrightarrow SbOCl + 2HCl$$

（4）分别取 1 mL 0.1 mol·L$^{-1}$ $Al_2(SO_4)_3$ 和 1 mL 0.1 mol·L$^{-1}$ $NaHCO_3$ 溶液于小试管中，并用 pH 试纸测试它们的 pH，写出它们的水解反应方程式。然后将 $NaHCO_3$ 慢慢滴入 $Al_2(SO_4)_3$ 中，观察有何现象？从水解平衡的移动解释所看到的现象。

**4. 沉淀的生成和溶解**

（1）在两支小试管中分别加入约 0.5 mL 饱和 $(NH_4)_2C_2O_4$ 溶液和 0.5 mL 0.1 mol·L$^{-1}$ $CaCl_2$ 溶液，观察白色 $CaC_2O_4$ 沉淀的生成。然后在一支试管内加入 2 mol·L$^{-1}$ HCl 溶液约 2 mL，搅拌，看沉淀是否溶解？在另一支试管中加入 2 mol·L$^{-1}$ HAc 溶液约 2 mL，沉淀是否溶解？加以解释。

（2）在两支试管中加入 1 mL 0.1 mol·L$^{-1}$ $MgCl_2$ 溶液，并逐滴加入 2 mol·L$^{-1}$ $NH_3·H_2O$ 至有白色 $Mg(OH)_2$ 沉淀生成，然后在第一支试管中加入 2 mol·L$^{-1}$ HCl 溶液，沉淀是否溶解？在第二支试管中加入饱和 $NH_4Cl$ 溶液，沉淀是否溶解？加入 HCl 和 $NH_4Cl$ 对下列反应

$$Mg(OH)_2 \Longrightarrow Mg^{2+} + 2OH^-$$

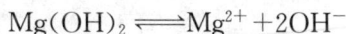

的平衡各有何影响？

（3）$Ca(OH)_2$，$Mg(OH)_2$ 和 $Fe(OH)_3$ 溶解度比较

①分别取约 0.5 mL 0.1 mol·L$^{-1}$ $CaCl_2$，$MgCl_2$ 和 $FeCl_3$ 溶液倒入三支小试管中，各加入 2 mol·L$^{-1}$ NaOH 溶液数滴，观察并记录三支试管中有无沉淀生成。

②分别取约 0.5 mL 0.1 mol·L$^{-1}$ $CaCl_2$，$MgCl_2$ 和 $FeCl_3$ 溶液倒入三支小试管中，各加入 2 mol·L$^{-1}$ $NH_3·H_2O$ 数滴，观察并记录三支试管中有无沉淀产生。

③分别取约 0.5 mL 0.1 mol·L$^{-1}$ $CaCl_2$，$MgCl_2$ 和 $FeCl_3$ 溶液倒入三支小试管中，分别加入 0.5 mL 饱和 $NH_4Cl$ 和 2 mol·L$^{-1}$ $NH_3·H_2O$ 混合溶液（体积比为 1:1），观察并记录三支试管中有无沉淀生成。

通过上述三个实验比较 $Ca(OH)_2$，$Mg(OH)_2$ 和 $Fe(OH)_3$ 溶解度的相对大小，并加以解释。

**5. 沉淀转化**

（1）在一支离心试管中加入 0.1 mol·L$^{-1}$ $Pb(NO_3)_2$ 溶液约 0.5 mL，再加入约 0.5 mL 0.1 mol·L$^{-1}$ $Na_2SO_4$，观察白色沉淀生成。离心分离，除去清液，在沉淀上滴加 0.1 mol·L$^{-1}$ $K_2CrO_4$ 溶液，搅拌观察白色 $PbSO_4$ 沉淀转化为黄色 $PbCrO_4$ 沉淀。写出反应式并根据溶度积原理解释。

（2）取数滴 0.1 mol·L$^{-1}$ $AgNO_3$ 于离心试管中，加入数滴 $K_2CrO_4$ 溶液，观察砖红色 $Ag_2CrO_4$ 沉淀生成。沉淀经离心，洗涤，然后加入 0.1 mol·L$^{-1}$ NaCl 溶液，观察砖红色沉淀转化为白色 AgCl 沉淀。写出反应式并解释。

## 思考题

1. 已知 $H_3PO_4$，$NaH_2PO_4$，$Na_2HPO_4$ 和 $Na_3PO_4$ 四种溶液的摩尔浓度相同，它们依

次分别显酸性，弱酸性，弱碱性和碱性。试解释之。

2. 加热对水解有何影响？为什么？

3. 将 10 mL 0.20 mol·L$^{-1}$ HAc 和 10 mL 0.10 mol·L$^{-1}$ NaOH 混合，则所得溶液是否有缓冲作用？这个溶液的 pH 在什么范围内？

4. 沉淀的溶解和转化的条件各有哪些？

# Exp 9　Ionization Equilibrium and Precipitate Reaction

## Objectives

1. To deepen the comprehension on the theories of ionization equilibrium, common ion effect and the hydrolysis of salts.

2. To study the preparation of buffer solutions as well as their buffer effects.

3. To understand the forming conditions and dissolving conditions of precipitate.

## Principles

The weak electrolytes (such as the weak acid and weak base) partly ionize in the aqueous solution and the ionized and unionized ions are in the balanced condition. Take HAc as an example:

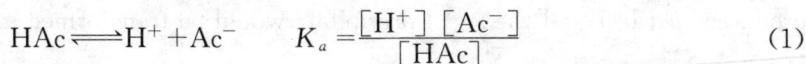

$$HAc \rightleftharpoons H^+ + Ac^- \qquad K_a = \frac{[H^+][Ac^-]}{[HAc]} \qquad (1)$$

If more $Ac^-$ ion or $H^+$ ion is added to the solution, the current equilibrium will be broken, and the direction of reaction will move to the left, decreasing the concentration of HAc. This phenomenon is called common ion effect.

When the concentration ($mol \cdot L^{-1}$) of $H^+$ in the solution is lower than 1, its acid degree is usually measured by pH, defined as:

$$pH = -lg[H^+]$$

In pure water or neutral solutions, $[H^+] = [OH^-] = 10^{-7}$ $mol \cdot L^{-1}$ (25℃). That's to say, pH = pOH = 7. While in the alkalinous solution, pH = 14 − pOH > 7; in acid, pH < 7.

In the condition that a weak acid and its salt are coexisting in one solution, HAc and NaAc, for instance, adding small amount of acid or base will not change its pH value much, for the newly added $H^+$ or $OH^-$ can be neutralized by $Ac^-$ or HAc respectively. This sort of solution is named buffer solution. Based on the same principle, a weak base and its salt can also form buffer solutions. The pH value of the buffer solution is as follows (taking HAc and NaAc as an example):

$$pH = pK_a - lg[acid]/[salt] = pK_a - lg[HAc]/[Ac^-]$$

All of the weak acid and strong base, or the weak base and strong acid, as well as the salts formed by the weak acid and weak base would always hydrolyze in aqueous solutions, for example:

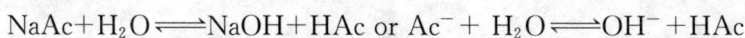

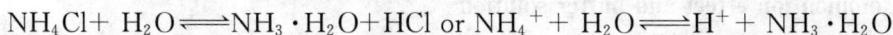

$$NaAc + H_2O \rightleftharpoons NaOH + HAc \text{ or } Ac^- + H_2O \rightleftharpoons OH^- + HAc$$

$$NH_4Cl + H_2O \rightleftharpoons NH_3 \cdot H_2O + HCl \text{ or } NH_4^+ + H_2O \rightleftharpoons H^+ + NH_3 \cdot H_2O$$

According to the theory of common ion effect, it would stop hydrolysis of some ions ($NH_4^+$ or $Ac^-$) when $H^+$ or $OH^-$ is added to the solution. Otherwise, heating the solution could promote the hydrolysis of the salts, so that the phenomenon of hydrolysis is a kind of endothermic reaction.

The indissoluble strong electrolytes would balance their ions in the saturated solutions at certain temperature. For example:

$$AgCl \rightleftharpoons Ag^+ + Cl^-$$

Its equilibrium constant is the cross product of the two ion concentrations in the saturated solution, called solubility product $K_{sp}$ (AgCl). As long as the cross product of the two ions exceeds the solubility product, the precipitate comes out. On the contrary, the precipitate would dissolve if you decline the concentration of some ions in the saturated solution to the degree of their product is less than the solubility product. For example, when adding $NH_3 \cdot H_2O$ to the solution mentioned above to transform $Ag^+$ into $Ag(NH_3)_2^+$, AgCl would be dissolved. On the basis of the same principles, when adding $I^-$ to the solution, it would combine with $Ag^+$ to form AgI whose solubility is much less. At this moment, the concentration of $Ag^+$ is lowed as a result, it is an unsaturated solution with regard to AgCl, but for AgI, it is oversaturated solution as long as there is enough $I^-$. The total consequence is: on one hand the AgCl is dissolving continuously, on the other hand, more AgI is produced. At last, all the AgCl precipitate would be transformed into AgI precipitate.

## Instruments and Reagents

pH meter, centrifugal machine, test tubes, beaker (100 mL).

NaAc(s), $NH_4Cl$(s), $Fe(NO_3)_3 \cdot 9H_2O$ (s), HCl (0. 1 mol·$L^{-1}$, 2 mol·$L^{-1}$, 6 mol·$L^{-1}$), HAc(0. 1 mol·$L^{-1}$, 2 mol·$L^{-1}$), $HNO_3$ (6 mol·$L^{-1}$), NaOH (0. 1 mol·$L^{-1}$, 2 mol·$L^{-1}$), $NH_3 \cdot H_2O$(0. 1 mol·$L^{-1}$, 2 mol·$L^{-1}$), $FeCl_3$(0. 1 mol·$L^{-1}$), $Pb(NO_3)_2$(0. 1 mol·$L^{-1}$), $Na_2SO_4$(0. 1 mol·$L^{-1}$), $K_2CrO_4$(0. 1 mol·$L^{-1}$), $AgNO_3$(0. 1 mol·$L^{-1}$), NaAc (0. 1 mol·$L^{-1}$), NaCl(0. 1 mol·$L^{-1}$), $NH_4Cl$ (0. 1 mol·$L^{-1}$, saturated), $Na_2CO_3$ (0. 1 mol·$L^{-1}$), $SbCl_3$(0. 1 mol·$L^{-1}$), $NH_4Ac$(0. 1 mol·$L^{-1}$), $(NH_4)_2C_2O_4$(saturated), $CaCl_2$ (0. 1 mol·$L^{-1}$), $MgCl_2$(0. 1 mol·$L^{-1}$), $NaHCO_3$(0. 1 mol·$L^{-1}$), $Al_2(SO_4)_3$(0. 1 mol·$L^{-1}$).

pH test paper, methyl orange, phenolphthalein.

## Experimental Procedures

### 1. The pH value of solution

Test the pH value of HCl, HAc, NaOH and $NH_3 \cdot H_2O$ respectively with pH test paper, and compare the results to the calculated value (both of the ionization constant of HAc and $NH_3 \cdot H_2O$ is 1. $8 \times 10^{-5}$).

### 2. Common ion effect and buffer solution

(1) Add about 2 mL 0. 1 mol·$L^{-1}$ HAc solution in a small test tube, and a drop of methyl orange. Observe the color of the solution. Add some solid of NaAc, and then observe the change of the color and explain it.

(2) Add about 2 mL 0. 1 mol·L$^{-1}$ NH$_3$·H$_2$O solution in a small test tube, and a drop of phenolphthalein. Observe the color of the solution. Then add a small solid of NH$_4$Cl, observe the change of the color and explain it.

(3) Mix 15 mL 0. 1 mol·L$^{-1}$ HAc and 15 mL 0. 1 mol·L$^{-1}$ NaAc in a small beaker. Test the pH value with pH meter or precise pH test paper when they are mixed evenly after stirring. And then divide the solution into two portions. Test the pH value of the first portion with pH meter or precise pH test paper after adding 1 mL 0. 1 mol·L$^{-1}$ HCl well-shakedly, test the pH value of the second portion with pH meter or precise pH test paper after adding 1 mL 0. 1 mol·L$^{-1}$ NaOH solution and shake well. Explain the observed phenomena.

(4) Add about 30 mL distilled water in a beaker and test its pH value. Divide it into two parts.

Add 1 mL 0. 1 mol·L$^{-1}$ HCl solution in the first part and test its pH value. Perform the same operation with the second part with 1 mL 0. 1 mol·L$^{-1}$ NaOH solution. Comparing the pH values to the experiment above, what conclusion can be drawn?

### 3.　The hydrolysis of the salts and its influencing factors

(1) Test the pH values of NaCl, NH$_4$Cl, Na$_2$CO$_3$ and NH$_4$Ac solution in the concentration of 0. 1 mol·L$^{-1}$ with the precise pH test paper. Explain the observed phenomena.

(2) Dissolve a small solid Fe(NO$_3$)$_3$·9H$_2$O (2 grains of a mung bean-like size) in 6 mL water. Divide it into three portions after observing its color. Make the first portion to be the control. Add several drops of 6mol·L$^{-1}$ HNO$_3$ into the second portion. Heat the third portion to boil. Observe the phenomena. The aqua ion of Fe$^{3+}$ is achromatic, but its solution appears brown after being hydrolyzed into various subsalts. What influence on the hydrolysis equilibrium will be brought when adding HNO$_3$ or heating? Explain it.

(3) Dilute 0. 5 mL SbCl$_3$ solution with water and observe whether the precipitate forms. Does the precipitate dissolve while adding 6 mol·L$^{-1}$ HCl? Dilute it again, whether the precipitate forms again? Explain it. The total reaction equation of the hydrolysis of SbCl$_3$ is as follows:

$$SbCl_3 + H_2O \rightleftharpoons SbOCl + 2HCl$$

(4) Add 1 mL 0. 1 mol·L$^{-1}$ Al$_2$(SO$_4$)$_3$ and 1 mL 0. 1 mol·L$^{-1}$ NaHCO$_3$ solution in two small test tubes, and test their pH values with pH test paper. Write down the hydrolysis reaction equation. Then pour NaHCO$_3$ solution into Al$_2$(SO$_4$)$_3$ solution, observe the phenomena. Explain it with the knowledge of hydrolysis balance.

### 4.　The formation and dissolving of the precipitate

(1) Add 0. 5 mL saturated solution of (NH$_4$)$_2$C$_2$O$_4$ and 0. 5 mL 0. 1 mol·L$^{-1}$ CaCl$_2$ in a test tube. Prepare another one as described above. Then, in one test tube, add about 2 mL 2 mol·L$^{-1}$ HCl, stir, and observe whether the precipitate dissolve or not. And in another test tube, add about 2 mL 2 mol·L$^{-1}$ HAc, stir, and observe whether the precipitate dis-

solve? Explain the phenomena.

(2) Add 1 mL 0.1 mol·L$^{-1}$ MgCl$_2$ solution in two test tubes respectively, and add 2 mol·L$^{-1}$ NH$_3$·H$_2$O in the solution till the white precipitate of Mg(OH)$_2$ is produced. Then add 2 mol·L$^{-1}$ HCl solution in one test tube and saturated NH$_4$Cl solution in the other, observe in which test tube the precipitate is dissolved. Explain the possible influences HCl or NH$_4$Cl has on the precipitate equilibrium.

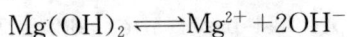

$$Mg(OH)_2 \rightleftharpoons Mg^{2+} + 2OH^-$$

(3) Compare the solubility of Ca(OH)$_2$, Mg(OH)$_2$ and Fe(OH)$_3$

In three test tubes, add 0.5 mL CaCl$_2$(0.1 mol·L$^{-1}$), MgCl$_2$(0.1 mol·L$^{-1}$) and FeCl$_3$ (0.1 mol·L$^{-1}$), add several drops of 2 mol·L$^{-1}$ NaOH respectively, and observe whether precipitate forms.

In three test tubes, add 0.5 mL CaCl$_2$(0.1 mol·L$^{-1}$), MgCl$_2$(0.1 mol·L$^{-1}$) and FeCl$_3$(0.1 mol·L$^{-1}$), add several drops of 2 mol·L$^{-1}$ NH$_3$·H$_2$O respectively, and observe whether precipitate forms.

In three test tubes, add 0.5 mL CaCl$_2$(0.1 mol·L$^{-1}$), MgCl$_2$(0.1 mol·L$^{-1}$) and FeCl$_3$ (0.1 mol·L$^{-1}$), add 5 mL of mixed solution of saturated NH$_4$Cl solution and 2 mol·L$^{-1}$ NH$_3$·H$_2$O (V:V=1:1) respectively, and observe whether precipitate forms.

Compare the solubility of Ca(OH)$_2$ and Mg(OH)$_2$ and Fe(OH)$_3$ in the above experiments, explain the reason.

### 5. Transformation of precipitate

(1) Mix about 0.5 mL 0.1 mol·L$^{-1}$ Pb(NO$_3$)$_2$ with about 0.5 mL 0.1 mol·L$^{-1}$ Na$_2$SO$_4$ in a test tube, observe the formation of the white precipitate. Then add about 0.5 mL 0.1 mol·L$^{-1}$ K$_2$CrO$_4$ in the solution, stir, and observe the phenomenon that the white precipitate of PbSO$_4$ gradually transforms into the yellow precipitate of PbCrO$_4$. Write down its reaction equation and explain the phenomenon based on the principle of solubility product.

(2) Add several drops of 0.1 mol·L$^{-1}$ AgNO$_3$ in a test tube, and a few drops of K$_2$CrO$_4$ solution. Observe the formation of the brown red precipitate of Ag$_2$CrO$_4$, centrifugate to separate the precipitate and wash it with water. Then add the precipitate in the 0.1 mol·L$^{-1}$ solution of NaCl, carefully observing the process that the brown red precipitate changes into the white precipitate of AgCl. Write down its reaction equation and explain.

## Questions

1. Explain the phenomenon of that of the four solutions—when H$_3$PO$_4$, NaH$_2$PO$_4$, Na$_2$HPO$_4$ and Na$_3$PO$_4$ solutions are in the same molarity, but they show acid, weak acid, weak alkalinous and alkalinous characteristic respectively.

2. What is the influence of heating on hydrolysis? Why?

3. Does it appear buffer effect when we mix 10 mL (0.20 mol·L$^{-1}$) HAc and 10 mL (0.10) NaOH? What is the range of the solution's pH value?

4. What are the conditions of the dissolving and transformation of the precipitate?

# 实验十　电离度和电离常数的测定

## 实验目的

1. 学习测定弱电解质的电离常数及电离度。
2. 加强对溶液电导等概念及其有关知识的理解。
3. 学习使用酸度计、电导率仪等仪器并初步掌握它们的使用技术。

## 酸度计及其有关技术

酸度计（也称 pH 计）是用来测量溶液 pH 的仪器。实验室常用的酸度计有雷磁 25 型，pHS-2 型和 pHS-3 型等。它们的原理相同，结构略有差别。下面介绍 pHS-2 型酸度计，其他型号酸度计的使用可查阅使用说明书。

### 1. 基本原理

酸度计测 pH 的方法是电位测定法。它除测量溶液的酸度外，还可以测量电池电动势（mV）。酸度计主要是由参比电极（甘汞电极）、测量电极（玻璃电极）和精密电位计三部分组成。

甘汞电极（图 10-1）：由金属汞、氯化亚汞（$Hg_2Cl_2$，即甘汞）和饱和氯化钾溶液组成的电极，内玻璃管封接着一根铂丝，铂丝插入纯汞中，纯汞下面有一层甘汞和汞的糊状物。外玻璃管中装入饱和 KCl 溶液，下端用素烧陶瓷塞塞住，通过素瓷塞的毛细孔，可使内外溶液相通。电极反应：$Hg_2Cl_2 + 2e^- = 2Hg + 2Cl^-$。甘汞电极的电极电势不随溶液 pH 变化而变化，在一定温度下有一定值。25℃时饱和甘汞电极电势为 0.245V。

玻璃电极（图 10-2）：它的主要部分是头部的球泡，由特殊的敏感玻璃薄膜构成。薄膜对氢离子有敏感作用，当它浸入被测溶液内，被测溶液的氢离子与电极球泡表面水化层进行离子交换，球泡内层也同样产生电极电势。由于内层氢离子浓度不变，而外层氢离子浓度在变化，因此内外层的电势差也在变化，所以该电极电势随待测溶液的 pH 不同而改变。

$$\varphi_{玻} = \varphi_{玻}^{\ominus} + 0.0592 \lg [H^+]$$
$$= \varphi_{玻}^{\ominus} - 0.0592 \, pH$$

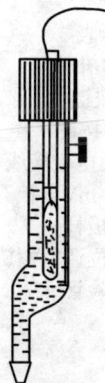

图 10-1　饱和甘汞电极　　　　　　图 10-2　玻璃电极

将玻璃电极和饱和甘汞电极一起浸在被测溶液中组成电池，并连接上精密电位计，即可测定电池电动势 $E$。在 25℃ 时，$E=\varphi_{正}-\varphi_{负}=\varphi_{甘汞}-\varphi_{玻}=0.245-\varphi_{玻}+0.0592\text{pH}$。

整理上式得：

$$\text{pH}=(E+\varphi_{玻}^{\ominus}-0.245)/0.0592$$

$\varphi_{玻}^{\ominus}$ 可用已知 pH 的缓冲溶液代替待测溶液而求得。为了省去计算过程，酸度计把测得的电池电动势直接用 pH 刻度值表示出来。因而从酸度计上可以直接读出溶液的 pH。

复合电极：为了使操作和管理更加方便，使用时不易损坏，现在酸度计大多配用 pH 复合电极，即把 pH 玻璃电极和外参比电极（如 Ag-AgCl 电极）以及外参比溶液（有的还有温度测定探头）一起装在一根电极塑管中，合为一体。底部露出玻璃球泡，加护罩保护。电极头还有一个带有保护液（饱和 KCl 溶液）的外套，pH 玻璃电极和外参比电极的引线用缆线及复合插头与测量仪器连接。结构如图 10-3 所示。

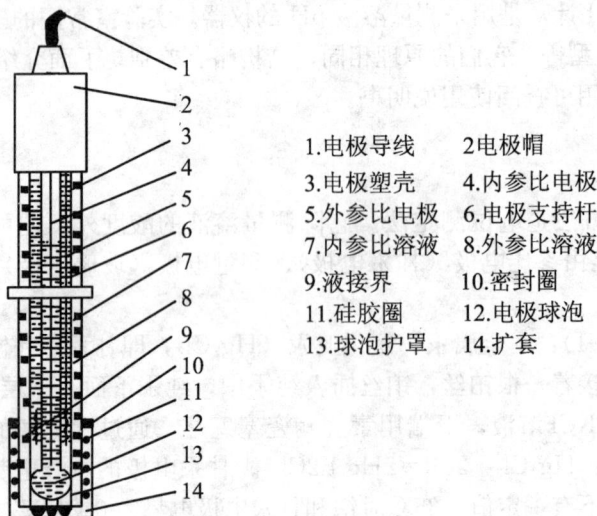

| | |
|---|---|
| 1.电极导线 | 2.电极帽 |
| 3.电极塑壳 | 4.内参比电极 |
| 5.外参比电极 | 6.电极支持杆 |
| 7.内参比溶液 | 8.外参比溶液 |
| 9.液接界 | 10.密封圈 |
| 11.硅胶圈 | 12.电极球泡 |
| 13.球泡护罩 | 14.扩套 |

图 10-3　pH 复合电极

### 2. pHS-2 型酸度计使用方法

（1）仪器的安装。如图 10-4，装好电极杆 13，接触电源。电源为交流电，电压必须符合标牌上所指明的数值，电压太低或电压不稳定影响使用。电源插头中的黑线表示接地线，不能与其他两根线错位。

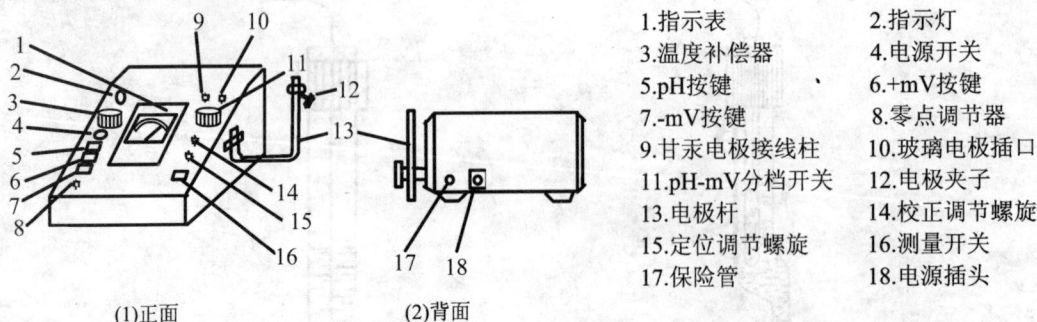

| | |
|---|---|
| 1.指示表 | 2.指示灯 |
| 3.温度补偿器 | 4.电源开关 |
| 5.pH按键 | 6.+mV按键 |
| 7.-mV按键 | 8.零点调节器 |
| 9.甘汞电极接线柱 | 10.玻璃电极插口 |
| 11.pH-mV分档开关 | 12.电极夹子 |
| 13.电极杆 | 14.校正调节螺旋 |
| 15.定位调节螺旋 | 16.测量开关 |
| 17.保险管 | 18.电源插头 |

(1)正面　　　(2)背面

图 10-4　pHS-2 型酸度计

（2）电极安装。先把电极夹子 12 夹在电极杆 13 上，然后将玻璃电极夹在夹子上，玻璃电极的插头插在电极插口 10 内，并将小螺丝旋紧。甘汞电极夹在另一夹子上，甘汞电极引线连接在接线柱 9 上。使用时应把上面的小橡皮塞和下端橡皮塞拔去，以保持液位压差，不用时要把它们套上。

（3）校正。如果测量 pH，先按下按键 5，但读数开关 16 保持不按下状态。左下角指示灯 2 应亮，为要保持仪表稳定，测量前要预热半小时以上。

①用温度计测量被测溶液的温度。

②调节温度补偿器到被测溶液的温度值。

③将分档开关 11 放在"6"，调节零点调节器 8 使指针指在 pH "1.00"上。

④将分档开关 11 放在"校"位置，调节校正调节器 14 使指针指在满刻度。

⑤将分档开关 11 放在"6"位置上，重复检查 pH 是否在"1.00"位置上。

⑥重复③和④两个步骤。

（4）定位。仪器附有三种标准缓冲溶液（pH 为 4.00，6.86，9.20），可选用一种与被测溶液的 pH 较接近的缓冲溶液对仪器进行定位。仪器定位操作步骤如下：

①向烧杯内倒入标准缓冲溶液，按溶液温度查出该温度时溶液的 pH。根据这个数值，将分档开关 11 放在合适的位置上。

②将电极插入缓冲溶液，轻轻摇动，按下读数开关 16。

③调节定位调节器 15 使指针指在缓冲溶液的 pH（即开档开关上的指示数加表盘上的指示数），至指针稳定为止。重复调节定位调节器。

④开启读数开关，将电极上移，移去标准缓冲溶液，用蒸馏水清洗电极头部，并用滤纸将水吸干。这时，仪器已定好位，后面测量时，不得再动定位调节器。

（5）测量

①放上盛有待测溶液的烧杯，移下电极，将烧杯轻轻摇动。

②按下读数开关 16，调节分档开关 11，读出溶液的 pH。如果指针打出左面刻度，则应减少分档开关的数值。如指针打出右面刻度，应增加分档开关的数值。

③重复读数，待读数稳定后，放下读数开关，移走溶液，用蒸馏水冲洗电极，将电极保存好。

④关上电源开关，套上仪器罩。

### 3. 仪器的维护技术

（1）玻璃电极的维护

①玻璃电极的主要部分为下端的玻璃泡，因球泡极薄，切忌与硬物接触，一旦发生破裂，则完全失效。取用和收藏时应特别小心。安装时，玻璃电极球泡下端应略高于甘汞电极的下端，以免碰到烧杯底。

②新的玻璃电极在使用前应在蒸馏水中浸泡 48 h 以上，不用时最好浸泡在蒸馏水中。

③在强碱溶液中应尽量避免使用玻璃电极。如果使用应迅速操作，测完后立即用水洗涤，并用蒸馏水浸泡（为什么？）。

④电极球泡有裂纹或老化（久放二年以上）则应调换，否则反应缓慢，甚至造成较大的测量误差。

（2）仪器的输入端（即玻璃电极插口）必须保持清洁，不用时将接续器插入，以防灰尘落入。在环境温度较高时，应把电极插子用干净的布擦干。

（3）在按下读数开关时，如果发现指针严重甩动，应放开读数开关，检查分档开关位置及其他调节器是否适当，电极头是否浸入溶液。

（4）转动温度调节按钮时，不要用力太大，防止移动紧固螺丝的位置，造成误差。

（5）当测定完毕后，必须先放开读数开关，再移去溶液，如果不放开读数开关就移去溶液，则指针甩动厉害，影响后面测定的准确性。

# 电导率仪及其有关技术

## 1. 基本原理

在一定的温度下，HAc 在水中的电离常数 $K_a$ 和 HAc 起始浓度 $c$ 及电离度 $\alpha$ 的关系为：

$$K_a = c\alpha^2/(1-\alpha) \tag{1}$$

而 HAc 的电离度 $\alpha$ 又等于浓度为 $c$ 时 HAc 溶液的摩尔电导率 $\Lambda_m$（HAc）和极限摩尔电导率 $\Lambda_m^\infty$（HAc）之比，即为：

$$\alpha = \Lambda_m(\text{HAc})/\Lambda_m^\infty(\text{HAc}) \tag{2}$$

将式（2）代入式（1）后得：

$$K_a = c(\Lambda_m)^2/[\Lambda_m^\infty(\Lambda_m^\infty - \Lambda_m)] \tag{3}$$

整理后为：

$$c\Lambda_m = [(\Lambda_m^\infty)^2 K_a/\Lambda_m] - \Lambda_m \times K_a \tag{4}$$

可见，若一系列不同浓度的 HAc 溶液的摩尔电导率，以 $c\Lambda_m$(HAc)对 $1/\Lambda_m$(HAc)作图应得一直线，其斜率为 $[\Lambda_m^\infty]^2 K_a$。如果知道 $\Lambda_m^\infty$(HAc)的数值，即可求得 $K_a$。$\Lambda_m^\infty$(HAc)可由文献中查出 $\Lambda_m(\text{H}^+)$ 和 $\Lambda_m(\text{Ac}^-)$ 加和得到，而 $\Lambda_m$(HAc)的数值可由实验测定。$\Lambda_m$ 与溶液浓度 $c$ 及电导率 $k$ 的关系为：

$$\Lambda_m = k/c \tag{5}$$

$$k = GL/A = K_{\text{cell}}G \tag{6}$$

其中，$L$ 是电导池两极间的距离，$A$ 是电极面积，$G$ 是所测溶液的电导，$K_{\text{cell}}$ 是电导池的电池常数。若将已知摩尔电导率的电解质溶液（通常用 KCl 溶液）放入电导池，测得其电导，代入（6）式，算出该电导池的电池常数。然后用该电导池测定浓度为 $c$ 的 HAc 溶液的电导，运用（6）式和（5）式算出该 HAc 溶液的摩尔电导率。

## 2. DDS-11A 型电导率仪及其使用技术

电导率仪即为测定液体电导率的仪器，在小烧杯内盛待测水样，插入电导电极，即可从表头读出电导率的值。下面介绍 DDS-11A 型电导率仪（图 10-5）的使用技术。

（1）检查、准备

①在接通电源前表头指针应指为零，若不在零位，可调节表头螺丝使指针指向零位。

②按说明书选择 DJS-I 型铂黑电极 1 支，用待测溶液润洗电极头 2 次，然后插入待测溶液（浸没铂片部分）。调节电极常数调节器 6，指向所用电极常数处（电极上已标明）。

③把量程选择开关 10 转到最高档（右旋到底）；"校正测量"开关 4 放在"校正"位置高、低周开关 3 放在"高周"位置。

④打开电源开关 1，指示灯 2 亮，预热 5~10 min。

（2）校正、测量

①将电极的插头插入电极插口 7，拧紧插口边的小螺丝。

②调节"校正调节器" 5，使指针满刻度。

③将 4 拨向"测量"，此时表头所指读数乘以量程选择开关的倍率，即为被测液实际电导率。测纯水的电导率时，将量程开关 10 选在 10 档，开关 3 放在"低周"；高于 30 ms·m$^{-1}$时，放在"高周"。

④重复（1）中②、③两步，取两个读数的平均值。

测量完毕将开关 10 还原到最高档，开关 4 拨向"校正"，关闭电源，拔下电极，用蒸馏水冲洗后，放回电极盒中。

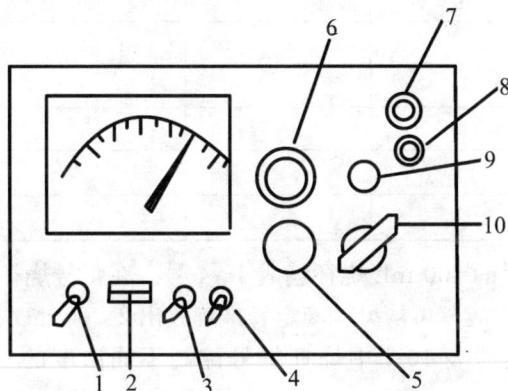

1.电源开关　2.指示灯　3.高、低周开关
4.校正测量工关　5.校正调节器
6.电极常数调节器　7.电极插口
8.电导池　9.电容补偿　10.量程开关

图 10-5　DDS-11A 型电导率仪

## 乙酸电离度和电离常数测定实验

### 1. 实验原理

乙酸（$CH_3COOH$ 或 HAc）是弱电解质，在水溶液中存在以下电离平衡：

$$HAc \rightleftharpoons H^+ + Ac^-$$

若 $c$ 为 HAc 的起始浓度，$[H^+]$，$[Ac^-]$，$[HAc]$分别为 $H^+$，$Ac^-$ 和 HAc 的平衡浓度，$\alpha$ 为电离度，$K_a$ 为电离常数。$H_2O$ 的电离可以忽略时，在纯的 HAc 溶液中$[H^+]=[Ac^-]$，$[HAc]=c(1-\alpha)$，$\alpha=[H^+]/c\times100\%$，则 $K_a=[H^+][Ac^-]/[HAc]=[H^+]^2/(c-[H^+])$。

当 $\alpha<5\%$ 时，$K_a=[H^+]^2/c$。

根据 pH $=-\lg[H^+]$可知，只要测定已知浓度 HAc 溶液的 pH，就可以计算它的电离度和电离常数。本实验用酸度计来测量 HAc 溶液的 pH。

### 2. 实验仪器与药品

酸度计，滴定管（配滴定台、夹），锥形瓶（250 mL），温度计，烧杯（100 mL）4 支，电导率仪，铂黑电导电极（DJS-I 型铂黑电极），移液管（50 mL）2 支，烧杯（高型，100 mL 和 150 mL）各 1 个，1 支玻璃搅拌棒。

NaOH（0.1 mol·L$^{-1}$）标准溶液，HAc（0.1 mol·L$^{-1}$），酚酞指示剂，标准缓冲溶液（pH=4.00）。

### 3. 实验内容

**(1) 酸度计法**

①乙酸溶液浓度的标定。从酸式滴定管中加入 25.00 mL 待标定的 HAc 溶液于锥形瓶中，加入 2~3 滴酚酞溶液，用标准 NaOH 溶液滴定至刚出现微红色摇动后约半分钟不再褪去为止。记下用去标准 NaOH 溶液的体积，重复上述测定，结果填入表 10-1。

表 10-1　乙酸溶液浓度测定数据

| 滴定序号 | | Ⅰ | Ⅱ | Ⅲ |
|---|---|---|---|---|
| 所取 0.10mol·L⁻¹HAc 溶液体积（mL） | | | | |
| 标准 NaOH 溶液的浓度（mol·L⁻¹） | | | | |
| 标准 NaOH 溶液的体积（mL） | | | | |
| 测得 HAc 溶液的浓度 | 测定值 | | | |
| | 平均值 | | | |

②配制不同浓度的乙酸溶液。将 4 支干燥的 100 mL 烧杯编成 1~4 号。在 1 号烧杯中，从滴定管准确放入 24.00 mL 已标定的 HAc 溶液，再从另一滴定管准确放出 24.00 mL 蒸馏水，用干燥的玻璃棒搅匀。用同样方法，按照表 10-2 中烧杯编号配制不同浓度的 HAc 溶液。

③测定乙酸溶液的 pH，并计算乙酸的电离度和电离常数。

按由浓到稀的次序（为什么？）在 pH 计上分别测定它们的 pH，记录数据和室温并填入表 10-2 中。计算电离度及电离常数。

表 10-2　HAc 电离常数和电离度的测定实验数据［室温（　℃）］

| 烧杯编号 | HAc 的体积（mL） | H₂O 的体积（mL） | [HAc]（mol·L⁻¹） | pH | [H⁺]（mol·L⁻¹） | α | $K_a$ |
|---|---|---|---|---|---|---|---|
| 1 | 24.00 | 24.00 | | | | | |
| 2 | 12.00 | 36.00 | | | | | |
| 3 | 6.00 | 42.00 | | | | | |
| 4 | 3.00 | 45.00 | | | | | |

**(2) 电导率仪法**

①接好 DDS-11A 型电导率仪测量线路。先将铂黑电极放在蒸馏水中浸泡数分钟。取出后用蒸馏水淋洗（不要直接冲铂黑），用滤纸吸干电极上的水（切勿触及铂黑）。

②测定 HAc 溶液的电导率，用移液管向干燥洁净的 150 mL 烧杯中加入 50.00 mL 已标定的 HAc 溶液，插入铂黑电极，测定其电导率值。

用另一支移液管向上述烧杯中，加入 50.00 mL 蒸馏水，搅拌均匀（不要触及铂黑电极），测其电导率值。

再用移液管从上述烧杯中吸出 50.00 mL HAc 溶液（移液管壁尽量不沾带出溶液）弃去，并补充 50.00 mL 蒸馏水，搅拌均匀，测其电导率值。

如此，再稀释三次，共测出六种不同浓度 HAc 溶液的电导率。测毕，以蒸馏水洗净铂黑电极，浸入蒸馏水中。

③测定蒸馏水的电导率，取 100 mL 烧杯，洗净后再以蒸馏水荡洗数次。测蒸馏水电

导率。

④数据记录和处理（表 10-3）

HAc 试液的浓度（mol·L$^{-1}$）_____；室温（℃）_____；

HAc 的极限摩尔电导率 $\Lambda_m^{\infty}$（室温）（S·m$^2$·mol$^{-1}$）_____。

**表 10-3　电导率法测定醋酸的电离常数数据**

| 实验序号 | 1 | 2 | 3 | 4 | 5 | 6 | 7 |
|---|---|---|---|---|---|---|---|
| $c$(HAc)(mol·L$^{-1}$) | | | | | | | |
| $k$(HAc. aq)(S·m$^{-1}$) | | | | | | | |
| $k$(HAc. aq)$-k$(H$_2$O)(S·m$^{-1}$) | | | | | | | |
| $k$(S·m$^{-1}$) | | | | | | | |
| $\Lambda_m$(HAc)(S·m$^2$·mol$^{-1}$) | | | | | | | |
| $1/\Lambda_m$(HAc)(S$^{-1}$·m$^{-2}$·mol) | | | | | | | |
| $c\Lambda_m$(HAc)(S·m$^{-1}$) | | | | | | | |

以 $c\Lambda_m$(HAc)对 $1/\Lambda_m$(HAc) 作图，并根据所得直线的斜率求出 $K_a$。

## 思考题

1. 本实验的原理是什么？为何测得 HAc 的起始浓度及其溶液的 pH，便可计算求得 HAc 的 $\alpha$ 和 $K_a$ 值？实验中[HAc]和[Ac$^-$]浓度是怎样测得的？要做好本实验，操作的关键是什么？

2. 改变所测 HAc 的浓度或温度，则电离度和电离常数有无变化？若有变化，会有怎样的变化？

3. 若所用 HAc 溶液的浓度极稀，是否还能用 $K_a=[H^+]^2/c$ 求电离常数？为什么？

4. 使用酸度计测量溶液 pH 的操作步骤有哪些？请写出各个操作步骤的要点。

5. 怎样正确使用玻璃电极？在实验中哪些不正确的操作会使玻璃电极损坏？

6. 什么是电导、电导率、摩尔电导率、极限摩尔电导率？电解质溶液导电的特点是什么？

7. 测定 HAc 溶液的电导率时，溶液的浓度为什么要由浓变稀？

8. 在计算 HAc 溶液的电导率时，为什么要考虑蒸馏水的电导率？测量蒸馏水的电导率时，动作缓慢，可以给实验带来什么误差？

9. 本实验忽略了温度的影响，请讨论这将给 HAc 电导率的测定带来什么影响？

10. 查文献可知，298K 时 HAc 的电离常数为 $1.754\times10^{-5}$，求实验测定值的相对误差，并分析误差原因。

11. 使用铂黑电极时应注意些什么？实验时如何保护好铂黑电极。

# Exp 10   Determining the Degree of Ionization and Ionization Constant

## Objectives

1. To determine the degree of ionization and ionization constant of some weak acid like acetic acid.

2. To consolidate understanding the concepts of conductance of solution and other related knowledge intensively.

3. To learn how to use acidometer, conductivity apparatus and other instruments, and master their using skill.

## pH meter and Some Related Skill

Acidometer (also named pH meter) is a kind of apparatus to determine the pH value of solution. Widely used acidometer in laboratory is Leici-25, pHS-2, pHS-3 and so on . Their principles are the same but have minor differences in structures, we will introduce pHS-2 type acidometer, and we can use other type of acid meter by reading their instructions.

### 1. Principles

The method of measuring pH value of solution by acidometer is electrode potential determination method. It not only determines the pH value of solution, but also determines the electromotive force of a cell (mV). Acidometer is mainly made up of three parts: reference electrode (calomel electrode), measuring electrode (glass electrode), and precise potentiometer.

calomel electrode (Fig. 10-1): it is made up of mercury, $Hg_2Cl_2$ and saturated KCl solution, a platinum wire is sealed in the inner glass tube, which inserts into the mercury, batter mixture of $Hg_2Cl_2$ and mercury is below the pure mercury. Outer glass tube is filled with saturated solution, the bottom is blocked by ceramic, the capillary holes make the inner and outer link with each other. Electrode reaction: $Hg_2Cl_2 + 2e^- = 2Hg + 2Cl^-$. Electrode potential of the calomel electrode does not vary with the change of the solution pH value, at a certain temperature, the pH is definite. At 25℃, the pH value is 0.245V.

The electrode potential of glass electrode (Fig. 10-2):

The main part of a glass electrode is a glass bulb in its upper part, which is composed of special sensible glass thin film. Thin film is sensible to hydrogen-ion, when it dips into solution to be determined, the ion exchange occurs between the hydrogen-ion in the solution to be determined and electrode bubble of surface hydration shell, and the inlet of bubble produce electric potential. Because the $H^+$ ion concentration inside is unchanged, while the outer layer's changes, the potentials difference of the ectonexine is of course changed. As a result, the electric potential varying by the value of pH of solution can be determined.

$$\varphi_{glass} = \varphi_{glass}^{\ominus} + 0.0592 lg[H^+] = \varphi_{glass}^{\ominus} - 0.0592pH$$

Put glass electrode and saturated calomel electrode into the solution to be determined to make up the cell, then link with precise potentiometer, the electromotive force of a cell E can be determined. At the temperature of 25℃, $E = \varphi_{positive} - \varphi_{negative} = \varphi_{calomel} - \varphi_{glass} = 0.245 - \varphi_{glass}^0 + 0.0592pH$.

To arrange the above equation:

$$pH = (E + \varphi_{glass}^0 - 0.245)/0.0592$$

$\varphi_{glass}^0$ can be calculated by using the solution to be determined to take the place of buffer solution with certain value of pH. In order to simplify the calculating procedures, acidometer directly demonstrate the electromotive force of a cell by the value of pH. As a result, we can read the value of the solution from the acidometer directly.

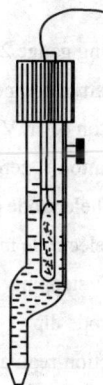

Fig. 10-1　saturated calomel electrode　　　　Fig. 10-2　glass electrode

Combination electrode: in order to operate and manage more conveniently, and the electrode is not prone to be destroyed, for most of the pH-meters employ pH combination electrode, that is to say, the combination electrode including following units: pH glass electrode, exo-reference electrode (e. g Ag—AgCl) and exo-reference solution (for certain electrode, there is temperature detection probe). At the bottom of the electrode, the glass bubble exposes , so a shroud is employed. There is also a overcoat (filled with saturated KCl) for protecting electrode tip. The fuse of the pH glass electrode and exo-reference electrode is linked with measurement apparatus by and combination plug. Its structure is showed in Fig. 10-3.

## 2. pHS-2 acidometer and its use skills

(1) Installation of the instrument

Set up the electrode pole 13 and turn on the power as Fig. 10-4. Power is alternating current, the voltage must accord with the numerical value that on the placard. It will interfere the usage if the voltage is too high or too low, the black wire represent link with floorwire and it must not be mistaken with the other two lines.

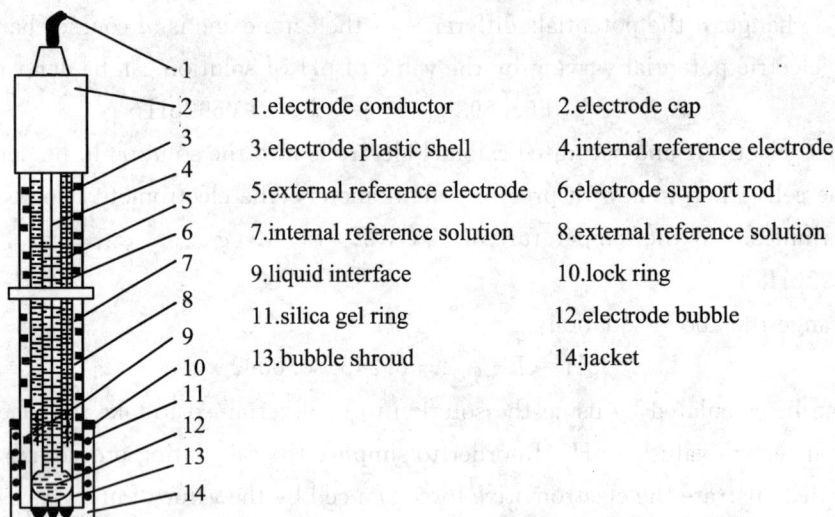

1. electrode conductor　　　　2. electrode cap

3. electrode plastic shell　　　4. internal reference electrode

5. external reference electrode　6. electrode support rod

7. internal reference solution　8. external reference solution

9. liquid interface　　　　　　10. lock ring

11. silica gel ring　　　　　　12. electrode bubble

13. bubble shroud　　　　　　14. jacket

Fig. 10-3　pH combination electrode

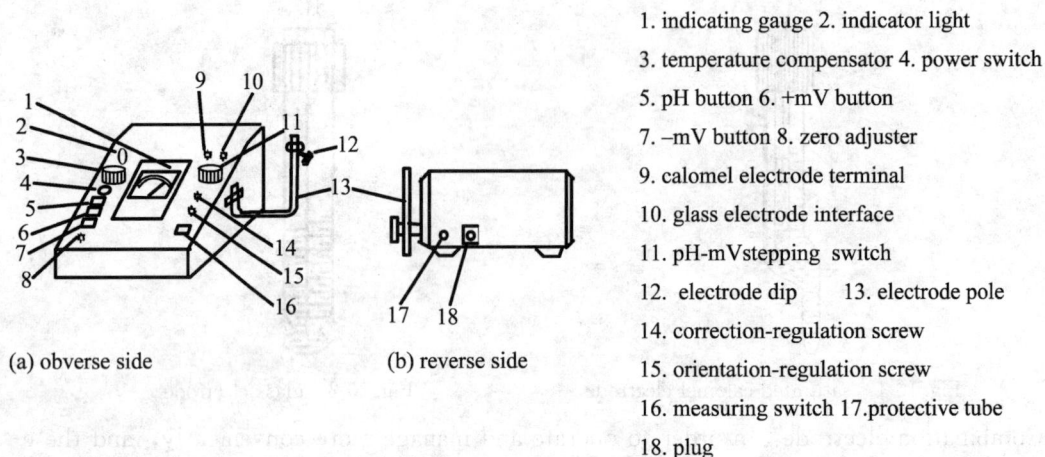

1. indicating gauge 2. indicator light

3. temperature compensator 4. power switch

5. pH button 6. +mV button

7. −mV button 8. zero adjuster

9. calomel electrode terminal

10. glass electrode interface

11. pH-mVstepping  switch

12. electrode dip　　　13. electrode pole

14. correction-regulation screw

15. orientation-regulation screw

16. measuring switch 17.protective tube

18. plug

(a) obverse side　　　　　(b) reverse side

Fig. 10-4　　pHS-2 acidometer

（2）Installation of the electrode

First, clip the electrode clip 12 onto electrode pole 13, and then clip the glass electrode to the clip. The plugs of glass electrode are plugged in electrode jack 10 and make the screw fasten. Clip calomel electrode to another clip, the terminal of calomel electrode connect with the terminal post 9. We should pick up the small rubber stopper in the upper and the rubber stopper in the inferior extremity when we use it. In order to keep the differential pressure of the liquid, sheathe them when are not in use.

（3）Correction

First, press the key 5, if determine the value of pH, keep the reading switch not being pressed. The indicator light on the lower left quarter should be bright. It is better to pre-heat before measuring in order to make the instrument stable.

①Determine the temperature of the solution determined by thermometer.

②Adjust the temperature compensator to make indicating temperature the same as the value of the solution to be determined.

③Put the switch 11 to "6", adjust zero adjuster 8 to make the pointer point to pH "1.00".

④Put the switch 11 to "adjust", adjust correction adjuster 14 make the pointer point to full scale.

⑤Put the switch 11 to "6", repeat checking the position pH "1.00".

⑥Repeat procedures ③ and ④.

(4) Fixing

The instrument have three standard buffer solutions (pH is 4.00, 6.68, 9.20), we can fix the instrument with the above buffer solutions which are approach of the solution pH value to be determined.

①Pour standard buffer solution to a beaker, to find out the value of pH at this temperature according to solution temperature. Put the switch 11 to suitable position on the basis of the data.

②Dip the electrode to buffer solution, shake slightly, press the reading switch 16.

③Adjust the fixing adjuster 15 to make the pointer point to the value of the pH of buffer solution (that is to say the indicator number on switch add the indicator number on the dial), until the pointer stable. Repeat fiting the fixing adjuster.

④Turn on the reading switch, move up the electrode, move away the standard buffer solution, clean the upper part of the electrode with distilled water, and blot the water with filter paper. The instrument has been fixed now. Don't adjust the fixing adjuster in the latter determination.

(5) Determination

①Place the electrode in the beaker in the solution to be determined, shake the beaker slightly.

②Press the reading switch 16; adjust the switch 11, read the pH value of the solution. We should decrease the numerical value of switch if the pointer goes beyond the scale on the left, and increase the numerical value of switch if the pointer goes beyond the scale on the right.

③Repeat reading, release the reading switch until the reading is stable, move away the solution, wash the electrode with distilled water, and preserve it well.

④Turn off the power switch; coat the instrument with cover.

## 3. Maintenance skill of the instrument

(1) Maintenance of glass electrode

①The main fraction part of a glass electrode is a glass bulb in its upper part. The bulb is so thin that make sure not to touch it with any hard substance in use. When the bulb is broken, this glass electrode will be invalid completely. So in the fixing procedures, please pay attention to keep the glass bulb a little higher than the bottom part of the calomel electrode to avoid being broken by the beaker.

②A new glass electrode should be dipped into distilled water no less than 48 hours before use.

③It should be avoided using glass electrode in the strong alkalinous solution. If we had to use it, we should operate quickly, wash it with water immediately, and soak it with distilled water (Why?).

④The crazing or aging electrode bulb (to place more than 2 years) should be abandoned otherwise it will response slowly, even cause obvious measuring error.

(2) The input end (that is the jack of electrode) should be kept clear, insert the connecter. When it is not used to prevent the dust from falling into it, we should wipe the electrode plug dry with clean cloth, when the temperature is too high.

(3) If the needle shakes seriously when the reading switch is pressed, we should release reading switch, check the position of switch and other adjuster to see whether they are suitable; and whether the electrode tip is dipped into the solution.

(4) Don't exert your strength to rotate the thermoregulation button, in order not to move the position of tightening screw and cause error.

(5) We must first loose the reading switch then move away the solution when measuring finishes. If not, the needle will shake seriously and bring harm to the correction of the latter determination.

## Conductivity apparatus and some related skill

### 1. Principles

At a certain temperature in water, the relationship among the ionization constant of the HAc, the original concentration and degree of ionization is as follow:

$$K_a = c\alpha^2/(1-\alpha) \tag{1}$$

And the degree of ionization of HAc, $\alpha$ is equal to the ratio of molar conductivity $\Lambda_m^\infty$ (HAc) to limit molar conductivity, that is

$$\alpha = \Lambda_m(\text{HAc})/\Lambda_m^\infty(\text{HAc}) \tag{2}$$

Substitute formula (2) to (1):

$$K_a = c(\Lambda_m)^2/[\Lambda_m^\infty(\Lambda_m^\infty - \Lambda_m)] \tag{3}$$

to arrange the above formula is

$$c\Lambda_m = [(\Lambda_m^\infty)^2 K_a/\Lambda_m] - \Lambda_m^\infty \times K_a \tag{4}$$

From the formula, we can see if a series of molar conductivity of HAc solutions with various concentrations are determined. $c\Lambda_m(\text{HAc})$ as Y-axis, $1/\Lambda_m(\text{HAc})$ as X-axis, draw graph to get a line, its slope rate is $[\Lambda_m^\infty(\text{HAc})]^2 K_a$. If the $\Lambda_m^\infty(\text{HAc})$ is known, $K_a$ can be calculated. We can make use of data of $\Lambda_m^\infty(\text{H}^+)$ and $\Lambda_m^\infty(\text{Ac}^-)$ in the documents to calculate $\Lambda_m^\infty(\text{HAc})$, while the value of $\Lambda_m(\text{HAc})$ can be determined by experiment.

The realtionship among $\Lambda_m^\infty$, $c$ and $k$ is: $\tag{5}$

$$\Lambda_m^\infty = k/c$$

$$k = GL/A = K_{cell}G \tag{6}$$

In the formula: $L$ represents the distance of two electrodes. $A$ represents the area of e-lectrode; $G$ represents the conductivity of solution. $K_{cell}$ represents the constant of conductant, if the known electrolyte solution (usually KCl) is put into conductance cell, the conductance can be determined. Its conductivity can be calculated by formula 5 and 6.

In this experiment, the conductivity of electrolyte can be directly determined by conductivity instrument.

### 2.  DDS-11A conductivity apparatus and its using skills

Conductivity instrument is the apparatus, which determines the conductivity of solution. First add the solution to be determined in a small beaker, and then insert the conductance electrode. We can read the value of conductivity from the gauge head. In this chapter, we will introduce using skills of DDS-11A conductivity apparatus.

(1) Checking, preparing

①Before the power is on, the needle should point to zero, we can adjust the screw on the gauge head to let it point to zero.

②Choose a DJS-Ⅰ Pt electrode matched with the instrument, and wash the electrode tip for 2 times by the solution to be determined.

Then insert it into the solution to be determined (immerse the part of Pt absolutely). Regulate the electrode constant adjuster 6 to point at the used electrode constant (already being labeled on the electrode).

③Rotate the range selection switch to the highest grade (on the very right), put "correction, determination" switch 4 to the position of "correction", high/low cycle switch 3 to the position of "high cycle".

④Turn on the power switch 1, the indicator light 2 brightens; preheat for 5~10 min.

(2) Correction, determination

①Insert the plug of the electrode to electrode jack 7; make the small screw near the jack fasten.

②Adjust "correction adjuster" 5, make the pointer full scale.

③Pluck 4 to "determination", the reading, which the gauge head points plus multiply factor of range switch, is the factual conductivity of the solution to be determined. When measuring the conductivity of purified water, we should put range switch 10 to 10th, switch 3 to 'low cycle', when conductivity is higher than 30 ms/m, switch 3 should be put to ' high cycle'.

④Repeat step ②and ③, get the average value of two readings.

Put the switch 10 to the highest step, put the switch 4 to 'correction', turn off the power, pull out the electrode, wash it with distilled water and put it back to the electrode box.

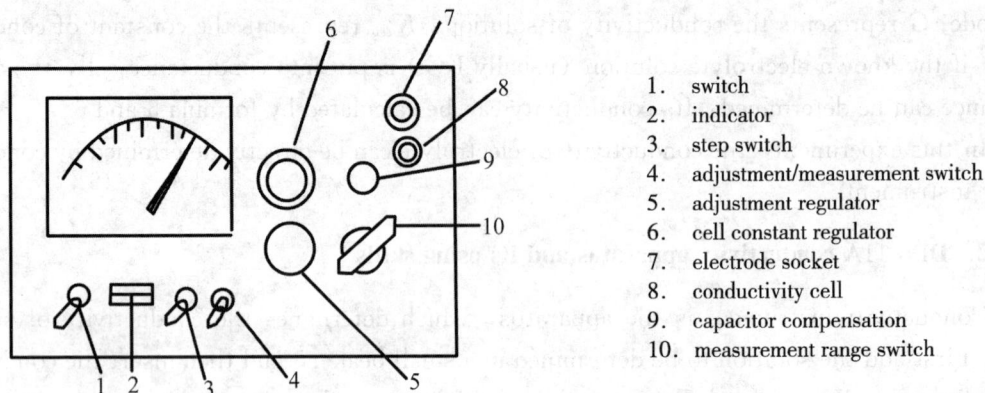

1. switch
2. indicator
3. step switch
4. adjustment/measurement switch
5. adjustment regulator
6. cell constant regulator
7. electrode socket
8. conductivity cell
9. capacitor compensation
10. measurement range switch

Fig. 10-5   DDS-11A conductivity apparatus

# Experiment of determining the degree of ionization and ionization constant

## 1. Principles

Acetic acid ($CH_3COOH$ or HAc) is a weak electrolyte. There exists ionization equilibrium in aqueous solution:

$$HAc \rightleftharpoons H^+ + Ac^-$$

In this reaction, $c$ represents the starting concentration of HAc, $[H^+]$, $[Ac^-]$ represent the equilibrium concentration of $H^+$ and $Ac^-$, $\alpha$ represents degree of ionization, $K$ represents ionization constant. In pure HAc solution, $[H^+] = [Ac^-]$, $[HAc] = c(1-\alpha)$, $\alpha = [H^+]/c \times 100\%$, $K_a = [H^+][Ac^-]/[HAc] = [H^+]^2/(c-[H^+])$, when $\alpha < 5\%$, $K_a = [H^+]^2/c$.

According to the formula $pH = -\lg[H^+]$, we can calculate its degree of ionization and ionization constant on the basis of measuring the pH value of HAc solution which has certain concentration. This experiment determines the pH value of HAc solution by acidometer.

## 2. Instrument and Reagents

acidometer, burette (matching with titration desk, clamp), dropping pipette, conical flask (250 mL), thermometer, four beakers (100 mL), conductivity apparatus, Pt conductance electrode, (DJS-I Pt electrode), two transfer pipettes (50 mL), beaker (high type, 100 mL and 150 mL), glass rod.

NaOH (0.1 mol·$L^{-1}$) standard solution, HAc (0.1 mol·$L^{-1}$), phenolphthalein indicator, standard buffer solution (pH=4.00).

## 3. Experimental Procedures

(1) By acidometer

①Standardization of the acetic acid solution. Transfer 25.00 mL HAc solution which will be standardized to a conical flask by an acid burette, then add 2~3 drops of phenol-phthalein indicator. Titrate it with standard NaOH solution until red color appears and doesn't fade after shaking constantly for about half a minute. Record the volume of standard NaOH solution, repeat the titration and fill in Tab 10-1.

**Tab 10-1   Determining data of the concentration of acetic acid solution**

| titration number | I | II | III |
|---|---|---|---|
| the volume of 0.10 mol·L$^{-1}$ HAc solution (mL) | | | |
| the concentration of standard NaOH solution (mol·L$^{-1}$) | | | |
| the volume of standard NaOH solution (mL) | | | |
| concentration of HAc solution by measuring — measuring value | | | |
| concentration of HAc solution by measuring — average value | | | |

②Prepare acetic acid solution with various concentrations. Label four dry 100 mL beakers with numbers from 1 to 4. Transfer 24.00 mL standardized HAc solution to No.1 beaker by a burette, then add 24.00 mL distilled water by another burette, stir it constantly with a dry glass rod. Use the above method to prepare acetic acid solutions with various concentrations according to the beaker labeled in Tab 10-2.

③Determine the pH value of acetic acid solution and calculate the degree of ionization and ionization constant of acetic acid.

Determine their pH values by pH meter as the concentration is decreasing (why?). Record the data at room temperature, and then fill in the blanks. Calculate the degree of ionization and ionization constant.

**Tab 10-2   The experimental data of degree of ionization and ionization constant of HAc [r. t ( ℃)]**

| No. | volume of HAc (mL) | volume of H$_2$O (mL) | [HAc] (mol·L$^{-1}$) | pH | [H$^+$] (mol·L$^{-1}$) | $\alpha$ | $K_a$ |
|---|---|---|---|---|---|---|---|
| 1 | 24.00 | 24.00 | | | | | |
| 2 | 12.00 | 36.00 | | | | | |
| 3 | 6.00 | 42.00 | | | | | |
| 4 | 3.00 | 45.00 | | | | | |

(2) By conductivity apparatus

①Link the circuit of the DDS-11A conductivity apparatus well. First dip the Pt electrode into the distilled water for several minutes, use distilled water to elute after taking out of it (Don't elute Pt directly), then blot the water onto the electrode by filter paper (Don't touch the Pt).

②Determine the conductivity of HAc solution using HAc. Add 50 mL standardized HAc solution to a dry, clean 150 mL beaker by a transfer pipet, insert Pt electrode, and determine its value of conductivity.

Add 50 mL distilled water to the above beaker by another transfer pipet, stir it well-distributed (Don't touch the Pt electrode), determine its value of conductivity.

Take out 50 mL HAc solution from the above beaker by the first transfer pipet (Don't take out the solution on the wall of transfer pipet as far as possible) and throw away, add 50 mL distilled water instead, stir constantly, determine its value of conductivity.

Repeat the above operation, dilute three times, and determine the value of conductivity of six solutions with different concentrations. After that, wash the Pt electrode with distilled water, dip it into the distilled water.

③To determine the conductivity of distilled water, fetch a clean 100 mL beaker, and wash it with distilled water for several times. Determine the conductivity of distilled water.

④Data record and data process (Tab 10-3).

The concentration of HAc test solution (mol·L$^{-1}$)

room temperature ( ℃);

the limited molar conductivity of HAc $\Lambda_m^{\infty}$ (r. t) (S·m$^2$·mol$^{-1}$)

Tab 10-3　The experimental data of degree of ionization and ionization constant of HAc by conductivity method

| serial number | 1 | 2 | 3 | 4 | 5 | 6 | 7 |
|---|---|---|---|---|---|---|---|
| $c$(HAc)(mol·L$^{-1}$) | | | | | | | |
| $k$(HAc. aq)(S·m$^{-1}$) | | | | | | | |
| $k$(HAc. aq)$-k$(H$_2$O)(S·m$^{-1}$) | | | | | | | |
| $k$(S·m$^{-1}$) | | | | | | | |
| $\Lambda_m$(HAc)(S·m$^2$·mol$^{-1}$) | | | | | | | |
| $1/\Lambda_m$(HAc)(S$^{-1}$·m$^{-2}$·mol) | | | | | | | |
| $c\Lambda_m$(HAc)(S·m$^{-1}$) | | | | | | | |

Draw a line according to $c\Lambda_m$ and $1/\Lambda_m$(HAc) and get the $K_a$ according to the slope rate of the straight line.

## Questions

1. What's the principle of this experiment? Why we can calculate the value of α and $K_a$ of HAc by first measuring the initial concentration and its pH value of the HAc? How to determine the concentration of HAc and Ac$^-$? What's the key to this experiment in order to perform it successfully?

2. Whether the degree of ionization and ionization constant changed when the temperature and the concentration of the HAc change? If it is changed, how does it change?

3. When the concentration of the HAc solution is rather dilute, can we get ionization constant by using this formula $K_a=[H^+]^2/c$? Why?

4. What are the operation steps when we use the acidometer? Please write down the main points of each operation step.

5. How to use the glass electrode correctly? What wrong operations will lead to the damage of glass electrode?

6. What's the meaning of conductance, conductivity, and molar conductivity, limited molar conductivity? What's the characteristic of electrolyte to conduct?

7.  When we determine the conductivity of HAc solution, why the concentration of the solution should be decreased?

8.  As we calculate the conductivity of HAc solution, why it is needed to consider the conductivity of distilled water? If we determine the conductivity of distilled water slowly, what's the error it will lead to?

9.  This experiment neglects the influence on the temperature; please discuss what will it lead to on the determining of the conductivity of HAc?

10.  According to the references, at the temperature of 298K, the ionization constant of HAc is $1.754 \times 10^{-5}$, please calculate the relative error of the experimental determine value, and explain the reason of the error.

11.  What's the caution when we use the Pt electrode? How to protect the Pt electrode in the experiment?

# 实验十一　氧化还原反应和电化学

## 实验目的

1. 试验并掌握电极电势与氧化还原反应方向的关系，以及介质和反应物浓度对氧化还原反应的影响。

2. 定性观察并了解化学电池的电动势，氧化态或还原态浓度变化对电极电势的影响。

3. 试验并了解电解反应。

## 基本原理

氧化还原过程也就是电子的转移过程。氧化剂在反应中得到了电子，还原剂失去了电子。这种得、失电子能力的大小或者说氧化、还原能力的强弱，可用它们的氧化态－还原态（例如 $Fe^{3+}-Fe^{2+}$，$I_2-I^-$，$Cu^{2+}-Cu$）所组成的电对的电极电势的相对高低来衡量。一个电对的电极电势（以还原电势为准）代数值越大，其氧化态的氧化能力越强，还原态的还原能力越弱。反之亦然。所以根据其电极电势（$\varphi$）的大小，便可判断一个氧化还原反应的进行方向。例如 $\varphi^{\ominus}(I_2/I^-)=+0.535V$，$\varphi^{\ominus}(Fe^{3+}/Fe^{2+})=+0.771V$，$\varphi^{\ominus}(Br_2/Br^-)=+1.08V$，所以对于下列两反应：

$$2Fe^{3+}+2I^- \rlap{=\!=\!=\!=} \quad I_2+2Fe^{2+} \tag{1}$$

$$2Fe^{3+}+2Br^- \rlap{=\!=\!=\!=} \quad Br_2+2Fe^{2+} \tag{2}$$

式（1）应向右进行，式（2）应向左进行，也就是说 $Fe^{3+}$ 可以氧化 $I^-$ 而不能氧化 $Br^-$。反过来说，$Br_2$ 可以氧化 $Fe^{2+}$，而 $I_2$ 则不能。总之氧化态的氧化能力 $Br_2>Fe^{3+}>I_2$，还原态的还原能力 $I^->Fe^{2+}>Br^-$。

298K 时浓度与电极电势的关系可用奈斯特方程式表示：

$$\varphi=\varphi^{\ominus}+\frac{0.0592}{n}\lg\frac{[氧化型]}{[还原型]}$$

例如以 $Fe^{3+}-Fe^{2+}$ 电对为例：

$$\varphi_{(Fe^{3+}/Fe^{2+})}=\varphi^{\ominus}_{(Fe^{3+}/Fe^{2+})}+\frac{0.0592}{1}\lg\frac{c(Fe^{3+})/c^{\ominus}}{c(Fe^{2+})/c^{\ominus}}$$

由此可知，$Fe^{3+}$ 或 $Fe^{2+}$ 浓度的变化都会改变其电极电势数值。特别是有沉淀剂（包括 $OH^-$）或络合剂的存在，能够大大减少溶液中某一种离子浓度，甚至可以改变反应的方向。

有些反应特别是含氧酸根离子参加的氧化还原反应中，经常有 $H^+$ 参加，介质的酸度也对 $\varphi$ 值产生影响。例如对于半电池反应：

$$MnO_4^-+8H^++5e^- \rlap{\rightleftharpoons} \quad Mn^{2+}+4H_2O$$

$$\varphi=\varphi^{\ominus}(MnO_4^-/Mn^{2+})+\frac{0.0592}{5}\lg\frac{[MnO_4^-]\cdot[H^+]^8}{[Mn^{2+}]}$$

$H^+$ 增大可使 $MnO_4^-$ 氧化性增强。

单独的电极电势是无法测量的，只能从实验中测量两个电对组成的原电池的电动势。因为在一定条件下一个原电池的电动势 $E^{\ominus}$ 为正、负电极的电极电势之差：

$$E^{\ominus}=\varphi_{+}^{\ominus}-\varphi_{-}^{\ominus}$$

所以先规定在一定大气压，298 K 和 $H^+$ 浓度为 1.00 mol·$L^{-1}$ 的条件下 $\varphi^{\ominus}$（$H^+/H_2$）为零，然后测定一系列原电池（包括氢电池或其他参比电极）的电动势，从而直接或间接测出其他电极的 $\varphi^{\ominus}$。准确的电动势是用对消法在电位差计上测量。本实验中只是为了进行比较，只需知道其相对数值，所以在 pH 计上进行测量。

电流通过电解质溶液，在电极上引起的化学变化叫电解。电解时电极电势的高低，离子浓度的大小，电极材料等因素都可以影响两极上的电解产物。在本实验中电解 $Na_2SO_4$ 溶液时以铜作电极，其电极反应如下：

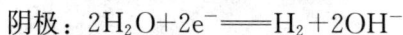

阴极：$2H_2O+2e^-\!=\!\!=\!\!=\!H_2+2OH^-$

阳极：$Cu-2e^-\!=\!\!=\!\!=\!Cu^{2+}$

## 仪器和药品

雷磁 pHS-3 型酸度计，烧杯（50 mL），小试管，盐桥，导线。

$H_2SO_4$（3 mol·$L^{-1}$，2 mol·$L^{-1}$），HAc（6 mol·$L^{-1}$），$Pb(NO_3)_2$（0.5 mol·$L^{-1}$），$CuSO_4$（1 mol·$L^{-1}$、0.5 mol·$L^{-1}$、0.1 mol·$L^{-1}$），KI（0.1 mol·$L^{-1}$），$FeCl_3$（0.1 mol·$L^{-1}$），KBr（0.1 mol·$L^{-1}$），$FeSO_4$（0.1 mol·$L^{-1}$），$KMnO_4$（0.01 mol·$L^{-1}$），$ZnSO_4$（1 mol·$L^{-1}$、0.5 mol·$L^{-1}$、0.1 mol·$L^{-1}$），$Na_2SO_4$（0.5 mol·$L^{-1}$），$CCl_4$，浓硫酸，浓氨水，碘水，溴水，酚酞，锌片，铜片，铅粒，砂纸，品红试纸。

## 实验内容

### 1. 电极电势与氧化还原反应的关系

（1）比较锌、铅和铜在电位序中的位置

在两只小试管中分别注入 0.5 mol·$L^{-1}$ 的 $Pb(NO_3)_2$ 和 0.5 mol·$L^{-1}$ 的 $CuSO_4$，各放入一块表面擦净的锌片，放置片刻，观察锌片表面有何变化。

用表面擦净的铅粒代替锌片，分别与 0.5 mol·$L^{-1}$ $ZnSO_4$ 和 0.5 mol·$L^{-1}$ $CuSO_4$ 溶液反应，观察铅粒表面有何变化。

写出反应式，说明电子转移方向，并确定锌、铜和铅在电位序中的相对位置。

（2）在小试管中将 3~4 滴 0.1 mol·$L^{-1}$ KI 溶液用蒸馏水稀释至 1 mL。加入 0.5 mL 四氯化碳，震荡，观察四氯化碳层中的颜色。加入 2 滴 0.1 mol·$L^{-1}$ $FeCl_3$，再充分振荡，观察四氯化碳液层的颜色有何变化（$I_2$ 溶于四氯化碳层显紫红色）。

（3）用 0.1 mol·$L^{-1}$ KBr 溶液代替 0.1 mol·$L^{-1}$ KI 溶液进行同样的实验，观察四氯化碳层的颜色（溴溶于四氯化碳中显棕黄色）。

根据（2）、（3）实验的结果，定性地比较 $Br_2-Br^-$，$I_2-I^-$ 和 $Fe^{3+}-Fe^{2+}$ 三个电对的电极电势的相对高低（即代数值的相对大小），并指出哪个电对的氧化态是最强的氧化剂，哪个电对的还原态是最强的还原剂。

（4）仿照上面实验，分别用碘水和溴水同 0.1 mol·$L^{-1}$ $FeSO_4$ 溶液作用，观察四氯化碳层的颜色，判断反应是否进行。写出有关的化学反应式。

根据（2），（3）和（4）的实验结果和上面比较得出的三个电对的电极电势的相对大小，说明电极电势与氧化还原反应方向的关系。

## 2. 酸度对氧化还原反应速度的影响

在两个各盛 $0.5$ mL $0.1$ mol·L$^{-1}$ KBr 溶液的试管中，分别加入 $0.5$ mL $3$ mol·L$^{-1}$ H$_2$SO$_4$ 溶液和 $6$ mol·L$^{-1}$ HAc 溶液，然后往两个试管中各加入 $2$ 滴 $0.01$ mol·L$^{-1}$ KMnO$_4$ 溶液。观察并比较两个试管中的紫色溶液褪色的快慢。写出反应式，并加以解释。

## 3. 浓度对氧化还原反应的影响

往两个分别盛有 $2$ mol·L$^{-1}$ H$_2$SO$_4$ 和浓 H$_2$SO$_4$ 的试管加入一片擦去表面氧化膜的铜片，稍加热，观察所发生的现象。

在盛有浓 H$_2$SO$_4$ 的试管，用润湿的品红试纸检验气体（若品红褪色表示有 SO$_2$ 产生），写出有关方程式，并加以解释。

## 4. 浓度对电极电势的影响

（1）在 $100$ mL 烧杯中加入 $30$ mL $1$ mol·L$^{-1}$ CuSO$_4$，在另一个 $100$ mL 烧杯中加入 $30$ mL $1$ mol·L$^{-1}$ ZnSO$_4$ 溶液，然后在 CuSO$_4$ 溶液内放一铜片，在 ZnSO$_4$ 溶液内放一锌片，组成两个电极。用一个盐桥将它们连接起来，通过导线将铜电极接入酸度计的正极，把锌电极通过"接续头"插入酸度计的负极插孔，测定其电势差。

（2）取下盛 CuSO$_4$ 溶液的烧杯，在其中加浓氨水，搅拌，至生成的沉淀完全溶解，形成了深蓝色的溶液：

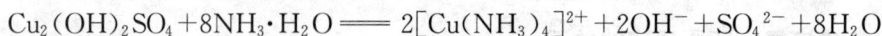

$$SO_4{}^{2-}+2Cu^{2+}+2NH_3·H_2O \Longrightarrow Cu_2(OH)_2SO_4\downarrow+2NH_4{}^+$$
$$Cu_2(OH)_2SO_4+8NH_3·H_2O \Longrightarrow 2[Cu(NH_3)_4]^{2+}+2OH^-+SO_4{}^{2-}+8H_2O$$

测量电势差，观察有何变化，这种变化是怎样引起的？

（3）再在 ZnSO$_4$ 溶液中加浓氨水至生成的沉淀完全溶解：

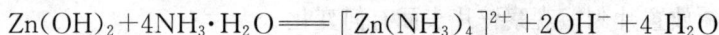

$$Zn^{2+}+2NH_3·H_2O \Longrightarrow Zn(OH)_2\downarrow+2NH_4{}^+$$
$$Zn(OH)_2+4NH_3·H_2O \Longrightarrow [Zn(NH_3)_4]^{2+}+2OH^-+4H_2O$$

测量电势差，其值又有何变化，试解释上面的实验结果。

## 5. 测定下列浓差电池的电动势

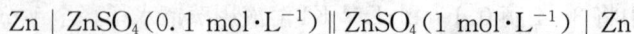

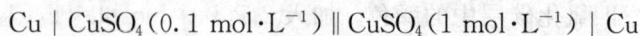

$$Zn \mid ZnSO_4(0.1 \text{ mol·L}^{-1}) \parallel ZnSO_4(1 \text{ mol·L}^{-1}) \mid Zn$$
$$Cu \mid CuSO_4(0.1 \text{ mol·L}^{-1}) \parallel CuSO_4(1 \text{ mol·L}^{-1}) \mid Cu$$

运用奈斯特方程式计算上面浓差电池的电动势，并与实验值比较。

## 6. 电解

往一只小烧杯中加入 $50$ mL $0.5$ mol·L$^{-1}$ 的 ZnSO$_4$ 溶液，在其中插入锌片；往另一只小烧杯中加入 $50$ mL $0.5$ mol·L$^{-1}$ CuSO$_4$ 溶液，在其中插入铜片，按图 11-1 把线路连接好。把两根分别连接锌片和铜片的铜线的另一端插入装有 $50$ mL $0.5$ mol·L$^{-1}$ Na$_2$SO$_4$ 溶液和三滴酚酞的小烧杯中，观察连接锌片的那根铜线周围的 Na$_2$SO$_4$ 溶液有何变化？试加以解释。

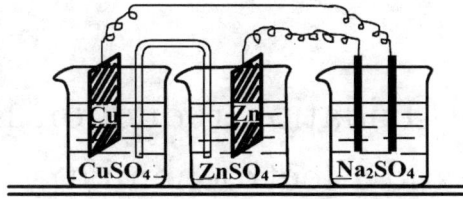

图 11-1　电解装置

## 思考题

1. 怎样利用 pH 计测定电极电势？

2. 原电池的正极同电解池的阳极以及原电池的负极同电解池的阴极，其电极上的反应本质是否相同？

3. 电解硫酸钠水溶液时，为什么在阴极上得不到金属钠？用石墨作电极和以铜作电极，在阳极上的反应是否相同？为什么？

# Exp 11　Oxidation-reduction Reaction and Electrochemistry

## Objectives

1. To test and grasp the relationship between electrode potential and the direction of oxidation-reduction reaction, the effect of medium and concentration of the reactant on oxidation-reduction reaction.

2. Qualitatively observe and know about the effect of EMF (electromotive force), how the change of concentration of oxidation state or reduction state influence the electrode potential.

3. To test and know the reaction of electrolysis.

## Principles

Oxidation-reduction process is namely electron transfer process. Oxidant obtains electron and reductant loses electron in the reaction. The ability of obtaining or losing electron, in other words, the ability of reducibility or oxidizability can be scaled by relative discretion of electrode potential of redox couple (for example $Fe^{3+} - Fe^{2+}$, $I_2 - I^-$, $Cu^{2+} - Cu$). The more algebra value electrode potential of redox couple (according to deoxidize electrode potential), the stronger oxidizabiity of oxidation states are, the weaker reducibility of reduction states, vice versa. So the direction of oxidation-reduction reaction can be judged according to the value of electrode potential ($\varphi$). For example $\varphi^\ominus(I_2/I^-) = +0.535V$, $\varphi^\ominus(Fe^{3+}/Fe^{2+}) = +0.771V$, $\varphi^\ominus(Br_2/Br^-) = +1.08V$, so the formula (1) shall process rightward and the formula (2) shall process leftward in the following two reactions. Namely $Fe^{3+}$ can oxidize $I^-$ but can't oxidize $Br^-$, on the contrary, $Br_2$ can oxidize $Fe^{2+}$, $I_2$ but can't oxidize $Fe^{2+}$. In a word, oxidizabiity sequence of oxidation states is $Br_2 > Fe^{3+} > I_2$, reducibility sequence of reduction states is $I^- > Fe^{2+} > Br^-$.

$$2Fe^{3+} + 2I^- = I_2 + 2Fe^{2+} \tag{1}$$
$$2Fe^{3+} + 2Br^- = Br_2 + 2Fe^{2+} \tag{2}$$

The relationship between concentration and electrode potential can be expressed with Nernst equation as follows:

$$\varphi = \varphi^\ominus + \frac{0.0592}{n} \lg \frac{[\text{oxi}]}{[\text{red}]}$$

For example redox couple of $Fe^{3+} - Fe^{2+}$:

$$\varphi(Fe^{3+}/Fe^{2+}) = \varphi^\ominus(Fe^{3+}/Fe^{2+}) + \frac{0.0592}{1} \lg \frac{c(Fe^{3+})/c^\ominus}{c(Fe^{2+})/c^\ominus}$$

So the variation of concentration of $Fe^{3+}$ or $Fe^{2+}$ can change the value of electrode potential $\varphi$. Concentrations of some ions will greatly reduce, which can change the direction of reaction especially in the presence of deposition reagent or coordination reagent.

In some oxidation-reduction reaction joined oxyacid radical $H^+$ often attends the reaction, so the acidity of medium will affect $\varphi$ value. For example for half-electrode reaction:

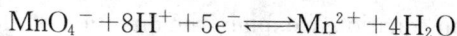

$$MnO_4^- + 8H^+ + 5e^- \rightleftharpoons Mn^{2+} + 4H_2O$$

$$\varphi = \varphi^{\ominus}(MnO_4^-/Mn^{2+}) + \frac{0.0592}{5} \lg \frac{[MnO_4^-][H^+]^8}{[Mn^{2+}]}$$

The greater $[H^+]$ will make oxidisability of $MnO_4^-$ stronger.

Single electrode potential can't be measured; we can measure electromotive force of primary cell composed by two redox couple, because electromotive force of primary cell $E$ is the difference of electrode potential of positive and negative electrode:

$$E^{\ominus} = \varphi_+^{\ominus} - \varphi_-^{\ominus}$$

Standard hydrogen electrode $\varphi_{H^+/H_2}$ is defined as zero in the condition of an atmosphere press, 25℃ and $\alpha(H^+) = 1 \text{ mol} \cdot L^{-1}$ in advance. Then measure a series of electromotive force of primary cells (including hydrogen or other reference electrode), accordingly measure $\varphi^{\ominus}$ straightly or indirectly. Exact electromotive force shall be measured using 'oppsite elimination law' in potentiometer. Electromotive force is measured in pH meter in this experiment because its relative value is needed for comparison.

When the current gets across electrolyte solution, the chemical change brought in electricity board is named electrolysis. Electrolysis production on the two poles can be effected by the factors of discretion of electrode potential, ion concentration, pole materials and so on. Copper acts as the electrode in electrolyzing $Na_2SO_4$ solution in this experiment, its electrolysis reaction is as follows:

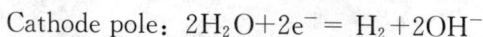

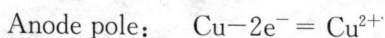

Cathode pole: $2H_2O + 2e^- = H_2 + 2OH^-$

Anode pole:   $Cu - 2e^- = Cu^{2+}$

## Instruments and Reagents

Leici pHS-35 pH meter, beaker (50 mL), small test tubes, salt bridge, wire.

$H_2SO_4$ (3 mol$\cdot$L$^{-1}$, 2 mol$\cdot$L$^{-1}$), HAc(6 mol$\cdot$L$^{-1}$), Pb(NO$_3$)$_2$(0.5 mol$\cdot$L$^{-1}$), CuSO$_4$(1 mol$\cdot$L$^{-1}$, 0.5 mol$\cdot$L$^{-1}$, 0.1 mol$\cdot$L$^{-1}$), KI(0.1 mol$\cdot$L$^{-1}$), FeCl$_3$(0.1 mol$\cdot$L$^{-1}$), KBr(0.1 mol$\cdot$L$^{-1}$), FeSO$_4$(0.1 mol$\cdot$L$^{-1}$), KMnO$_4$ (0.01 mol$\cdot$L$^{-1}$), ZnSO$_4$ (1 mol$\cdot$L$^{-1}$, 0.5 mol$\cdot$L$^{-1}$, 0.1 mol$\cdot$L$^{-1}$), Na$_2$SO$_4$ (0.5 mol$\cdot$L$^{-1}$), CCl$_4$, concentrated sulfuric acid, concentrated ammonia water, iodine water, bromine water, phenolphthalein, zinc sheet, copper sheet, plumbum grain, sand paper, fuchsine test paper.

## Experimental Procedures

### 1. The relationship between electrode potential and oxidation-reduction reaction

(1) Compare the position of Zn, Pb and Cu in electrode potential sequence

Add 0.5 mol$\cdot$L$^{-1}$ Pb(NO$_3$)$_2$ and 0.5 mol$\cdot$L$^{-1}$ CuSO$_4$ into two small test tubes, put a piece of scraped zinc sheet respectively, stand for a minute. Observe the change on surface of zinc sheet.

Add $0.5 \text{ mol} \cdot \text{L}^{-1}$ $Pb(NO_3)_2$ and $0.5 \text{ mol} \cdot \text{L}^{-1}$ $CuSO_4$ to two small test tubes, put a grain of lead scraped respectively. Observe the change on surface of lead grain.

Write down the reaction equations; explain the direction of electron transfer. Determine the relative position in electrode potential sequence of Zn, Pb and Cu.

(2) Dilute 3~4 drops of $0.1 \text{ mol} \cdot \text{L}^{-1}$ KI in a small test tube to 1 mL with distilled water. Add 0.5 mL $CCl_4$, shake and observe the color of $CCl_4$. Then add 2 drops of $0.1 \text{ mol} \cdot \text{L}^{-1}$ $FeCl_3$, shake well. Observe the color change of $CCl_4$ after shaking ($I_2$ dissolved in $CCl_4$ appears mauve).

(3) Repeat the same experiment by replacing $0.1 \text{ mol} \cdot \text{L}^{-1}$ KI with $0.1 \text{ mol} \cdot \text{L}^{-1}$ KBr, observe the color change of $CCl_4$ ($Br_2$ dissolved in $CCl_4$ shows brown-yellow).

Qualitatively compare relative sequence of electrode potential of $Br^- - Br_2$, $I^- - I_2$, $Fe^{2+} - Fe^{3+}$ three redox couples (namely relative magnitude of algebra value) according to the results of experiment (2) and (3). Point out which of the oxidation states of the redox couples is the best oxidant, and which of the reduction states of the redox couple is the best reductant.

(4) Follow the above experiments, let iodine water and bromide water react with $0.1 \text{ mol} \cdot \text{L}^{-1}$ $FeSO_4$ solution respectively. Observe the color of $CCl_4$, judge whether the reaction carry out or not. Write down the related reaction equations.

Explain the relationship between electrode potential and oxidation-reduction reaction according to the experimental results of experiments(2),(3)and(4)and explain the relationship between electrode potential and direction of oxidation-reduction reaction.

## 2. The effect of acidity on the velocity of reaction

Add 0.5 mL $3 \text{ mol} \cdot \text{L}^{-1}$ $H_2SO_4$ solution and $6 \text{ mol} \cdot \text{L}^{-1}$ HAc solution to two test tubes which containing 0.5 mL $0.1 \text{ mol} \cdot \text{L}^{-1}$ KBr solution respectively. Then add 2 drops of $0.01 \text{ mol} \cdot \text{L}^{-1}$ $KMnO_4$ solution into them. Observe and compare the fading speed of amaranth solution in the two test tubes. Write down the reaction equation and explain it.

## 3. The effect of concentration on reaction

Add $2 \text{ mol} \cdot \text{L}^{-1}$ $H_2SO_4$ and concentrated $H_2SO_4$ into the two test tubes, add a sheet of copper with surface oxidation film scraped to them respectively. Heat slightly and observe the phenomena.

Test the gas produced by fuchsine test paper in which concentrated $H_2SO_4$ is added (if fuchsine fades, it indicates $SO_2$ gas has produced), write down the related reaction equation and explain it.

## 4. The effect of concentration on electrode potential

(1) Add 30 mL $1 \text{ mol} \cdot \text{L}^{-1}$ $CuSO_4$ into a 100 mL beaker, add 30 mL $1 \text{ mol} \cdot \text{L}^{-1}$ $ZnSO_4$ into the other 100 mL beaker, then put a copper sheet into $CuSO_4$ solution, put a zinc sheet

into $ZnSO_4$ solution, thus forms two electrodes. Connect them with a salt bridge, copper e-lectrode meets the anode of acidity meter using wire, zinc electrode meets the cathode jack of acidometer though "pigtail implement", measure its electric potential difference.

(2) Take the beaker containing $CuSO_4$ solution and add concentrated ammonia, mix round till the precipitate formed dissolves completely, and then deep blue solution is got.

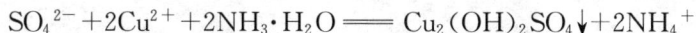

$$SO_4^{2-} + 2Cu^{2+} + 2NH_3 \cdot H_2O \Longrightarrow Cu_2(OH)_2SO_4 \downarrow + 2NH_4^+$$

$$Cu_2(OH)_2SO_4 + 8\ NH_3 \cdot H_2O \Longrightarrow 2[Cu(NH_3)_4]^{2+} + 2OH^- + SO_4^{2-} + 8H_2O$$

Measure its electric potential difference and observe what the change is. What factors are brought for the change?

(3) Add concentrated ammonia into $ZnSO_4$ solution, till to the precipitate formed dis-solves completely.

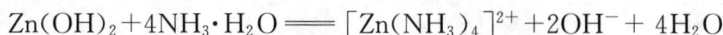

$$Zn^{2+} + 2NH_3 \cdot H_2O \Longrightarrow Zn(OH)_2 \downarrow + 2NH_4^+$$

$$Zn(OH)_2 + 4NH_3 \cdot H_2O \Longrightarrow [Zn(NH_3)_4]^{2+} + 2OH^- + 4H_2O$$

Measure its electric potential difference, how does its value change? Try to explain the above experimental results.

### 5.  Measure the electromotive force of the above following concentration difference cell

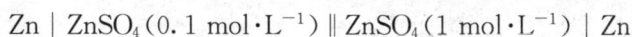

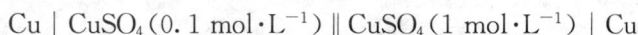

$$Zn \mid ZnSO_4\,(0.\,1\ mol \cdot L^{-1}) \parallel ZnSO_4\,(1\ mol \cdot L^{-1}) \mid Zn$$

$$Cu \mid CuSO_4\,(0.\,1\ mol \cdot L^{-1}) \parallel CuSO_4\,(1\ mol \cdot L^{-1}) \mid Cu$$

Calculate the electromotive force of the above concentration difference cell using Nernst equation and compare with the experimental values.

### 6.  Electrolysis

Add 50 mL 0. 5 $mol \cdot L^{-1}$ $ZnSO_4$ solution into a small beaker, in which a zinc sheet is plugged. Add 50 mL 0. 5 $mol \cdot L^{-1}$ $CuSO_4$ solution to a small beaker, in which a copper sheet is plugged. Connect the circuitry according to the chart (Fig. 11-1). The other end of copper line connected with zinc sheet and copper sheet respectively, plug it into a small beaker added 50 mL 0. 5 $mol \cdot L^{-1}$ $Na_2SO_4$ solution is added and three drops of phenolphthale-in, observe what change will happen in the $Na_2SO_4$ solution around the steel wire connected with zinc sheet. Try to explain it.

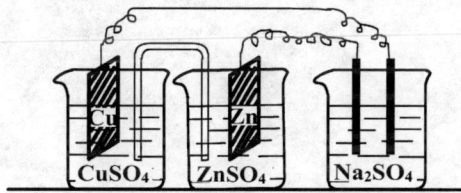

Fig. 11-1 electrolysis equipment

## Questions

1.  How to determine the electrode potential by pH meter?

2. Whether the nature of electrode reaction is the same between positive electrode of primary cell and anode electrolytic cell and between negative electrode of primary cell and cathode electrolytic cell?

3. Why can't we get metal sodium on cathode when we electrolyze sodium sulfate solution? Whether the reaction on cathode is the same as using the graphite or copper as electrode? Why?

# 实验十二　氧化还原原理实验

## 实验目的

1. 加深对电极电势与氧化还原反应方向关系的认识。
2. 进一步了解浓度、介质酸度对电极电势及氧化还原反应的影响。
3. 学习用酸度计测定原电池电动势及电极电势的方法。

## 实验仪器、药品与材料

$Pb(NO_3)_2$（$0.1\ mol \cdot L^{-1}$，$0.5\ mol \cdot L^{-1}$，$1.0\ mol \cdot L^{-1}$），$H_2SO_4$（$0.5\ mol \cdot L^{-1}$，$1\ mol \cdot L^{-1}$，$3\ mol \cdot L^{-1}$），$HAc$（$1\ mol \cdot L^{-1}$，$3\ mol \cdot L^{-1}$，$6\ mol \cdot L^{-1}$），$CuSO_4$（$0.5\ mol \cdot L^{-1}$），$ZnSO_4$（$0.5\ mol \cdot L^{-1}$），$HNO_3$（$0.5\ mol \cdot L^{-1}$，浓），$KI$（$0.1\ mol \cdot L^{-1}$），$KBr$（$0.1\ mol \cdot L^{-1}$），碘水，溴水，$CCl_4$，浓 $NH_3 \cdot H_2O$，饱和 $KNO_3$，$FeCl_3$（$0.1\ mol \cdot L^{-1}$），$FeSO_4$（$0.1\ mol \cdot L^{-1}$），$(NH_4)_2Fe(SO_4)_2$（$0.1\ mol \cdot L^{-1}$），$NH_4F$（$10\%$），$Na_2SO_4$（$0.1\ mol \cdot L^{-1}$），$KMnO_4$（$0.01\ mol \cdot L^{-1}$，$0.1\ mol \cdot L^{-1}$），$NaOH$（$6\ mol \cdot L^{-1}$，$40\%$），$K_2CrO_4$（$0.1\ mol \cdot L^{-1}$），$Na_3AsO_3$（$0.1\ mol \cdot L^{-1}$），$Na_3AsO_4$（$0.1\ mol \cdot L^{-1}$），奈斯勒试剂，$Na_2SiO_3$（$d = 1.06$）溶液。

50 mL 烧杯 4 只、试管、盐桥、酸度计、甘汞电极、铜电极、锌电极、铁电极、石墨电极（2 支）、锌粒、铜片、铅片、锌片（3 块）、导线、砂纸、琼脂。

## 实验内容

### 1. 电极电势与氧化还原反应

（1）现有浓度均为 $0.5\ mol \cdot L^{-1}$ 的 $Pb(NO_3)_2$，$CuSO_4$ 及 $ZnSO_4$ 溶液和铅片、锌片及铜片，请设计实验确定 Zn，Cu 和 Pb 在电极电势次序中的先后位置。

（2）在试管中加入 0.5 mL KI 溶液和 0.5 mLCCl$_4$，震荡，观察 CCl$_4$ 层的颜色。再加入 2 滴 $FeCl_3$ 溶液，充分振荡，观察 CCl$_4$ 层的颜色有何变化？写出反应方程式。

用 $0.1\ mol \cdot L^{-1}$ KBr 溶液代替 $0.1\ mol \cdot L^{-1}$ KI 溶液进行相同的实验，能否发生反应？为什么？

分别用碘水和溴水与 $FeSO_4$ 溶液作用，观察有何现象？

根据以上实验的结果，定性地比较 $Br_2/Br^-$，$I_2/I^-$ 和 $Fe^{3+}/Fe^{2+}$ 三个电对的电极电势的相对高低（或代数值的相对大小），指出哪个是最强的氧化剂，哪个是最强的还原剂。说明电极电势与氧化还原反应方向的关系。

### 2. 浓度和酸度对电极电势的影响

（1）浓度的影响

在两只 50 mL 小烧杯中分别加入 $ZnSO_4$ 溶液和饱和 $KNO_3$ 溶液各 20 mL。将 Zn 电极插入 $ZnSO_4$ 溶液，甘汞电极插入 $KNO_3$ 溶液，两烧杯间放入盐桥便构成了原电池，如图 12-1

所示。因甘汞电极的电极电势已知，若测得该原电池的电动势，便可知道锌电极的电极电势。

盐桥的制法：称取 1 g 琼脂，放在 100 mL 饱和的 KCl 溶液中浸泡一会，加热煮成糊状，趁热倒入 U 形玻管中（里面不能留有气泡），冷却后即成。制好的盐桥不用时应浸在饱和 KCl 溶液中保存。KCl 在不同温度下的饱和溶解度为表 12-1 所示：

**表 12-1　KCl 在不同温度下的饱和溶解度（g/100 g 水）**

| $t$（℃） | 10 | 20 | 30 | 40 | 50 |
|---|---|---|---|---|---|
| 溶解度 | 20.9 | 31.6 | 45.7 | 63.9 | 85.5 |

电池电动势可按下法测得：给酸度计接通电源，调节零点至 0 mV，将甘汞电极插入酸度计插口 9（图 12-2），在插口 10 插入"接续器"，锌电极导线接在"接续器"上，按下"+ mV"按键 6，再按下读数开关 16 便可测得原电池电动势 $E$，记下溶液温度和 $E$，求出锌电极电势。

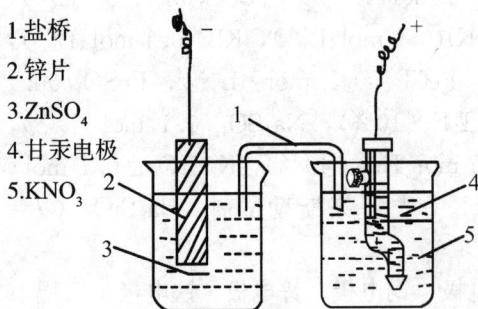

1.盐桥
2.锌片
3.ZnSO$_4$
4.甘汞电极
5.KNO$_3$

图 12-1　$\varphi_{Zn}$ 测定装置　　　　　　图 12-2　pHS-2 型酸度计

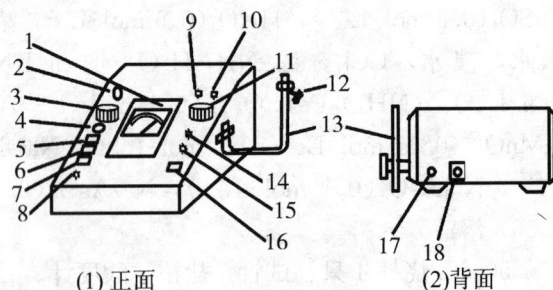

再往 ZnSO$_4$ 溶液中加入浓氨水至生成的沉淀溶解为止，因生成四氨合锌（Ⅱ）配离子而使溶液中锌离子浓度降低：

$$Zn^{2+} + 4NH_3 = [Zn(NH_3)_4]^{2+}$$

观察电极电势有何变化，解释现象。

将烧杯中 ZnSO$_4$ 用同浓度 CuSO$_4$ 溶液代替，Zn 电极换成 Cu 电极，按同样操作步骤，测定铜电极的电极电势，这时铜电极和甘汞电极哪个是正极，哪个是负极？

在 CuSO$_4$ 溶液中加入浓氨水至生成的沉淀溶解为止，形成深蓝色的溶液：

$$Cu^{2+} + 4NH_3 = [Cu(NH_3)_4]^{2+}$$

观察电极电势有何变化，利用奈斯特方程式来解释实验现象。

（2）酸度的影响

在两只 50 mL 的小烧杯中分别加入 30 mL FeSO$_4$ 和 K$_2$Cr$_2$O$_7$ 溶液。在 FeSO$_4$ 溶液中插入铁电极，K$_2$Cr$_2$O$_7$ 溶液中插入石墨电极，将铁电极和石墨电极通过导线分别与酸度计的负极和正极相接，中间以盐桥相通，测量两极之间的电势差。

在 K$_2$Cr$_2$O$_7$ 溶液中慢慢加入 1 mol·L$^{-1}$ 的 H$_2$SO$_4$ 溶液，观察溶液颜色以及电动势有何变化？再在 K$_2$Cr$_2$O$_7$ 溶液中逐滴加入 6 mol·L$^{-1}$ 的 NaOH 溶液，观察溶液的颜色以及电动势有何变化？解释现象。

**3. 浓度和酸度对氧化还原反应速度和产物的影响**

（1）浓度的影响

①往两个各盛一粒锌粒的试管中，分别加入浓 $HNO_3$ 和 $0.5\ mol\cdot L^{-1}HNO_3$ 观察所发生的现象。

不同浓度的硝酸与锌的反应速度和反应产物有何不同？待反应进行片刻后，取加稀 $HNO_3$ 的试管中的试液于白色点滴板上，加 1 滴奈斯勒试剂，生成红棕色沉淀，表示有 $NH_4^+$（$NH_4^+$ 浓度小时仅呈棕黄色溶液）。奈斯勒试剂是 $K_2HgI_4$ 的 KOH 溶液，它与 $NH_3$ 可发生如下反应：

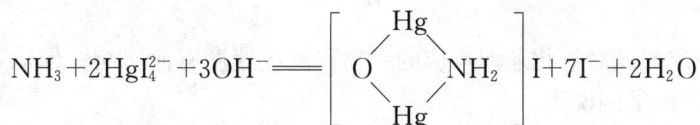

$$NH_3+2HgI_4^{2-}+3OH^- =\!=\!= \left[ O \begin{array}{c} Hg \\ \\ Hg \end{array} NH_2 \right] I+7I^-+2H_2O$$

$NH_3$ 的浓度低时，没有沉淀产生，但溶液呈黄色或棕色。

②往三支粗试管中分别装入 1 mL 浓度各为 0.1、0.5 和 1.0 $mol\cdot L^{-1}$ 的 $Pb(NO_3)_2$ 溶液，再各加入浓度为 1 $mol\cdot L^{-1}$ 的 HAc 溶液 10 mL，振荡混匀后缓慢加入密度 $d=1.06$ 的硅酸钠溶液 10 mL，搅匀后在 90℃ 左右的热水浴加热至形成乙酸铅凝胶。在三支试管中分别插入表面积相同的锌片，观察三支试管中铅树生长的速度有何不同？解释现象。

（2）酸度对氧化还原反应的影响

①在两个各盛有 0.5 mL KBr 溶液的试管中，分别加入 0.1 mL 3 $mol\cdot L^{-1}$ 的 $H_2SO_4$ 和 HAc；然后往两个试管中各加入 2 滴 0.01 $mol\cdot L^{-1}KMnO_4$ 溶液，观察并比较两个试管中紫色溶液褪色的快慢，写出反应式，并加以解释。

②在三支试管中，各加入 0.5 mL $Na_2SO_3$ 溶液，分别加入 1 $mol\cdot L^{-1}$ 的 $H_2SO_4$ 溶液，蒸馏水和 1 $mol\cdot L^{-1}$ NaOH 溶液各 4 滴，然后往三支试管各滴几滴 0.01 $mol\cdot L^{-1}$ 的 $KMnO_4$ 溶液，观察反应产物有何不同，写出反应式。

### 4. 浓度和酸度对氧化还原反应方向的影响

（1）浓度的影响

$Fe^{3+}$ 与 $I^-$ 发生如下反应：

$$2Fe^{3+}+2I^- =\!=\!= 2Fe^{2+}+I_2$$

①往盛有 2 mL 水和 1 mL$CCl_4$ 的试管中，滴加 2 mL（$NH_4$）$_2Fe(SO_4)_2$ 溶液，振荡后观察 $CCl_4$ 层的溶液。

②往盛有 2 mL 水和 1 mL $CCl_4$ 的试管中，滴加 2 mL（$NH_4$）$_2Fe(SO_4)_2$ 溶液，振荡，观察 $CCl_4$ 的颜色，再滴加 2 mL KI 溶液，振荡后观察 $CCl_4$ 的颜色与上面实验中的 $CCl_4$ 层颜色有无改变。

③往盛有 2 mL（$NH_4$）$_2Fe(SO_4)_2$ 溶液中滴加 1 mL $CCl_4$ 震荡后观察 $CCl_4$ 层的颜色。滴加 2 mL KI 溶液到试管中，振荡，观察 $CCl_4$ 的颜色。再往试管中注入 2 mL $NH_4F$ 溶液，用力振荡试管，观察 $CCl_4$ 层颜色。用化学平衡移动的观点解释。

（2）酸度的影响

$AsO_4^{3-}$ 与 $I^-$ 发生如下反应：

$$AsO_4^{3-}+2I^-+2H^+ =\!=\!= AsO_3^{3-}+I_2+H_2O$$

将 10 mL $Na_3AsO_3$ 和 10 mL $Na_3AsO_4$ 混和在一小烧杯中，另一烧杯中混合 10 mL KI 溶液和 10 mL 碘水。每一烧杯中各插一炭棒，以盐桥连通，用导线把原电池与酸度计连接。

利用指针的偏转，了解化学反应方向的改变。在 $Na_3AsO_3$ 和 $Na_3AsO_4$ 的混合液中逐滴加入浓盐酸，观察微安表指针的移动；再在该混合溶液中滴入 $40\%NaOH$ 溶液，观察电流方向的改变。

## 思考题

1. 电极电势适用于什么场合？有何用途？它的值受哪些因素所影响？其中哪些在本实验中得到验证。

2. 浓度和介质酸度对氧化还原反应的产物和反应速度有何影响？催化剂对氧化还原反应速率有无影响？试举例说明。

3. 将实验内容2.(2)的电池用电池符号表示出来。其 $K_2Cr_2O_7$ 溶液中加入 $H_2SO_4$ 溶液时电池符号是否作相应改变？若用 $NaOH$ 溶液代替 $H_2SO_4$ 时呢？

4. 在实验内容1.(2)和4.(1)中加入 $CCl_4$ 的目的是什么？

5. 为什么实验内容3.(1).②中所制的凝胶称为醋酸铅凝胶而不称其为硝酸铅凝胶？

# Exp 12　Principle of Oxidation-reduction Reaction

## Objectives

1. To deepen the comprehension of the relationship between electrode potential and the direction of oxidation-reduction reaction.

2. To know the effect of concentration, medium acidity on electrode potential and oxidation-reduction reaction.

3. To learn the method of measuring electromotive force of primary cell and electrode potential by acidometer.

## Instruments, Reagents and Materials

$Pb(NO_3)_2$(0.1 mol·$L^{-1}$, 0.5 mol·$L^{-1}$, 1.0 mol·$L^{-1}$), $H_2SO_4$(0.5 mol·$L^{-1}$, 1mol· $L^{-1}$, 3 mol·$L^{-1}$), HAc(1mol·$L^{-1}$, 3mol·$L^{-1}$, 6 mol·$L^{-1}$), $CuSO_4$(0.5 mol·$L^{-1}$), $ZnSO_4$(0.5 mol·$L^{-1}$), $HNO_3$(0.5 mol·$L^{-1}$, concentrated), KI(0.1 mol·$L^{-1}$), KBr(0.1 mol·$L^{-1}$), iodine water, bromine water, $CCl_4$, concentrated $NH_3$·$H_2O$, saturated $KNO_3$, $FeCl_3$(0.1 mol·$L^{-1}$), $FeSO_4$(0.1 mol·$L^{-1}$), $(NH_4)_2Fe(SO_4)_2$(0.1 mol·$L^{-1}$), $NH_4F$ (10%), $Na_2SO_4$(0.1 mol·$L^{-1}$), $KMnO_4$ (0.01mol·$L^{-1}$, 0.1 mol·$L^{-1}$), NaOH (6 mol· $L^{-1}$, 40%), $K_2CrO_4$(0.1 mol·$L^{-1}$), $Na_3AsO_3$(0.1 mol·$L^{-1}$), $Na_3AsO_4$(0.1 mol·$L^{-1}$), Nessler reagent, $Na_2SiO_3$ ($d$=1.06) solution.

four beakers (50 mL), test tube, salt bridge, acidometer, calomel electrodg, raphite electrode copper electrode, zinc electrode, iron electrode, graphite electrode, forceps, microammeter, two black lead electrodes, zinc grain, copper sheet, lead sheet, wire, sand paper, agar.

## Experimental Procedures

### 1. Electrode potential and the oxidation-reduction reaction.

(1) 0.5 mol·$L^{-1}$ $Pb(NO_3)_2$, $CuSO_4$, $ZnSO_4$ solutions and lead sheet, zinc sheet, copper sheet have been offered. Please design a scheme to determine the position of Zn, Cu and Pb in electrode potential sequence.

(2) Add 0.5 mL KI and 0.5 mL $CCl_4$ to the test tube, shake, observe the color of $CCl_4$ layer; 2 drops of $FeCl_3$ solution are then added, shake fully, observe the color change of $CCl_4$. Write down the reaction equation.

Repeat the same experiment by replacing 0.1 mol·$L^{-1}$ KI with 0.1 mol·$L^{-1}$ KBr, whether will the reaction happen or not? Why?

Replace iodine water and bromine water with 0.1 mol·$L^{-1}$ $FeSO_4$ solution respectively. Observe the phenomena.

Qualitatively compare relative order of electrode potential of three redox couples ($Br^- -$ $Br_2$, $I^- -I_2$ and $Fe^{2+} -Fe^{3+}$, namely relative magnitude of algebra value) according to the results of the above experiments. Point out which is the strongest oxidant and the strongest reductant respectively. Explain the relationship between electrode potential and oxidation-reduction reaction.

## 2. The effect of concentration and acidity on electrode potential

(1) The effect of concentration on electrode potential

Add 20 mL $ZnSO_4$ and saturated $KNO_3$ solution into two 50 mL beakers respectively. Then put a zinc electrode into $ZnSO_4$ solution, put a calomel electrode into $KNO_3$ solution, connect them with a salt bridge between two breakers, thus a primary cell is formed (Fig. 12-1). If electromotive force of the primary cell is measured, electrode potential of zinc electrode can be obtained because electrode potential of calomel electrode is known.

Preparation of salt bridge: soak 1 g agar in 100 mL saturated KCl solution for a minute, heat till it is in pasty shape, pour it into U shape glass tube while it is hot, salt bridge is prepared after cooling down. Keep the salt bridge by soaking in saturated KCl solution. Saturated solubilities of KCl at different temperatures are listed in Tab 12-1 (g/100 g water):

**Tab 12-1   Solubility of KCl at different temperature (g/100 g $H_2O$)**

| $t$ (℃) | 10 | 20 | 30 | 40 | 50 |
|---|---|---|---|---|---|
| solubility | 20.9 | 31.6 | 45.7 | 63.9 | 85.5 |

Electromotive force of primary cell can be measured according to the following method: turn on the acidimeter, adjust zero point to 0 mV, plug calomel electrode into faucet 9 of acidometer (Fig. 12-2). Plug "connector" into faucet 10 of acidometer. Zinc electrode lead meets the "connector". Electromotive force of primary cell can be measured when pressing "+mV" key-press 6 and reading on-off 16, note down solution temperature and $E$, try to obtain electrode potential of zinc electrode.

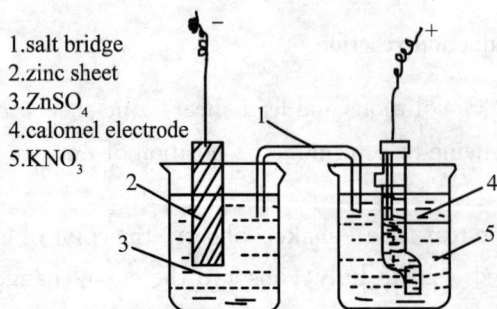

1.salt bridge
2.zinc sheet
3.$ZnSO_4$
4.calomel electrode
5.$KNO_3$

Fig. 12-1   Measuring instrument of $\varphi_{Zn}$

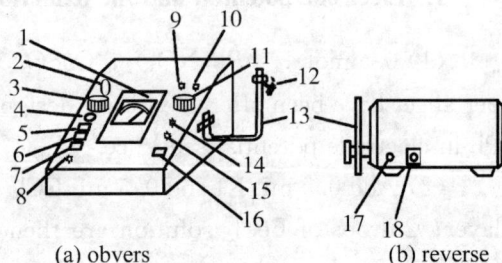

Fig. 12-2   pHS-2 acidometer

(a) obvers          (b) reverse

Add concentrated ammonia water to $ZnSO_4$ solution till the precipitate is dissolved completely, which will make concentration of zinc ion reduce for forming $[Zn(NH_3)_4]^{2+}$.

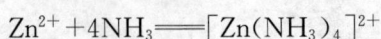

$$Zn^{2+} +4NH_3 = [Zn(NH_3)_4]^{2+}$$

Observe the change of electrode potential and explain the phenomenon.

Measure the electrode potential of copper electrode as the same operation process in condition of $ZnSO_4$ replaced with $CuSO_4$ solution, Zn electrode replaced with Cu one. Which is the anode in copper electrode and calomel electrode? Which is the cathode?

Add concentrated ammonia into $CuSO_4$ solution till to the precipitate built dissolve entirely, form a deep blue solution.

$$Cu^{2+} + 4NH_3 = [Cu(NH_3)_4]^{2+}$$

Observe the change of electrode potential and explain the experimental phenomenon using Nernst equation.

(2) The effect of acidity on electrode potential

Add 30 mL $FeSO_4$ and $K_2Cr_2O_7$ solution into two 50 mL beakers respectively. Then put an iron electrode into $FeSO_4$ solution, put a charcoal stick electrode into $K_2Cr_2O_7$ solution. Connect iron electrode with cathode of acidometer, charcoal stick electrode with anode of acidometer, use a salt bridge between the two beakers. Measure the electrode potential difference between the electrodes.

Slowly add 1 mol·$L^{-1}$ $H_2SO_4$ solution to $K_2Cr_2O_7$ solution, observe the color of the solution and the change of electromotive force, then gradually add 6 mol·$L^{-1}$ NaOH solution to $K_2Cr_2O_7$ solution. Observe the color of the solution and the change of electromotive force. Explain the phenomenon.

## 3. The effect of concentration and acidity on the velocity and production of oxidation-reduction reaction

(1) The effect of concentration

①Add concentrated $HNO_3$ and 0.5 mol·$L^{-1}$ $HNO_3$ in two test tubes containing a grain of zinc respectively. Observe the phenomenon.

What difference is the velocity and products of reaction $HNO_3$ with of different concentrations of $HNO_3$ and zinc? Take a sample from the solution which dilute $HNO_3$ is added, then add to white dropping board for some moments. Add a drop of Nessler reagent, if umber precipitate is formed, it shows the existence of $NH_4^+$ (palm yellow solution is possible when $NH_4^+$ concentration is dilute). nessler reagent is the solution of $K_2[HgI_4]$ and KOH, which can react with $NH_3$, the reaction equation is as follows:

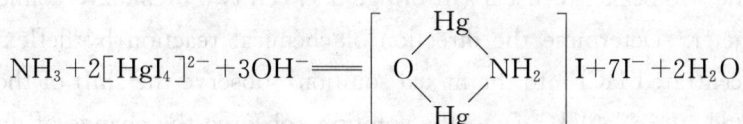

$$NH_3 + 2[HgI_4]^{2-} + 3OH^- = \left[ O \underset{Hg}{\overset{Hg}{\diagdown \diagup}} NH_2 \right] I + 7I^- + 2H_2O$$

It is possible that precipitate can not be formed and palm yellow solution is obtained when $NH_4^+$ concentration is low.

②Add 1 mL 0.1, 0.5 and 1.0 mol·$L^{-1}$ $Pb(NO_3)_2$ into three wide mouthed test tubes respectively, then add 10 mL 1 mol·$L^{-1}$ HAc solution, slowly add 10 mL sodium silicate of $d = 1.06$ after shaking fully. Heat it to form lead acetate gel in hot water bath at 90℃ after mixing round. Plug three zinc sheets possessing the same surface into the three test tubes, respectively, observe the different velocities of lead tree growing. Explain the phenomenon.

(2) The effect of acidity on oxidation-reduction reaction

①Add 0. 1 mL 3 mol·L$^{-1}$ H$_2$SO$_4$ and HAc into two test tube containing 0. 5 mL KBr solution respectively. Then add 2 drops of 0. 01 mol·L$^{-1}$ KMnO$_4$ solution into them, observe and compare the fading speed of amaranth solution in two test tubes. Write down the reaction equation and explain it.

②Add 0. 5 mL Na$_2$SO$_3$ solution to three test tubes respectively, add 4 drops of 1 mol·L$^{-1}$ H$_2$SO$_4$, distilled water and 6mol·L$^{-1}$ NaOH solution to them respectively. Then add a few drops of 0. 01 mol·L$^{-1}$ KMnO$_4$ solution into the three test tubes, observe products of the reaction. Write down the reaction equation.

What differences have among the products of the reaction.

### 4. The effect of concentration and acidity on the direction of oxidation-reduction reaction

(1) The effect of concentration

The reaction of Fe$^{3+}$ and I$^-$ is as the follows:

$$2Fe^{3+} + 2I^- = 2Fe^{2+} + I_2$$

①Add 2 mL water and 1 mL CCl$_4$ into a test tube, then drop 2 mL (NH$_4$)$_2$Fe(SO$_4$)$_2$ solution, observe the color of CCl$_4$ layer after shaking.

②Add 2 mL water and 1 mL CCl$_4$ into the test tube, then drop 2 mL (NH$_4$)$_2$Fe(SO$_4$)$_2$ solution, shake, observe the color of CCl$_4$ layer. After that, 2 mL KI are added, observe whether the color of CCl$_4$ layer changes.

③Add 2 mL (NH$_4$)$_2$Fe(SO$_4$)$_2$ solution and 1 mL CCl$_4$ to the test tube, observe the color of CCl$_4$ layer after shaking. 2 mL KI are added, shake and observe the color, then add 2 mL NH$_4$F solution, observe the color of CCl$_4$ layer after shaking fully.

Explain the phenomenon with knowledge of shift of chemical equilibrium.

(2) The effect of acidity

The reaction of AsO$_4$$^{3-}$ and I$^-$ is as the follows:

$$AsO_4{}^{3-} + 2I^- + 2H^+ = AsO_3{}^{3-} + I_2 + H_2O$$

Mix about 10 mL Na$_3$AsO$_3$ and 10 mL Na$_3$AsO$_4$ solution in a small beaker, mix round 10 mL KI solution and 10 mL iodine water in another small beaker. Then put a graphite electrode into the two beakers, use a salt bridge between two breakers, connect the primary cell and acidometer. Determine the direction of chemical reaction by deflexion of needle. Slowly add concentrated HCl into the mixed solution, observe the shift of the needle of microammeter. Add 40% NaOH(aq) to the solution, observe the change of direction of current.

## Questions

1. Where is electrode potential applicable? What is the use? What factors will affect the value of electrode potential? What have been confirmed in this experiment?

2.   What effects are concentration and acidity on the product and reaction velocity of oxidation-reduction reaction? What effect is catalyst on reaction velocity of oxidation-reduction reaction? Try to explain through illustration.

3.   Express the cell of experiment 2. (2) using cell formula.  Whether will cell formula have the corresponding change when adding $H_2SO_4$ solution into $K_2Cr_2O_7$ solution? What is the change when $H_2SO_4$ is replaced by NaOH?

4.   What are the purposes of adding $CCl_4$ in experimental procedures 1. (2)and 4. (1)?

5.   Why the prepared gel is lead acetate not lead nitrate in experimental procedures 3. (1). ②?

# 实验十三　碘酸铜溶度积的测定

## 实验目的

1. 测定碘酸铜溶度积，加深对溶度积概念的理解。
2. 练习目视比色（或光电比色）测定溶液浓度的方法。

## 721 型分光光度计及其有关技术

### 1. 仪器工作原理

分光光度计的基本工作原理是基于物质对光（对光的波长）的吸收具有选择性，不同的物质都有各自的吸收光带，所以当光色散后的光谱通过某一溶液时，其中某些波长的光线就会被溶液吸收。在一定的波长下，溶液中物质的浓度与光能量减弱的程度有一定的比例关系，也即符合于比色原理—比尔定律（图 13-1）：

$$T = I/I_o$$
$$\lg(I_0/I) = \varepsilon c b$$

式中，$T$ 为透射率，$I_o$ 为入射光强度，$I$ 为透射光强度，$A$ 为吸光度，$\varepsilon$ 为吸收系数，$b$ 为溶液的光径长度（液层厚度），$c$ 为溶液的浓度。

从以上公式可以看出，当入射光、吸收系数和溶液液层厚度一定时，透光率是根据溶液的浓度而变化的。

721 型分光光度计允许的测定波长范围在 260～800 nm，其构造比较简单，测定的灵敏度和精密度较高，应用比较广泛。

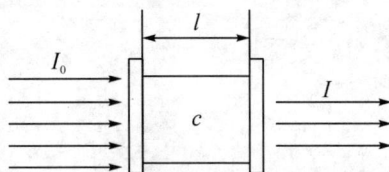

图 13-1　比尔定律原理示意图

### 2. 仪器的基本结构

721 型分光光度计的仪器构造见图 13-2。从光源灯发出的连续辐射光线，射到聚光透镜上，会聚后，再经过平面镜转角 90°，反射至入射狭缝。由此入射到单色器内，狭缝正好位于准光镜的焦面上，当入射光线经过准直物镜反射后，就以一束平行光射向棱镜。光线进入棱镜后，进行色散。色散后回来的光线，再经过准直镜反射，就会聚在出光狭缝上，再通过聚光镜后进入比色皿，光线一部分被吸收，透过的光进入光电管，产生相应的光电流，经放大后在微安表上读出。

### 3. 操作和使用方法

（1）首先接通电源，打开电源开关 7（图 13-3），指示灯亮，打开比色皿暗箱盖 8。预热

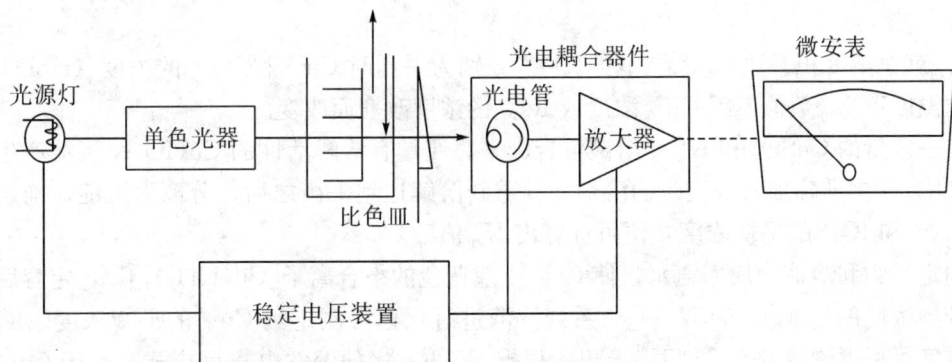

图 13-2　721 型分光光度计的基本结构示意图

20 min。

（2）波长选择旋钮 6，选择所需的单色光波长，用灵敏度旋钮 2 选择所需的灵敏档。

（3）放入比色皿，旋转零位旋钮 5 调零，将比色皿暗箱盖合上，推进比色皿拉杆 3，使参比比色皿处于空白校正位置，使光电管见光，旋转透光率调节旋钮 4，使微安表指针 9 准确处于 100％。按上述方法连续几次调整零位和 100％位，即可进行测定工作。

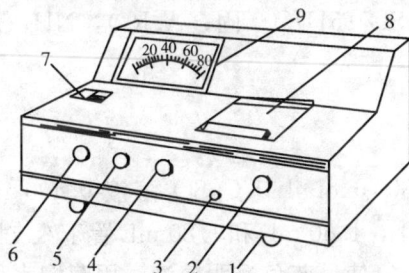

图 13-3　721 型分光光度计

**4. 仪器使用和维护中的注意事项**

（1）连续使用仪器的时间不应超过 2 h，最好是间歇 0.5 h 后，再继续使用。

（2）比色皿每次使用完毕后，要用蒸馏水洗净并倒置晾干后存放在比色皿盒内。在日常使用中应注意保护比色皿的透光面，使其不受损坏或产生划痕，以免影响透光度。

（3）仪器不能受潮。在日常使用中，应经常注意单色器上的防潮硅胶（在仪器的底部）是否变红，如硅胶的颜色已变红，应立即取出烘干或更换。

（4）在托运或移动仪器时，应注意小心轻放。

## 碘酸铜溶度积的测定实验

**1. 实验原理**

碘酸铜是难溶强电解质。在其水溶液中，已溶解的 $Cu^{2+}$ 和 $IO_3^-$ 与未溶解的 $Cu(IO_3)_2$ 之间，在一定温度下可达到动态平衡：

$$Cu(IO_3)_2 \rightleftharpoons Cu^{2+} + 2IO_3^-$$

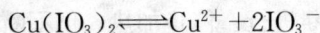

碘酸铜是难溶强电解质。在一定温度下，在其饱和水溶液中，$Cu^{2+}$ 浓度与 $IO_3^-$ 浓度（更确切地说应是活度）平方的乘积是一个常数：

$$K_{sp} = [Cu^{2+}][IO_3^-]^2$$

$K_{sp}$ 就是溶度积常数，$[Cu^{2+}]$ 和 $[IO_3^-]$ 分别为平衡时 $Cu^{2+}$ 和 $IO_3^-$ 的浓度（$mol \cdot L^{-1}$）。在一定温度下 $K_{sp}$ 数值不因 $Cu^{2+}$ 浓度或 $IO_3^-$ 的浓度改变而改变。

将一定量的 $CuSO_4$ 和 $KIO_3$ 溶液混合起来，便具有溶解活性的 $Cu(IO_3)_2$ 沉淀产生。在此 $Cu(IO_3)_2$ 中准确加入一定量的溶剂使其重新溶解达到平衡之后，分离去沉淀，确定出溶液中 $Cu^{2+}$ 和 $IO_3^-$ 的平衡浓度，便可计算出 $K_{sp}$ 值。

向定量饱和溶液中加入氨水，使 $Cu^{2+}$ 生成蓝色的络合离子 $[Cu(NH_3)_4]^{2+}$，定容后与已知其准确浓度的标准 $[Cu(NH_3)_4]^{2+}$ 系列溶液进行比色可测定出 $Cu^{2+}$ 的平衡浓度。$IO_3^-$ 的平衡浓度可根据测得的 $Cu^{2+}$ 间接求出：根据 $Cu(IO_3)_2$ 的溶解电离反应式，由于 $Cu(IO_3)_2$ 沉淀溶解电离出的 $IO_3^-$ 浓度是 $Cu^{2+}$ 浓度的二倍，测定出饱和溶液中的 $Cu^{2+}$ 便可计算出 $IO_3^-$ 的平衡浓度。

### 2. 仪器和药品

烧杯（50 mL），比色管（50 mL），刻度移液管（1.00 mL，2.00 mL 和 20.00 mL 刻度），定量滤纸，漏斗。

$KIO_3$（固，A.R.），$CuSO_4 \cdot 5H_2O$（固，A.R.），$NH_3 \cdot H_2O$（2 $mol \cdot L^{-1}$），$CuSO_4$（0.0800 $mol \cdot L^{-1}$）。

### 3. 标准色阶

用刻度移液管吸取 0.0800 $mol \cdot L^{-1}$ $CuSO_4$ 溶液 0.400 mL，0.600 mL，0.800 mL，1.00 mL，1.20 mL，1.40 mL，1.60 mL 和 1.80 mL 置于八支同规格的 25.00 mL 的比色管中，各加 2 $mol \cdot L^{-1}$ $NH_3 \cdot H_2O$ 使呈蓝色透明溶液，用蒸馏水稀释至刻度（氨水加入过量，以保证用水稀释时溶液不易发生浑浊）。

按上述条件配制的色阶分别为：

$1.28 \times 10^{-3}$　$1.92 \times 10^{-3}$　$2.56 \times 10^{-3}$　$3.20 \times 10^{-3}$
$3.84 \times 10^{-3}$　$4.48 \times 10^{-3}$　$5.12 \times 10^{-3}$　$5.78 \times 10^{-3}$

### 4. 实验步骤

（1）制备碘酸铜新鲜固相。

① 称取 4.2 g $KIO_3$ 固体溶于 85 mL 蒸馏水中，加热使其完全溶解（必要时进行过滤），得 $KIO_3$ 溶液。

② 称取 2.5 g $CuSO_4 \cdot 5H_2O$ 溶于 15 mL 蒸馏水中，加热溶解，得 $CuSO_4$ 溶液。

③ 将 $KIO_3$ 溶液缓慢倒入热的 $CuSO_4$ 溶液中，搅拌，然后在搅拌条件下自然冷却至室温（约半小时），以使其沉淀完全，且保持沉淀有一定溶解活性。用定量滤纸在减压过滤装置上过滤，并用少量水洗涤沉淀 5~6 次，最后将沉淀抽干。

（2）用玻璃小匙取约黄豆粒大碘酸铜晶体置于干燥洁净的 50 mL 烧杯中，用移液管加入 20.00 mL 蒸馏水在石棉网上小火加热，并不断搅拌，至温度 60~70℃时维持 1~2 min。为加速达到溶解平衡，搅拌应不断进行。停止加热，仍不断搅拌下自然缓慢冷却至室温。以干燥慢速定量滤纸在干燥洁净的漏斗中将溶液过滤，滤液用另一只干燥洁净小烧杯收集。用移液管量取滤液 10.00 mL。置于 25.00 mL 与标准色阶规格相同的比色管中，用于配制标准色阶完全相同的方法配制此未知样溶液，摇匀后以目视比色法（或光度法）与标准系列色阶比较，找出与未知样相同浓度的标准色阶浓度（$Cu^{2+}$）色阶，浓度值可以正好是标准色阶上所标之浓度，也可以是两个标准色阶之间的平均值，按实际比色结果而定。记下室温。

　　按上述方法再重复两次实验，但是 20.00 mL 蒸馏水一次改为 19.00 mL 蒸馏水和 1.00 mL 0.0800 $mol \cdot L^{-1}$ $CuSO_4$ 溶液，另一次改为 18.00 mL 蒸馏水和 2.00 mL 0.800 $mol \cdot L^{-1}$ $CuSO_4$ 溶液，即总体积仍保持 20.00 mL，通过比色找出各自相应的标准色阶浓度。

### 5. 实验数据处理

数据处理方法举例：

**表 13-1　碘酸铜溶度积测定数据　　　　（$t=15℃$）**

| 实验编号 | 1 | 2 | 3 |
|---|---|---|---|
| 加入已知 $Cu^{2+}$ 的浓度<br>$a=[Cu^{2+}] \times \dfrac{V_{Cu^{2+}}}{V_{平衡}}$ | 0.00 | $0.0800 \times \dfrac{1.00}{20.00}$ | $0.0800 \times \dfrac{2.00}{20.00}$ |
| 比色测定 $Cu^{2+}$ 的浓度<br>$b=[Cu^{2+}] \times \dfrac{V_2}{V_1}$ | $1.60 \times 10^{-3} \times \dfrac{25.0}{10.0}$<br>$=4.00 \times 10^{-3}$ | $2.88 \times 10^{-3} \times \dfrac{25.0}{10.0}$<br>$=7.20 \times 10^{-3}$ | $4.48 \times 10^{-3} \times \dfrac{25.0}{10.0}$<br>$=11.2 \times 10^{-3}$ |
| 平衡时由 $Cu(IO_3)_2$ 溶解产生的 $Cu^{2+}$ 的浓度 $(b-a)$ | $4.00 \times 10^{-3}$ | $3.20 \times 10^{-3}$ | $3.20 \times 10^{-3}$ |
| 平衡时由 $Cu(IO_3)_2$ 溶解产生 $IO_3^-$ 的浓度 $2(b-a)$ | $8.00 \times 10^{-3}$ | $6.40 \times 10^{-3}$ | $6.40 \times 10^{-3}$ |
| 溶度积常数<br>$K_{sp}=b \times [2(b-a)]^2$ | $2.56 \times 10^{-7}$ | $2.95 \times 10^{-7}$ | $4.59 \times 10^{-7}$ |

表中，$V_{Cu^{2+}}$ 表示每次实验中加入已知浓度的 $CuSO_4$ 溶液体积；$V_{平衡}$ 表示固液平衡时溶液总体积（本实验固定为 20.00 mL）；$[Cu^{2+}]$ 表示已知浓度的 $CuSO_4$ 溶液（本实验为 0.0800 $mol \cdot L^{-1}$）；$[Cu^{2+}]$ 色阶表示与未知浓度相等的标准色阶浓度；$V_1$ 表示比色用未知液体积（本实验为 10.00 mL）；$V_2$ 表示比色未知液稀释后的体积（本实验为 25.00 mL）。

　　由表 13-1 可知，平衡时 $IO_3^-$ 的浓度是用间接方法求出的，即根据平衡时由 $Cu(IO_3)_2$ 溶解产生的 $IO_3^-$ 的浓度应为 $Cu^{2+}$ 浓度的两倍。但实际比色的 $Cu^{2+}$ 浓度 $b$ 包括外加已知的 $Cu^{2+}$ 浓度 $a$（对未加入者 $a=0.00$），所以平衡时，由 $Cu(IO_3)_2$ 溶解产生的 $Cu^{2+}$ 浓度为 $(b-a)$，则 $IO_3^-$ 浓度为 $2(b-a)$。

## 思考题

1. 假如 $Cu(IO_3)_2$ 固体透过滤纸，对实验结果将会产生什么样的影响？

2. 为什么重复两次实验时要加入不同量的 $CuSO_4$ 溶液？

3. 估计一下本实验方法测得的 $Cu(IO_3)_2$ 溶度积常数值与文献中记载的数值的偏差，造成这种偏差的主要因素有哪些？

# Exp 13　Determination of the Solubility Product Constant of $Cu(IO_3)_2$

## Objectives

1. To consolidate the comprehension of the concept of solubility product by determing the $K_{sp}$ of $Cu(IO_3)_2$.

2. To practice measuring the concentrations of solutions by visualizing colorimetry (photoelectric colorimetry).

## Introduction of 721 Spectrophotometer and Relative Technologies

### 1. Principles

The fundamental working principle of spectrophotometer is based on the selectivity of materials absorption of light (selectivity of the wavelength of light). As different material absorbs light with different wavelength, when the light spectrum produced after dispersion is passing through solution, light with certain wavelengths would be absorbed. And with certain wavelengths, the concentration of the solution is proportionally related to the decrease of light energy, that's the shade selection principle—Bill theory (Fig. 13-1).

$$T = I/I_0$$
$$\lg(I_0/I) = \varepsilon cb$$

In the formula, $T$ is transmissivity, $I_0$ is figured as the strength of light shot in, $I$ is characterized with the strength of light shot out, $A$ means extinction value (absorbance), $\varepsilon$ indicates the absorption constant, $b$ is known as the length of light's pass way through the solution, while $c$ is defined as the concentration of the solution.

From the formula above, we could know that when the wavelength of light shot in, the absorption constant and the thickness of the solution are fixed, the transmissivity is proportionally changed with the concentration of the solution.

The effective measuring content of 721 spectrophotometer is from 260 nm to 800 nm. It's simple in structure, and high in accuracy and sensitivity. So, it has been used widely.

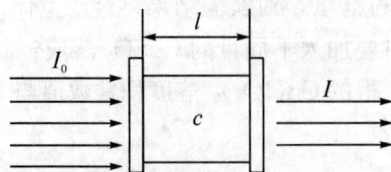

Fig. 13-1　The introduction of Bill theory

### 2. Basic structure of the apparatus

The structure of 721 spectrophotometer is shown in Fig. 13-2. The unremitting radio light from the bulb shoot on the condenser lens, focuses and then reflects to the shooting in slot by a plane mirror which makes the light turn 90° in its passing way. Through the slot, the light enters the monochromator. While the slot is just on the focus of sphere collimation objective, after reflection by it, the light shot would shoot in to prism in the form of a branch of parallel light. Then it would be dispersion, and the light produced by it will be reflected by collimating mirror again and focused on the shooting out slot. After that, the light will enter cell, through a condenser lens. The light will be partly absorbed by the solution in cell, the rest will enter photoelectric tube to produce correspondent light-electronic flow, whose strength is enlarged, could be shown on a microammeter.

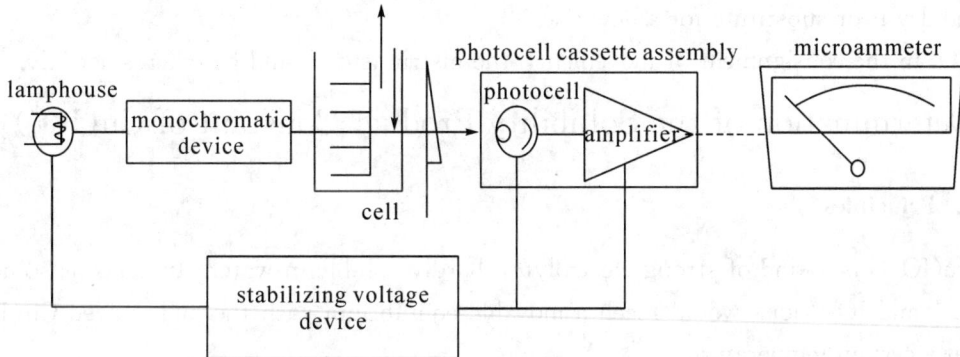

Fig. 13-2　The introduction of the basic structure of 721 spectrophotometer

## 3. Usage

(1) First, turn the power on (button 7, Fig. 13-3), then the indicating lamp turns on. Open the cell camera below cover 8. Preheating the instrument for 20 min.

(2) Select needed wavelength of monochrome light with the wavelength selecting turn button (turn button 6), and select needed sensitivity with the sensitivity turn button (turn button 2).

(3) Put cells in the instrument, zero the instrument with the zero knob by turning button 5, then cover the cell camera below cover. Push in the cell draw-bar 3 to adjust the control cell zero. Expose the photocell to light, turning the transmittancy adjusting turn button 4 to make the microammeter 9 point to 100%. Repeat the operation above several times to adjust zero and 100% for the instrument. Then measurement could be performed.

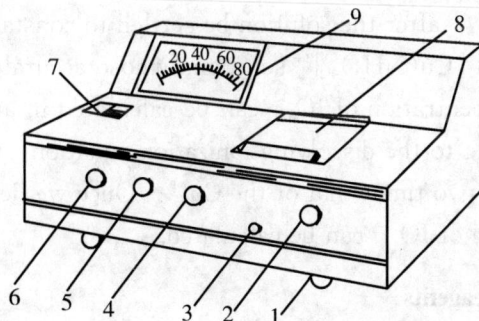

Fig. 13-3　721 Spectrophotometer

### 4. Notice

(1) Do not continuously use the instrument over 2 h. You'd better rest the instrument for 0.5 h before next usage.

(2) After using, cell should be washed with pure water and toppled to dry under room temperature, then restore them in the specialized cell box. In usage, pay attention to protect the lucency face of cell from damage or scoring.

(3) Keep the instrument dry. In daily use, attentions should be paid to the self-indicating silica gel (in the bottom of the monochromator). If it turns red, immediately take it out and dry it or substitute for a new one.

(4) In the consignment or movement, the instrument should be taken smoothly.

# Determination of the Solubility Product Constant of $Cu(IO_3)_2$

### 1. Principles

$Cu(IO_3)_2$ is a kind of strong electrolyte, hardly soluble in water. In aqueous solution, the $Cu^{2+}$ and $IO_3^-$ ions would reach a movable equilibrium with the undissolved $Cu(IO_3)_2$ solid at a certain temperature:

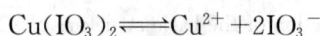

$$Cu(IO_3)_2 \rightleftharpoons Cu^{2+} + 2IO_3^-$$

When the solution is saturated, the equilibrium is reached. Under certain temperature, the multiple of the concentration of $Cu^{2+}$ and the square of the concentration (activity, in accuracy) of $IO_3^-$ is a constant in the saturated solution:

$$K_{sp} = [Cu^{2+}][IO_3^-]^2$$

$K_{sp}$ is the solubility product constant, while $[Cu^{2+}]$, $[IO_3^-]$ are the concentrations of $Cu^{2+}$ and $IO_3^-$ at equilibrium $(mol \cdot L^{-1})$. At certain temperature, the value of $K_{sp}$ won't change with $[Cu^{2+}]$ or $[IO_3^-]$.

If we mix solutions of $CuSO_4$ and $KIO_3$, precipitated $Cu(IO_3)_2$ with solubility activity will form. Add certain quantity of solvent to make the precipitate dissolve again, separate the precipitate, determine the current concentration of $Cu^{2+}$ and $IO_3^-$, the $K_{sp}$ value is available.

$NH_3 \cdot H_2O$ is added to certain quantity of saturated the solution to produce the blue complex ions of $[Cu(NH_3)_4]^{2+}$, after the solution be settled to constant volume, then compare its color with the standard $[Cu(NH_3)_4]^{2+}$ solutions whose accurate concentrations are definite. The equilibrium concentration of $IO_3^-$ can be calculated in an indirect method according to the $Cu^{2+}$. According to the dissolving-ionization equation, the concentration of $IO_3^-$ produced by $Cu(IO_3)_2$ are two times that of the $Cu^{2+}$. Once we determine the concentration of $Cu^{2+}$, the concentration of $IO_3^-$ can be calculated.

### 2. Instruments and Reagents

beaker (50 mL), color comparison tube (50 mL), scales dropping piette (1.00 mL,

2. 00 mL, 20. 00 mL), quantitative filter paper, funnel.

$KIO_3$(solid, A. R. ), $CuSO_4 \cdot 5H_2O$ (solid, A. R. ), $NH_3 \cdot H_2O$ (2 mol·$L^{-1}$), $CuSO_4$ (0. 800 mol·$L^{-1}$).

### 3. Standard color

Measure 0. 400 mL, 0. 600 mL, 0. 800 mL, 1. 00 mL, 1. 20 mL, 1. 40 mL, 1. 60 mL, 1. 80 mL, 0. 0800 mol·$L^{-1}$ $CuSO_4$ solution with scale dropping pipette, put them in eight 25. 00 mL color comparison tubes with the same standard respectively and then add 2 mol· $L^{-1}$ $NH_3 \cdot H_2O$ respectively in every tube to produce a kind of translucent, blue solution. Dilute those solutions with distilled water to the standards of tubes (notice: $NH_3 \cdot H_2O$ should be over amounted to prevent the solution from being cloudy in diluting. ).

Following the above method, the order color rules are:

　　1. 28×$10^{-3}$　　1. 92×$10^{-3}$　　2. 56×$10^{-3}$　　3. 20×$10^{-3}$
　　3. 84×$10^{-3}$　　4. 48×$10^{-3}$　　5. 12×$10^{-3}$　　5. 78×$10^{-3}$

### 4. Experimental Procedures

(1) Preparation of fresh solid copper iodate

①Weigh 4. 2 g $KIO_3$(s), and dissolve it in 85 mL distilled water, heat and filter if necessary, thus translucent $KIO_3$ solution is prepared.

②Weigh 2. 5 g $CuSO_4 \cdot 5H_2O$, and dissolve it in 15 mL distilled water, heat to dissolve, thus $CuSO_4$ solution is prepared.

③Slowly add the $KIO_3$ solution to the warm solution of $CuSO_4$ with continuous stirring, and cool it to room temperature (about 30 min, keep stirring) naturally, so as to make it precipitate completely and ensure the precipitate is resolvable. Filter with quantative filter paper by vacuum filtration, wash the precipitate 5~6 times with small quantity of water, and then transfer the precipitate to filter paper, drying it.

(2) Take out the wet solid of copper iodate crystal as big as a soybean by a glass spoon, put it in a clean and dry beaker (50 mL), then add 20. 00 mL distilled water by dropping pipette. Slowly heat the beaker on a asbestos net, keep stirring, and keep heating for 1~2 min at 60~70℃. Stirring should not be stopped in order to reach the equilibrium faster. After that, stop heating , and cool it at room temperature, keep stirring. Filter the solution with dry slow-speed quantitative filter paper in a clean and dry funnel with long stem funnel, then collect the filtrate with a clean and dry small beaker. Measure 10. 00 mL filtrate with piette, then add it in a 25. 00 mL color comparison tube, which is the same as that in the standard color concentrations. Prepare the unknown solution in the same way as preparing the standard color concentrations, and compare it with the standard color concentrations by eyes (or photometry) to find out the standard color concentration ($Cu^{2+}$) whose color is the same as or close to the unknown solution. The concentration of the unknown solution may be just the concentration of a certain standard color concetration or the average concentrations of two standard color concentrations, all being determined by experiment. Finally,

record the room temperature.

Repeat the experiment twice with the same method mentioned above, yet substitute the 20.00 mL distilled water with 19.00 mL distilled water plus 1.00 mL 0.0800 mol·L$^{-1}$ CuSO$_4$ and 18.00 mL pure water plus 2.00 mL 0.800 mol·L$^{-1}$ CuSO$_4$, keep the total volume as 20.00 mL. Then determine the correspondent concentration by colorimetry.

### 5. Data process

Data collected in experiments should be analyzed as Table 13-1:

**Table 13-1　Experiment data of copper iodate $K_{sp}$　($t = 15\,^{\circ}\!C$)**

| No. | 1 | 2 | 3 |
|---|---|---|---|
| The primary concentration of Cu$^{2+}$ in solutions $a = [Cu^{2+}] \times \dfrac{V_{Cu^{2+}}}{V_{eq}}$ | 0.00 | $0.08 \times \dfrac{1.00}{20.00}$ | $0.08 \times \dfrac{2.00}{20.00}$ |
| The concentration of Cu$^{2+}$ measured by shade selection $b = [Cu^{2+}] \times \dfrac{V_2}{V_1}$ | $1.60 \times 10^{-3} \times \dfrac{25.0}{10.0} = 4.00 \times 10^{-3}$ | $2.88 \times 10^{-3} \times \dfrac{25.0}{10.0} = 7.20 \times 10^{-3}$ | $4.48 \times 10^{-3} \times \dfrac{25.0}{10.0} = 11.2 \times 10^{-3}$ |
| The equilibrium concentration of Cu$^{2+}$ produced by the dissolving of Cu(IO$_3$)$_2$ $(b-a)$ | $4.00 \times 10^{-3}$ | $3.20 \times 10^{-3}$ | $3.2 \times 10^{-3}$ |
| The equilibrium concentration of IO$_3^-$ produced by the dissolving of Cu(IO$_3$)$_2$ $2(b-a)$ | $8.00 \times 10^{-3}$ | $6.40 \times 10^{-3}$ | $6.40 \times 10^{-3}$ |
| The solubility product constant $K_{sp} = b \times [2(b-a)]^2$ | $2.56 \times 10^{-7}$ | $2.95 \times 10^{-7}$ | $4.59 \times 10^{-7}$ |

In the abore table, $V_{Cu^{2+}}$ is volume of the CuSO$_4$ solution with known concentration being added in every color comparison tube; $V_{eq}$ is the total volume of the solution in solid-liquid equilibrium (in this experiment, the $V_{eq}$ is defined as 20.00 mL); $[Cu^{2+}]$ is the concentration of CuSO$_4$ solution (in this experiment, it is defined as 0.0800 mol·L$^{-1}$); $[Cu^{2+}]$ level represents that the unknown concentration is the same as the standard color concentration. $V_1$ is the volume of the unknown solution used in shade selection (in this experiment, it's 10.00 mL); $V_2$ is the volume of the unknown solution used in colorimetry after dissolving (in this experiment, it's 25.00 mL).

As shown in Table 13-1, the equilibrium concentration of IO$_3^-$ is calculated in an indirect method. That's to say, on the basis of the formula Cu(IO$_3$)$_2$ = Cu$^{2+}$ + 2IO$_3^-$, every Cu$^{2+}$ in the solution is corresponded with two IO$_3^-$, thus the concentration of IO$_3^-$ should be twice as that of Cu$^{2+}$. But in experiments, the Cu$^{2+}$ concentration includes the Cu$^{2+}$ concentration ($a$) being added solutions. If no other solution is added in the experiment, $a = 0.00$. So in the equilibrium, concentration of Cu$^{2+}$ produced by the solution of Cu(IO$_3$)$_2$ should be $(b-a)$. While the concentration of IO$_3^-$ is $2(b-a)$.

## Questions

1. If the Cu(IO$_3$)$_2$(s) passes the filter paper in filtering, what effects it may have on

experiment results?

  2. Why will we add different volumes of $CuSO_4$ solutions in repeating the experiment?

  3. Estimate the derivation between the $K_{sp}$ values determined in this experiment and the records in literature, and explain the major influential factors causing the derivation.

# 实验十四　设计实验初步——氯化铵的提纯

## 实验内容

1. 自拟提纯粗 $NH_4Cl$ 固体制备试剂 $NH_4Cl$ 的实验方案。其中包括实验原理、操作步骤及操作条件。

2. 称取 10 g 粗 $NH_4Cl$ 固体（含泥沙等不溶性杂质和 $Ca^{2+}$，$Mg^{2+}$，$Fe^{2+}$，$K^+$，$SO_4^{2-}$ 等可溶性杂质），按照教师审阅后的自拟方案进行提纯。

3. 成品中 $Fe^{3+}$ 和 $SO_4^{2-}$ 的限量分析。

(1) $Fe^{3+}$ 的限量分析。称取 2.00 g $NH_4Cl$ 样品，放入 25 mL 比色管中，加 10 mL 蒸馏水溶解，再加 2.00 mL 10%磺基水杨酸溶液和 2.00 mL 10% $NH_3 \cdot H_2O$，用蒸馏水稀释至刻度，摇匀，与标准色阶（准备室给出）进行比色，确定 $NH_4Cl$ 试样的等级。

(2) $SO_4^{2-}$ 的限量分析。称取 1.00 g $NH_4Cl$ 样品，放入 25 mL 比色管中，加 10 mL 蒸馏水溶解，再加 5.00 mL 95%乙醇和 1.00 mL 3 mol·L$^{-1}$ HCl 溶液，在不断振荡下滴加 3.00 mL 25%$BaCl_2$ 溶液，用蒸馏水稀释至刻度，摇匀，与标准比浊液（准备室给出）进行比浊，确定试样等级。

## 提示

1. 方案应有计划、有步骤地除去不溶性和可溶性杂质，可以从形成难溶性物质考虑除去可溶性杂质，同时根据有关难溶盐的溶度积数据，选择适当的沉淀剂。杂质除尽后，要作必要的"中间控制"检验。

2. 为使沉淀颗粒较大，便于分离，可在溶液处于较高温度时逐滴加入沉淀剂，当加足沉淀剂后，要让溶液温热一段时间（陈化）。

3. 磺基水杨酸在微碱性溶液中（pH=9~11.5 氨性溶液）与 $Fe^{3+}$ 生成黄色的磺基水杨酸铁配合物，反应如下：

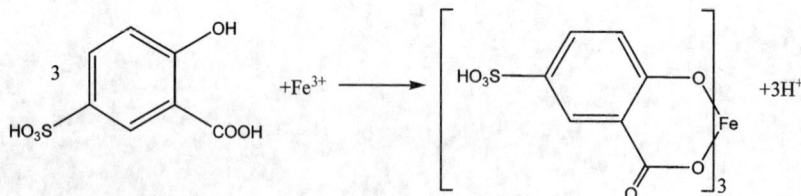

## 标准色阶的配制

1. 比色用标准色阶的配制。分别准确量取 0.40 mL、1.00 mL 和 2.00 mL 浓度为 0.01 mg·mL$^{-1}$ 的 $Fe^{3+}$ 溶液于 25 mL 比色管中，照本实验内容 3.(1) 的用量加入 10%磺基水杨酸和 10% $NH_3 \cdot H_2O$，并用水稀释至刻度，摇匀即可。

上述标准溶液内 $Fe^{3+}$ 含量依次为：0.0002%，0.0005%，0.001%。

各级 $NH_4Cl$ 中所允许的 $Fe^{3+}$ 最高含量为：优级纯不大于 0.0002%，分析纯不大于 0.0005%，化学纯不大于 0.001%。

2. 比浊用标准色阶的配制。分别量取 0.20 mL、0.50 mL 浓度为 0.1 mg·mL$^{-1}$ 的 SO$_4^{2-}$ 溶液于 25 mL 的比色管中，照本实验内容3.(2)中的用量加入 95% 乙醇、3 mol·L$^{-1}$ 的 HCl 和 25%BaCl$_2$ 溶液，用蒸馏水稀释至刻度，摇匀即可。

上述溶液内 SO$_4^{2-}$ 含量依次为 0.002%、0.005%。各级 NH$_4$Cl 中所允许的 SO$_4^{2-}$ 最高含量为：优级纯不大于 0.005%，分析纯不大于 0.01%，化学纯不大于 0.02%。

# Exp 14　Design Experiment—Purification of Ammonium Chloride

## Experimental Procedures

1. Design a experiment scheme for preparation of pure $NH_4Cl$ with crude $NH_4Cl$. Including the experiment principle, experimental procedures and operation.

2. Weigh 10 g crude $NH_4Cl$ solid (containing indissolvable impurity of mud sand and soluble impurity of $Ca^{2+}$, $Mg^{2+}$, $Fe^{2+}$, $K^+$, $SO_4^{2-}$), carry out the design after checked by instructor.

3. Analysis the dose limited of $Fe^{3+}$ and $SO_4^{2-}$ in end product

(1) Analysis dose limited of $Fe^{3+}$: weigh 2.00 g $NH_4Cl$ sample in a 25 mL color comparison tube, dissolve it with 10 mL distilled water, add 2.00 mL 10% 3-sulfosalicylic acid and 2.00 mL 10% $NH_3 \cdot H_2O$, continue to add distilled water to dilute it to the scale of 25 mL, shake it. Visual colorimetry with standard solution (given by preparative room), determine the quality grade of $NH_4Cl$ sample.

(2) Analysis dose limited of $SO_4^{2-}$: weigh 1.00 g $NH_4Cl$ sample in a color comparison test tube, dissolve it with 10 mL distilled water, 5.00 mL 95% of ethanol and 1.00 mL 3 $mol \cdot L^{-1}$ of HCl solution are added, then drop 3.00 mL 25% of $BaCl_2$ solution with continuous shaking. Continue to add distilled water to dilute it to the scale of 25 mL, shake it. Compare the color with standard color comparison solution given by preparative room to determine the quality grade of $NH_4Cl$ sample.

## Note

1. The project shall remove indissolvable and soluble impurity according to the corresponding plan and procedure. Remove dissolvable impurity in the form of indissolvable substance. Select proper precipitator based $K_{sp}$ data of related indissolvable salt. Note to have a necessary test "media control".

2. Add precipitant at a rather high temperature in order to make precipitate grain bigger and easily being separated. After adding enough precipitant, let the solution keep slightly hot for a period time.

3. Sulfonyl salicylic acid reacts with $Fe^{3+}$ to form yellow coordination compound of sulfur salicyl iron in alkaline solution (pH=9~11.5 ammonia solution), the reaction equation is as follows:

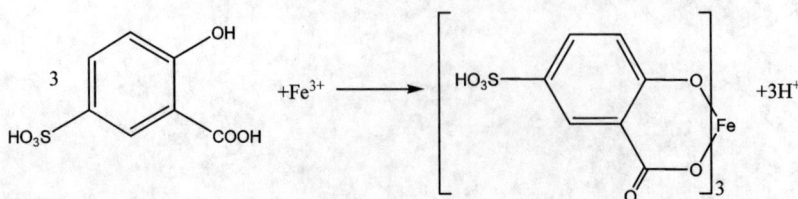

## Preparation standard color rank

1.  Preparation standard color rank for comparing color.  Measure 0. 40 mL, 1. 00 mL, 2. 00 mL (concentration 0. 01mg·mL$^{-1}$) $Fe^{3+}$ solution into three 25 mL color comparison tubes respectively.  Then add 10% 5-sulfosalicylic acid and 10% $NH_3·H_2O$ into it according to the dosage in experiment 3. (1), add distilled water to dilute it to the scale of 25 mL, shake.

$Fe^{3+}$ contents in standard solution are 0. 0002%, 0. 0005%, 0. 001% in order.

$Fe^{3+}$ content permitted in all levels are: excellent level not more than 0. 0002%, analytical level not more than 0. 0005%, chemical level not more than 0. 001%.

2.  Preparing standard color rank for turbidimetry.  Measure 0. 20 mL, 0. 50 mL (concentration is 0. 1 mg·mL$^{-1}$) $SO_4^{2-}$ solutions into three 25 mL color comparison tubes respectively, add 95% ethanol and 3 mol·L$^{-1}$ HCl solution according to the amount in experiment 3. (2), add distilled water to dilute it to the scale of 25 mL, shake it.

$SO_4^{2-}$ contents in standard solution are 0. 002%, 0. 005% in order.

$SO_4^{2-}$ contents permitted in all levels are: excellent level not more than 0. 005%, analytical level not more than 0. 01%, chemical level not more than 0. 02%.

# 实验十五　碱金属和碱土金属

## 实验目的

1. 试验并了解金属钠和过氧化钠的性质。
2. 了解钠、锂、钾盐的溶解性。
3. 试验并比较碱土金属氢氧化物的难溶性。
4. 试验碱土金属难溶盐的溶解性。
5. 学会焰色反应的操作。

## 实验仪器、药品及材料

仪器：小试管、小刀、镊子、研钵、坩埚、铂丝（或镍铬丝）、温度计、钴玻璃等
材料：pH 试纸、冰
固体药品：金属钠、钾、钙、镁，过氧化钠
液体药品：汞(水银)、$NaCl(1.0\ mol\cdot L^{-1})$、$KCl(1.0\ mol\cdot L^{-1})$、$LiCl(1.0\ mol\cdot L^{-1})$、$MgCl_2(0.5\ mol\cdot L^{-1})$、$CaCl_2(0.5\ mol\cdot L^{-1})$、$SrCl_2(0.5\ mol\cdot L^{-1})$、$BaCl_2(0.5\ mol\cdot L^{-1})$、$NaOH(2\ mol\cdot L^{-1})$、$NH_4Cl(饱和)$、$Na_2CO_3(0.5\ mol\cdot L^{-1})$、$HCl(2，6\ mol\cdot L^{-1})$、HAc $(6\ mol\cdot L^{-1})$、$HNO_3(浓)$、$Na_2SO_4(0.5\ mol\cdot L^{-1})$、$CaSO_4(饱和)$、$K_2CrO_4(0.5\ mol\cdot L^{-1})$、$KSb(OH)_6(饱和)$、$(NH_4)_2C_2O_4(饱和)$、$NaHC_4H_4O_6(饱和)$、$MnSO_4(0.1\ mol\cdot L^{-1})$、$H_2SO_4(浓)$、$NH_3\cdot H_2O(2\ mol\cdot L^{-1})$、酚酞

## 实验内容

### 1. 钠、钾与水的反应

用镊子各取绿豆大小一块金属钾和钠，用滤纸吸干表面的煤油，切去表面的氧化膜，立即将它们分别放入盛有水的 50 mL 的烧杯中，可将事先准备好的合适漏斗倒扣在烧杯上，以确保安全。观察两者与水反应的情况，并进行比较。反应终止后，检验溶液的酸碱性。根据反应进行的剧烈程度，说明钠、钾的金属活泼性。

### 2. 钠与空气中氧的反应和过氧化钠的性质

（1）钠与氧气的反应
用镊子取黄豆大小一块钠，用滤纸吸干表面的煤油，切去表面的氧化膜，立即置于坩埚中加热。当钠刚开始燃烧时，停止加热。观察反应情况和产物的颜色、状态，写出反应方程式。设计实验证明产物为过氧化钠（保留此产品供下面实验用）。
（2）过氧化钠的性质
①过氧化钠的碱性
取绿豆大小的过氧化钠固体，注入 1 mL 水，用冰冷却，并加以搅动，使其溶解。用 pH 试纸检验溶液的酸碱性。

②过氧化钠的分解

取黄豆大小的 $Na_2O_2$ 固体，注入 1 mL 水，微热，观察是否有气体放出？写出反应方程式。

### 3. 金属钠与汞反应

取一块绿豆大小的金属钠，擦干其表面的煤油，把它和 1 滴汞放在一起，在研钵中研磨，观察其反应现象和产物的颜色、状态。

将得到的钠汞齐转入盛有水的 50 mL 烧杯中，加入 1~2 滴酚酞，观察反应情况，写出反应方程式。将钠汞齐和水反应后产生的汞回收。

### 4. 钠、钾微溶盐的生成

（1）微溶性钠盐

往小试管中滴入 5 滴 1 mol·$L^{-1}$ NaCl 溶液，然后再滴入 5 滴饱和六羟基锑（V）酸钾 $KSb(OH)_6$ 溶液。如果无晶体析出，可用玻棒摩擦试管壁，然后放置一段时间。观察产物的颜色和状态，写出反应方程式。

（2）微溶性钾盐

往小试管中滴入 5 滴 1 mol·$L^{-1}$ KCl 溶液，接着滴入 5 滴饱和的酒石酸氢钠 $NaHC_4H_4O_6$ 溶液，如果无晶体析出，可用玻棒摩擦试管壁。观察反应产物的颜色和状态。写出反应方程式。

### 5. 镁、钙与水的反应

（1）取长为 1 cm 左右的镁条，用砂纸擦去表面的氧化膜，放入一支试管中，加入 1 mL 左右的水。观察有无反应，然后将试管加热，观察反应情况。加入几滴酚酞检验水溶液的碱性，写出反应方程式。

（2）用小刀切黄豆大小的一块钙，用滤纸吸干煤油后，放入盛有约 1 mL 水的试管中，观察反应情况，并检验溶液的 pH。

比较钙镁与水反应的情况，说明它们的金属活泼性顺序。

### 6. 氢氧化镁的生成和性质

（1）氢氧化镁的生成和性质

在三支试管中，都加入 5 滴 0.5 mol·$L^{-1}$ 的 $MgCl_2$ 溶液，然后各加入 5 滴 6 mol·$L^{-1}$ $NH_3·H_2O$。观察 $Mg(OH)_2$ 沉淀的生成。然后分别试验它们与饱和 $NH_4Cl$ 溶液，2 mol·$L^{-1}$ 的盐酸和 2 mol·$L^{-1}$ NaOH 溶液的反应情况。写出反应方程式。

（2）镁、钙、钡氢氧化物的难溶性

在三支试管中分别加入 5 滴 0.5 mol·$L^{-1}$ 的 $MgCl_2$，$CaCl_2$ 和 $BaCl_2$ 溶液，再加入等体积的新配制的 2 mol·$L^{-1}$ 氢氧化钠溶液，观察沉淀生成。

### 7. 碱土金属难溶性盐

（1）镁、钙、钡硫酸盐溶解性的比较

在三支试管中，分别盛有 5 滴 0.5 mol·$L^{-1}$ $MgCl_2$，$CaCl_2$ 和 $BaCl_2$ 溶液，然后分别注

入等量的 $0.5\ mol\cdot L^{-1}$ 硫酸钠溶液，观察现象。若 $MgCl_2$ 和 $CaCl_2$ 溶液加入硫酸钠溶液后无沉淀生成，可用玻棒摩擦试管壁，再观察有无沉淀生成，说明沉淀的情况，分别检验沉淀与浓 $H_2SO_4$ 的反应，写出反应方程式。

另外在两支试管中分别加入 5 滴 $0.5\ mol\cdot L^{-1}$ 的 $CaCl_2$ 和 $BaCl_2$ 溶液，各加入等体积饱和 $CaSO_4$ 溶液，观察沉淀生成的情况。

比较 $MgSO_4$，$BaSO_4$ 和 $CaSO_4$ 溶解度的大小。

（2）钙、钡铬酸盐的生成和性质

在两支试管中分别注入 5 滴 $0.5\ mol\cdot L^{-1}$ 的 $CaCl_2$ 和 $BaCl_2$ 溶液，再各注入 5 滴 $0.5\ mol\cdot L^{-1}$ 的铬酸钾溶液，观察现象。分别试验沉淀与 $6\ mol\cdot L^{-1}$ 的醋酸和 $2\ mol\cdot L^{-1}$ 的盐酸溶液反应，写出反应方程式。

（3）草酸钙的生成和性质

取 5 滴 $0.5\ mol\cdot L^{-1}CaCl_2$ 溶液于试管中，加入 5 滴饱和草酸铵溶液，观察反应产物的颜色和状态。把沉淀分成两份，分别试验它们与 $2\ mol\cdot L^{-1}$ 的盐酸和 $6\ mol\cdot L^{-1}$ 的醋酸反应。写出反应方程式。

### 8. 碱金属和碱土金属盐的焰色反应

取一支铂丝（或镍铬丝）蘸以 $6\ mol\cdot L^{-1}$ 盐酸溶液在氧化焰中烧至无色，再蘸上氯化锂溶液在氧化焰中灼烧，观察火焰颜色。依照此法，分别进行 NaCl，KCl，$CaCl_2$，$SrCl_2$ 和 $BaCl_2$ 溶液的焰色反应试验。每进行完一种溶液的焰色反应后，均需蘸 $6\ mol\cdot L^{-1}$ 盐酸溶液灼烧铂丝（或镍铬丝），观察钾盐的焰色反应时，为消除钠对钾焰色的干扰，一般需用蓝色的钴玻璃片滤光。

### 思考题

1. 试设计一个分离 $K^+$，$Mg^{2+}$ 和 $Ba^{2+}$ 的实验方案。
2. 若实验室中发生镁的燃烧事故，应用什么方法灭火？可否用水或二氧化碳来灭火？

## 附注

1. 金属钠、钾、钙平时应保存在煤油中或石蜡油中。取用时，可在煤油中用小刀切割，用镊子夹取，并用滤纸把煤油吸干。切勿与皮肤接触，未用完的钠屑不能乱丢，可放回原瓶中或放在少量酒精中，使其缓慢耗掉。

2. 汞盐和汞蒸气均有剧毒。使用汞时，应更加注意。由于汞的相对密度很大。用普通滴管吸取时容易自然下落。为了使汞不撒落在桌面或地面上，取汞的操作可在搪瓷盘上进行，如不慎将汞撒落时，一定要用滴管尽可能地将汞收回，然后在有可能残存汞的地方洒上一层硫磺粉。对撒在狭缝中的残汞应灌入熔化的硫磺。

3. 钠与汞形成钠汞齐时，若钠的含量很少时则呈粘液状态。钠的含量较多时，则呈固态，性脆。由于加入钠和汞的量不同，钠汞齐可以以不同的组成存在。

# Exp 15　Alkali Metals and Alkali Earth Metals

## Objectives

1. To test and understand the properties of metallic sodium and sodium peroxide.
2. To comprehend the solubility of salts of sodium, lithium, potassium.
3. To test and compare the insolubility of hydroxides of the alkali earth metals.
4. To test the insolubility of salts of alkali earth metals.
5. To master the operations of flame reaction.

## Instrument, Reagents and Materials

Instrument: small test tubes, knife, forceps, mortar box, melting pot, platinum filament (or nickel-chromium filament), thermometer, blue cobalt glass and so on

Materials: pH test paper, ice

Solid reagents: sodium, potassium, calcium, magnesium, sodium peroxide

Liquid reagents: mercury(liquid silver), $NaCl(1.0 \ mol \cdot L^{-1})$, $KCl(1.0 \ mol \cdot L^{-1})$, $LiCl(1.0 \ mol \cdot L^{-1})$, $MgCl_2(0.5 \ mol \cdot L^{-1})$, $CaCl_2(0.5 \ mol \cdot L^{-1})$, $SrCl_2(0.5 \ mol \cdot L^{-1})$, $BaCl_2(0.5 \ mol \cdot L^{-1})$, $NaOH(2 \ mol \cdot L^{-1})$, $NH_4Cl(saturated)$, $Na_2CO_3(0.5 \ mol \cdot L^{-1})$, $HCl(2, 6 \ mol \cdot L^{-1})$, $HAc(6 \ mol \cdot L^{-1})$, $HNO_3(concentrated)$, $Na_2SO_4(0.5 \ mol \cdot L^{-1})$, $CaSO_4(saturated)$, $K_2CrO_4(0.5 \ mol \cdot L^{-1})$, $KSb(OH)_6(saturated)$, $(NH_4)_2C_2O_4(saturated)$, $NaHC_4H_4O_6(saturated)$, $MnSO_4(1 \ mol \cdot L^{-1})$, $H_2SO_4(concentrated)$, $NH_3 \cdot H_2O(2 \ mol \cdot L^{-1})$, phenolphthalein

## Experimental Procedures

### 1. Reaction of sodium, potassium with water

Fetch small pieces of sodium and potassium of soybean size with forceps, its superficial kerosene is absorbed, remove the superficial oxide membrane, then immediately put them into beakers containing 50 mL water respectively and immediately. For our safety, put the suitable funnel which is prepared onto the beaker headstand, observe the phenomenon and compare. After the reaction, test the acidity or alkalinity of the solution. Explain the metals activity of the sodium, potassium according to the violent degree of the reactions.

### 2. Reaction of sodium with oxygen in air and property of sodium peroxide

(1) Reaction of sodium with oxygen

A piece of sodium of soybean size is cut with scissors, its superficial kerosene is absorbed, remove the fresh superficial membrane on metal sodium, then immediately put it into a crucible, heat, when the sodium begins burning, stop heating. Observe the phenomenon of the reaction, color and state of products. Write down the reaction equation, try to prove the product is the sodium peroxide (reserve the product for the following experiment).

(2) The properties of sodium peroxide

①The alkalinity of sodium peroxide

Take a piece of solid sodium peroxide of mung bean size, add 1 mL water, the test tube must be placed in cold water, dissolve it by stirring, detect whether the solution is alkaline with pH test paper.

②The decomposition of sodium peroxide

On a piece of solid sodium peroxide of soybean size, add 1 mL water, heat slightly. Whether the gas will be given out? Write down the reaction equation.

### 3. Reaction of sodium with mercury

A piece of sodium of mung bean size, its superficial kerosene is absorbed, put it together with one drop of mercury, put them into a mortar box to grind. Observe the phenomena of the reaction and the color and state of products, transfer the sodium amalgam to a 50 mL beaker, add 1~2 drops of phenolphthalein, then observe the phenomena of the reaction, and write down the reaction equations. Recycle the residual mercury of reaction product of sodium amalgam with water.

### 4. The formation of slightly soluble salts of sodium, potassium

(1) Slightly soluble sodium salts

Add 5 drops of 1 $mol \cdot L^{-1}$ NaCl solution to a test tube, then add 5 drops of saturated $KSb(OH)_6$ solution. If there is no crystal, rub the inside of the test tube with a glass rod. Hold on for a moment. Observe the color and state of the product. Write down the reaction equation.

(2) Slightly soluble potassium salts

Add 5 drops of 1 $mol \cdot L^{-1}$ KCl solution to a test tube, then add 5 drops of saturated $NaHC_4H_4O_6$ (saturated) solution. If there is no crystal, rub the inside of the test tube with a glass rod and hold on for a moment. Observe the color and state of product. Write down the reaction equation.

### 5. The reaction of calcium, magnesium with water

(1) Take magnesium of 1 cm length, reveal the fresh superficial membrane with sand paper, put it into a test tube, and add about 1 mL water. Observe whether the reaction happens, then heat the tube, observe the phenomena of the reaction. Examine whether the solution is alkaline by adding a few drops of phenolphthalein. Write down the reaction equation.

(2) A piece of calcium of soy bean size is cut with scissors, its superficial kerosene is absorbed. Put it into a test tube and add 1 mL water; observe the phenomena and test the

pH value of the solution.

Compare the reaction of calcium, magnesium with water. Explain and conclude the metals activity in proper order.

### 6. The properties and formation of magnesium hydroxide

(1) The properties and formation of magnesium hydroxide

Add 5 drops of 0.5 mol·L$^{-1}$ MgCl$_2$ solution to three test tubes respectively, then add 5 drops of 6 mol·L$^{-1}$ NH$_3$·H$_2$O, observe the formation of the precipitate of Mg(OH)$_2$. Observe the phenomena that precipitate reacts with saturated NH$_4$Cl solution, 2 mol·L$^{-1}$HCl, 2 mol·L$^{-1}$NaOH respectively, write down the reaction equation.

(2) The insolubility of hydroxide of calcium, magnesium, barium

Add 5 drops of 0.5 mol·L$^{-1}$ MgCl$_2$, CaCl$_2$ and BaCl$_2$ solution to three test tubes respectively, then add the equivalent volume of fresh 2 mol·L$^{-1}$ NaOH solution to the three test tubes. Observe the formation of precipitate.

### 7. The insoluble salts of alkali earth metals

(1) Compare the solubility of sulfate salts of calcium, magnesium, barium

Add 5 drops of 0.5 mol·L$^{-1}$ of MgCl$_2$, CaCl$_2$ and BaCl$_2$ solution to three test tubes respectively, then add the equivalent quantity of 0.5 mol·L$^{-1}$ Na$_2$SO$_4$ solution respectively, observe the phenomena. If there is no precipitate in MgCl$_2$, CaCl$_2$ solutions, add more Na$_2$SO$_4$ solution, rub the inside of the test tube with a glass rod, then observe the phenomena, is there some crystal in solution? Test the product reacting with concentrated H$_2$SO$_4$, write down the reaction equation.

In another two tubes, 5 drops of 0.5 mol·L$^{-1}$ CaCl$_2$, BaCl$_2$ are added first, then equivalent volume of CaSO$_4$ solution are added, observe the formation of precipitate.

Compare the solubility of MgSO$_4$, BaSO$_4$ and CaSO$_4$.

(2) The properties, formation of calcium chromate, barium chromate

Add 5 drops of 0.5 mol·L$^{-1}$ of CaCl$_2$, BaCl$_2$ solutions to two different test tubes respectively, and then 5 drops of 0.5 mol·L$^{-1}$ potassium chromate solution to each one, observe the phenomena, test the precipitate reacting with 6 mol·L$^{-1}$ HAc, 2 mol·L$^{-1}$HCl, write down the reaction equations.

(3) The properties, formation of CaC$_2$O$_4$

In solution of 5 drops of CaCl$_2$, 5 drops of (NH$_4$)$_2$C$_2$O$_4$ are added, observe the color and state of product. Divide the product into two portions, test the reaction between the product and 2 mol·L$^{-1}$ or 6 mol·L$^{-1}$ HAc. Write down the reaction equations.

### 8. Flame tests of alkali metals, alkali earth metals

To clean the wire, dip it into the test tube of 6 mol·L$^{-1}$ HCl and heat the wire on the oxidizing flame until the flame appears colorless. When the platinum wire is clean, dip the wire in the test tube containing LiCl solution and hold it on the hottest part of the flame.

Observe the color of flame, then test with NaCl, KCl, CaCl, $SrCl_2$ and $BaCl_2$, every time the wire must be cleaned. As we observe the color of the flame of the potassium ion, a piece of blue cobalt glass should be used.

## Questions

1. Design a scheme to separate $K^+$, $Mg^{2+}$ and $Ba^{2+}$.

2. If there is a fire accident caused by magnesium in lab, how shall we put out fire? Can we use the water or carbon dioxide to put it out?

## Notes

1. Sodium, potassium and calcium should be stored in kerosene or liquid paraffine. As we use them, we should cut them into small ones in the kerosene. Fetch it with forceps, and absorb the kerosene with filter paper. Make sure not to contact with skin. Residual sodium should not be thrown over casually, you may add a small amount of anhydrous ethanol to make it decompose slowly or recycle.

2. Salts of mercury and gaseous mercury are both poisonous. More care should be taken in use of mercury for it has relatively heavier density. It is not easy to drop naturally by common dropping pipette. To avoid leaving mercury on desk or floor, it's better to use it on enamel plate. If mercury is sprinkled, it should be reclaimed by dropping pipette and sprinkle sulfur powder on it. For residual mercury in narrow crevice, sulfur powder should be poured.

3. When sodium and mercury form sodium amalgam, if the content of sodium is low, it presents myxoid. But when the content of sodium is high, it presents solid appearance, frail. For the quantity of sodium and mercury dissimilarity, sodium amalgam presents different existence.

# 实验十六　卤　素

## 目的要求

1. 掌握卤素的氧化性和卤素离子的还原性。
2. 掌握次卤酸盐及卤酸盐的氧化性。
3. 了解卤素的歧化反应。
4. 试验实验室中制备卤化氢的方法及试验它们的性质。
5. 了解某些金属卤化物的性质。

## 预习与思考

1. 预习有关氯、溴、氢氟酸和氯酸钾的使用安全知识。
2. 预习有关卤族元素的 $\varphi^{\ominus} - n$ 图。
3. 进行卤素离子还原性实验时应注意哪些安全问题？
4. 如何区别次氯酸钠溶液和氯酸钠溶液？如何比较次氯酸钠和氯酸钾的氧化性？
5. 为什么用 $AgNO_3$ 检出卤素离子时，要先用 $HNO_3$ 酸化溶液，再用 $AgNO_3$ 检出？向一未知溶液中加入 $AgNO_3$ 时如果不产生沉淀，能否认为溶液中不存在卤素离子？

## 仪器和药品

玻璃片，分液漏斗，铅皿，带支管的大试管，氯气发生器装置（公用）。

氯水，溴水，碘水，$H_2S$（饱和水溶液），四氯化碳，品红溶液，淀粉溶液，浓氨水，红磷，石蜡。

$I_2(s)$，$KCl(s)$，$KBr(s)$，$KI(s)$，$KClO_3(s)$，$CaF_2(s)$，$NaCl(s)$，$H_3PO_4$（浓），$H_2SO_4$（浓，$3\ mol \cdot L^{-1}$，$6\ mol \cdot L^{-1}$），$KBrO_3$（饱和），$KClO_3$（饱和），$KIO_3(0.1\ mol \cdot L^{-1})$，$KBr(0.1\ mol \cdot L^{-1}$，$0.5\ mol \cdot L^{-1})$，$NaF(0.1\ mol \cdot L^{-1})$，$NaCl(0.1\ mol \cdot L^{-1})$，$Na_2S_2O_3(0.5\ mol \cdot L^{-1}$，$0.1\ mol \cdot L^{-1})$，$KI(0.01\ mol \cdot L^{-1}$，$0.1\ mol \cdot L^{-1})$，$KOH(6\ mol \cdot L^{-1}$，$2\ mol \cdot L^{-1})$，$FeCl_3(0.1\ mol \cdot L^{-1})$，$HCl$（浓，$2\ mol \cdot L^{-1}$），$MnSO_4(0.1\ mol \cdot L^{-1})$，$Ca(NO_3)_2(0.1\ mol \cdot L^{-1})$，$AgNO_3(0.1\ mol \cdot L^{-1})$，$HNO_3(2\ mol \cdot L^{-1})$，$NH_3 \cdot H_2O(2\ mol \cdot L^{-1})$，$Na_2SO_3(0.1\ mol \cdot L^{-1})$。

碘化钾-淀粉试纸，pH 试纸，醋酸铅试纸。

含 $Cl^-$，$Br^-$ 与 $I^-$ 离子的混合液，失落标签的 $KClO$，$KClO_3$ 和 $KClO_4$ 试剂。

## 实验内容

### 1. 卤素单质在不同溶剂中的溶解性

分别试验并观察少量的氯、溴、碘在水、四氯化碳和碘化钾水溶液中的溶解情况，以表格形式写出实验结果，并作理论解释。

**2. 卤素的氧化性**

（1）分别以 $0.1\ mol \cdot L^{-1}$ KBr，$0.1\ mol \cdot L^{-1}$ KI，$CCl_4$，氯水和溴水等试剂，设计一系列试管实验，说明氯、溴、碘的置换次序。记录有关实验现象，写出反应式。

（2）氯水、溴水、碘水氧化性差异的比较

分别向氯水、溴水、碘水溶液中滴加 $0.1\ mol \cdot L^{-1}$ $Na_2S_2O_3$ 溶液及饱和硫化氢水溶液，观察现象，写出反应式。

（3）氯水对溴、碘离子混合溶液的氧化顺序

在试管内加入 $0.5\ mL$（约 10 滴）$0.1\ mol \cdot L^{-1}$ KBr 溶液及 2 滴 $0.01\ mol \cdot L^{-1}$ KI 溶液，然后再加入 $0.5\ mL$ 四氯化碳，逐滴加入氯水，仔细观察四氯化碳液层颜色的变化，写出有关反应式。

通过以上实验说明卤素氧化性的递变顺序。

（4）请利用卤素置换次序的不同制作"保密信"一封。

**3. 卤素离子还原性（通风橱内进行）**

（1）分别向三支盛有少量（绿豆大小）KCl，KBr 和 KI 固体的试管中加入约 $0.5\ mL$ 浓硫酸。观察现象并选用合适的试纸或试剂检验各试管中逸出的气体产物。提供选择的试纸或试剂分别有醋酸铅试纸、碘化钾－淀粉试纸、pH 试纸和浓氨水。该实验说明了卤素离子的什么性质？写出反应式。

（2）$Br^-$、$I^-$ 还原性的比较。分别利用 KBr，KI 和 $FeCl_3$ 溶液之间的反应，说明 $Br^-$，$I^-$ 还原性的差异，写出反应式。

通过以上实验比较卤素离子还原性的相对强弱。

**4. 氯的歧化反应（在通风橱内进行）**

取氯水 $10\ mL$ 逐滴加入 $2\ mol \cdot L^{-1}$ KOH 至溶液呈弱碱性（用 pH 试纸检验）。将溶液分成四份，第一份溶液与 $2\ mol \cdot L^{-1}$ HCl 反应，选择合适的试纸检验气体产物，写出有关反应式；另外三份留作次氯酸钾氧化性实验用。

另取 $5\ mL$ $6\ mol \cdot L^{-1}$ KOH 溶液，水浴加热溶液至近沸下通入氯气。待有晶体析出后，用冰水冷却试管，滤去溶液，观察产物色态。晶体留作氯酸钾氧化性实验用。写出氯气在热碱溶液中歧化的反应式。

**5. 次卤酸盐及卤酸盐的氧化性**

（1）次氯酸钾的氧化性

由上述实验内容 4 制得的三份次氯酸钾溶液分别与 $0.1\ mol \cdot L^{-1}$ $MnSO_4$ 溶液、品红溶液及用 $H_2SO_4$ 酸化了的碘化钾－淀粉溶液反应。观察现象，写出反应式。

（2）氯酸钾的氧化性

①取少量由实验内容 4 制得的 $KClO_3$ 晶体置于试管中，加入少许浓盐酸，注意逸出气体的气味，检验气体产物，写出反应式，并作出解释。

②分别试验由实验室配制的饱和 $KClO_3$ 溶液与 $0.1\ mol \cdot L^{-1}$ $Na_2SO_3$ 溶液在中性及酸性条件下（用什么酸酸化？）的反应，用 $AgNO_3$ 验证反应产物，试用该实验说明 $KClO_3$ 的氧

化性与介质酸碱性的关系。

③取少量 $KClO_3$ 晶体，用 $1\sim2$ mL 水溶解后，加入少量四氯化碳及 $0.1$ mol·$L^{-1}$ KI 溶液数滴，摇动试管，观察试管内水相及有机相有什么变化？再加入 $6$ mol·$L^{-1}$ $H_2SO_4$ 酸化溶液又有什么变化？写出反应式。能否用 $HNO_3$ 或盐酸来酸化溶液？为什么？

（3）溴酸钾的氧化性（在通风橱内进行）

①饱和溴酸钾溶液经 $H_2SO_4$ 酸化后分别与 $0.5$ mol·$L^{-1}$ KBr 溶液及 $0.5$ mol·$L^{-1}$ KI 溶液反应，观察现象并检验反应产物，写出反应式。

②试验 $KBrO_3$ 溶液与 $Na_2SO_3$ 溶液在中性及酸性条件下的反应，记录现象，写出反应式。

（4）碘酸盐的氧化性

$0.1$ mol·$L^{-1}$ $KIO_3$ 溶液经 $3$ mol·$L^{-1}$ $H_2SO_4$ 酸化后加入几滴淀粉溶液，再滴加 $0.1$ mol·$L^{-1}$ $Na_2SO_3$ 溶液，观察现象，写出反应式。若体系不酸化，又有什么现象？改变加入试剂顺序（先加 $Na_2SO_3$ 最后滴加 $KIO_3$），又会观察到什么现象？

（5）溴酸盐与碘酸盐的氧化性比较

往少量饱和 $KBrO_3$ 溶液中加入少量浓 $H_2SO_4$ 酸化后再加入少量碘片，振荡试管，观察现象，写出反应式。

通过以上实验总结氯酸盐、碘酸盐、溴酸盐的氧化性。

**6. 卤化氢的制备与性质（通风橱内进行）**

（1）氟化氢的制备与性质

在一块涂有石蜡的玻璃片上，用小刀刻下字迹。在铅皿或塑料瓶盖上放入约 $1$ g 固体 $CaF_2$，加入几滴水调成糊状后，滴入 $1\sim2$ mL 浓 $H_2SO_4$，立即用刻有字迹的玻璃片覆盖。$1\sim2$ h 后，用水冲洗玻璃片并刮去玻璃片上的石蜡后可清晰地看到玻璃片上的字迹。解释现象，写出反应式。

（2）分别试验少量固体 NaCl，KBr，KI 与浓 $H_3PO_4$ 反应，适当微热，观察现象并与 3.(1)实验比较，写出反应式。

（3）碘化氢的制备与性质（通风橱内进行）

在干燥的大试管内装有粉状的碘及在干燥器干燥过的红磷（$I_2$:P$\approx$1:6），稍微加热试管，从分液漏斗中滴加少量水，反应生成的气体由导气管导入一支干燥的小试管中。将烧红了的玻璃棒插入收集碘化氢的试管中，观察现象，写出反应式。

**7. 金属卤化物的性质**

（1）卤化物的溶解度比较

①分别向盛有 $0.1$ mol·$L^{-1}$NaF，NaCl，KBr 以及 KI 溶液的试管中滴加 $0.1$ mol·$L^{-1}$ $Ca(NO_3)_2$ 溶液，观察现象，写出反应式。

②分别向盛有 $0.1$ mol·$L^{-1}$ NaF，NaCl，KBr 以及 KI 溶液的试管中滴加 $0.1$ mol·$L^{-1}$ $AgNO_3$ 溶液，制得的卤化银沉淀经过离心分离后分别与 $2$ mol·$L^{-1}$ $HNO_3$，$2$ mol·$L^{-1}$ $NH_3$·$H_2O$ 及 $0.5$ mol·$L^{-1}$ $Na_2S_2O_3$ 溶液反应，观察沉淀是否溶解？写出反应式。解释氧化物与其卤化物溶解度的差异及变化规律。

（2）卤化银的感光性

将制得的 AgCl 沉淀均匀地涂在滤纸上，滤纸上放上一把钥匙，光照约十多分钟后取出钥匙，可清楚地看到钥匙的轮廓。卤化银见光分解以氯化银较快，碘化银最慢。

### 8. 小设计

（1）混合液中含 $Cl^-$，$Br^-$ 和 $I^-$ 离子，试设计分离检出方案。

（2）有三瓶无色液体试剂失去了标签，它们分别是 KClO，$KClO_3$ 和 $KClO_4$。请设计实验方案加以鉴别。

## 安全知识

1. 氯气有毒和刺激性，少量吸入人体会刺激鼻咽部，引起咳嗽和喘息。大量吸入会导致严重损害，甚至死亡。因此，进行有关氯气的实验必须在通风橱内进行。

2. 溴蒸气对气管、肺部、眼鼻喉都有强烈的刺激作用。进行有关溴的实验，应在通风橱内进行，不慎吸入溴蒸气时，可吸入少量氨气和新鲜空气解毒。液态溴具有很强的腐蚀性，能灼烧皮肤，严重时会使皮肤溃烂。移取液态溴时，需戴橡皮手套。溴水的腐蚀性虽比液溴弱些，但在使用时，也不允许直接由瓶内倒出，而应用滴管移取，以防溴水接触皮肤。如果不慎把溴水溅在手上，应及时用水冲洗，再用以稀硫代硫酸钠溶液充分浸透的绷带包扎处理。

3. 氟化氢气体有剧毒和强腐蚀性。主要对骨骼、造血系统、神经系统、牙齿及皮肤黏膜造成伤害。吸入人体内会使人中毒，氢氟酸能灼烧皮肤。因此，在使用氢氟酸和进行有关氟化氢气体的实验时，应在通风橱内进行，在移取氢氟酸时，必须戴上皮手套，用塑料管吸取。

4. 氯酸钾是强氧化剂，保存不当时容易引起爆炸，它与硫、磷的混合物是炸药，因此，绝对不允许将它们混在一起。氯酸钾容易分解，不宜大力研磨，烘干或烤干。在进行有关氯酸钾的实验时，如同进行其他有强氧化性物质实验一样，应将剩下的试剂倒入回收瓶内回收处理，一律不准倒入废酸缸中。

# Exp 16　Halogen

## Objectives

1. To master the oxidizability of halogen and reducibility of halogen ion.
2. To master the oxidizability of hypohaloate .
3. To understand the disproportionation reaction of halogen.
4. To know the method of preparing hydrogen halides in laboratory and test their properties.
5. To understand the properties of some metal halides.

## Previewing and Thinking

1. Preview the safety knowledge of hydrochloride, hydrobromide, hydrofluoride and potassium chlorate.
2. Preview the $\varphi^\Theta$-n figure of halogen.
3. What problems should you note when you carry out the reducibility experiment of halogen ion?
4. How to distinguish sodium chlorate from sodium hypochlorite solution? How to compare the oxidizability of sodium chlorate and sodium hypochlorite?
5. When detecting halogen ion, we usually use $HNO_3$ to acidify the solution first , then use $AgNO_3$ solution, why? Add $AgNO_3$ to some unknown solutions, if no precipitate is produced, can you draw a conclusion that no halogen ions exist?

## Instruments and Reagents

glass plate, separating funnel, lead watch-glass, big test tube with branch, producing instrument of chlorine (for public use).

chlorine water, bromine water, iodine water, $H_2S$ solution (saturated), tetrachloride, fuchsine solution, starch solution, concentrated ammonia water, amorphous phosphorus, paraffin wax.

$I_2(s)$, $KBr(s)$, $KI(s)$, $KCl(s)$, $KClO_3(s)$, $CaF_2(s)$, $NaCl(s)$, $H_3PO_4$(concentrated), $H_2SO_4$(concentrated, $3\ mol \cdot L^{-1}$, $6\ mol \cdot L^{-1}$), $KBrO_3$(saturated), $KClO_3$(saturated), $KIO_3(0.1\ mol \cdot L^{-1})$, $KBr(0.1\ mol \cdot L^{-1}$, $0.5\ mol \cdot L^{-1})$, $NaF(0.1\ mol \cdot L^{-1})$, $NaCl$ $(0.1\ mol \cdot L^{-1})$, $Na_2S_2O_3(0.5\ mol \cdot L^{-1}$, $0.1\ mol \cdot L^{-1})$, $KI(0.01\ mol \cdot L^{-1}$, $0.1\ mol \cdot L^{-1})$, $KOH(6\ mol \cdot L^{-1}$, $2\ mol \cdot L^{-1})$, $FeCl_3(0.1\ mol \cdot L^{-1})$, $KOH(6\ mol \cdot L^{-1}$, $2\ mol \cdot L^{-1})$, $FeCl_3(0.1\ mol \cdot L^{-1})$, $HCl$(concentrated, $2\ mol \cdot L^{-1})$, $MnSO_4(0.1\ mol \cdot L^{-1})$, $Ca(NO_3)_2(0.1\ mol \cdot L^{-1})$, $AgNO_3$ $(0.1\ mol \cdot L^{-1})$, $HNO_3(2\ mol \cdot L^{-1})$, $NH_3 \cdot H_2O(2\ mol \cdot L^{-1})$.

KI-starch test paper, pH test paper, lead acetate test paper.

Mixed solution containing $Cl^-$, $Br^-$ and $I^-$, $KClO$, $KClO_3$ and $KClO_4$ reagents without labels.

## Experimental Procedures

### 1. Solubility of small quantities of halogen in different solvents

Test and observe the solubility of small quantities of chlorine, bromine and iodine in water, carbon tetrachloride and KI solution, write out the results in form and explain.

### 2. Oxidizability of halogen

(1) Design a series of test tube experiments, using $0.1\ mol \cdot L^{-1}$ KBr, $0.1\ mol \cdot L^{-1}$ KI, $CCl_4$, chlorine water, bromine water to illustrate the substitution sequence. Record the related experiment phenomenon and write down the reaction equations.

(2) Comparing oxidizability differences in chlorine, bromine, iodine water

Add $0.1\ mol \cdot L^{-1}$ $Na_2S_2O_3$ solution and saturated $H_2S$ solution to chlorine water, bromine water, iodine water. Observe the phenomenon and write down the reaction equations.

(3) Oxidizability sequence of chlorine water for bromine ion, iodine ion in mixture

Add about $0.5$ mL (about 10 drops) $0.1\ mol \cdot L^{-1}$ KBr solution and two drops of $0.01 mol \cdot L^{-1}$ KI solution, then $0.5$ mL $CCl_4$ is added, chlorine water is added dropwise, observe the color change of $CCl_4$ layer. Write down the reaction equations.

Explain the oxidizability sequence of halogen.

(4) Make use of the substitution sequence to prepare a "secret letter".

### 3. Reducibility of halogen ions (carrying out in fume hood)

(1) Add about $0.5$ mL $H_2SO_4$ to three test tubes containing KCl, KBr, KI respectively. Observe the phenomenon and choose the suitable test paper or reagents for the gas produced. You can choose test paper as follows: lead acetate test paper, KI-starch test paper, pH test paper or concentrated ammonia water. What properties do these experiments illustrate? Write down the reaction equation.

(2) Comparing the reducibility of $Br^-$, $I^-$

Using the reactions between KCl, KI and $FeCl_3$, illustrate the differences of reducibility of $Br^-$, $I^-$. Write down the reaction equations.

Compare the relative of halogen ions through above experiments.

### 4. Disproportionate reaction of halogen (carrying out in fume hood)

Add 10 mL chlorine water to $2\ mol \cdot L^{-1}$ KOH dropwise, until the solution is weak alkaline. Divide the solution into four portions, the first portion reacts with HCl, choose suitable test paper to test the gas produced. Write down the reaction equations. The other three portions are used in following experiments.

Fetch 5 mL $6 mol \cdot L^{-1}$ KOH solution, heat in water bath, gaseous $Cl_2$ is bubbled when the solution nearly boils. When the crystal produces, cool down the test tube in ice water, filter, observe the color and state of the product. The crystal is conserved for later use. Write down the disproportionate reaction equation of $Cl_2$ in hot alkalinous solution.

### 5. Oxidizability of hypochlorite and chlorate

(1) Three portions of potassium hypochlorite produced in procedure 4 react with 0.1 $mol \cdot L^{-1}$ $MnSO_4$, fuchsine solution and KI-starch solution acidifying with $H_2SO_4$ respectively. Observe the phenomenon and write down the reaction equations.

(2) Oxidizability of potassium chlorate

①Add a small quantity of $KClO_3$ produced in procedure 4 in the test tube, then a small quantity of concentrated hydrochloric acid is added. Observe the gas produced, test the gas, write down the reaction equations, and explain the phenomenon.

②Test the reaction between saturated $KClO_3$ produced in laboratory and 0.1 $mol \cdot L^{-1}$ $Na_2SO_3$ solution under neutral acid medium. Test the product produced by $AgNO_3$. What conclusion can you draw on the relationship between the oxidizability of potassium chlorate in acid or alkaline properties of medium?

③Fetch a small quantity of crystal $KClO_3$, dissolve it with $1{\sim}2$ drops of water, add a small quantity of $CCl_4$ and 0.1 $mol \cdot L^{-1}$ KI solution, shake the test tube, observe the phenomena between organic phase and water phase. Add 6 $mol \cdot L^{-1}$ $H_2SO_4$ solution, what change does it happen? Write down the reaction equation. Can you acidify the solution with $HNO_3$? Why?

(3) Oxidizability of potassium bromate (carrying out in fume hood)

① Acidify the saturated potassium bromate solution in two test tubes, then add 0.5 $mol \cdot L^{-1}$ KBr solution and 0.5 $mol \cdot L^{-1}$ KI solution respectively, observe and test the product produced, write down the reaction equations.

②Test the reaction between $KBrO_3$ solution and $Na_2SO_3$ solution in neutral and acid medium. Record the phenomenon and write down the reaction equation.

(4) Oxidizability of iodate

0.1 $mol \cdot L^{-1}$ $KIO_3$ solution acidified with 3 $mol \cdot L^{-1}$ $H_2SO_4$ is added to several drops of starch solution, then 0.1 $mol \cdot L^{-1}$ $Na_2SO_3$ solution orderly. Observe the phenomenon and write down the reaction equation. If the mixture isn't acidified, what phenomenon will we observe? Change the sequence when you add the reagents. What phenomenon will be observed?

(5) Compare the oxidizability of bromate and iodate

Add a small quantity of concentrated sulfuric acid to a small quantity of saturated $KBrO_3$ solution, then add a small quantity of iodine(s), shake the test tube. Observe the phenomenon and write down the reaction equation

Summarize the oxidizability sequence of chlorate, bromate and iodate.

### 6. Preparation and properties of hydrogen halide (carrying out in fume hood)

(1) Preparation and properties of hydrogen fluoride

Inscribe handwriting on a glass plate being overlayed with paraffin wax. Place about 1 g $CaF_2$ in a plastic coveror, add several drops of water and blend to paste, then drop 1~2 concentrated $H_2SO_4$, covering with glass plate inscribed with handwriting. After 1~2 h, wash the glass plate and scrape the paraffin wax. Then we can see clear handwriting on the glass plate. Explain the phenomenon and write down the reaction equations.

(2) Test the reaction between NaCl(s), KBr(s), KI(s) and concentrated $H_3PO_4$ respectively, heat them slightly. Observe and compare the phenomenon with that in procedure 3. (1) and write down the reaction equation.

(3) Preparation and properties of hydrogen iodide (carrying out in fume hood)

Add powdered iodine and dried red phosphorus to a dry test tube ($I_2 : P \approx 1 : 6$), heat the test tube, add a small quantity of water by separating funnel, ventilate the produced gas to another small test tube. Insert the burnt glass rod to the test tube which has collected HI. Observe the phenomenon and write down the reaction equation.

### 7. The properties of metal halide

(1) Comparison of halide's solubility

①Drop 0.1 mol·$L^{-1}$ Ca($NO_3$)$_2$ solution to 0.1 mol·$L^{-1}$ NaF, NaCl, KBr, KI respectively, Observe the phenomenon, write down the reaction equations.

②Drop 0.1 mol·$L^{-1}$ $AgNO_3$ solution to the test tube with 0.1 mol·$L^{-1}$ NaF, KBr, KI, $AgNO_3$ precipitate produced is centrifugalzed, then make precipitate react with 2 mol·$L^{-1}$ $HNO_3$, 2 mol·$L^{-1}$ $NH_3·H_2O$, 0.5 mol·$L^{-1}$ $Na_2S_2O_3$ solution, observe the dissolving of precipitate. Write down the reaction equations. Explain the differences of solubility between fluoride and their halides and summarize the changing law.

(2) Light sensitiveness of silver halides

Overlay produced AgCl on filter paper, put a key on the filter paper, illuminate about ten minutes, clear adumbration can be seen. AgCl decomposes most quickly, while AgI does slowly.

### 8. Design

(1) For mixture contains $Cl^-$, $Br^-$ and $I^-$, design a separation and identification scheme.

(2) Three bottles of colorless liquid reagent missing labels, they are $KClO$, $KClO_3$ and $KClO_4$. Design an identification scheme.

## Safety Notes

1. Chlorine gas is toxic and irritate, breathing in a small quantity can stimulate nose and throat, and cause cough and gasping. Too much breathing will cause to death. So the experiments related chlorine must be carried out in fume hood.

2. Gaseous bromine is stimulus to trachea, bellows, eyes, nose and larynx. The experiments related to bromine should be carried out in fume hood. If you breathe in bromine

carelessly, you should breathe in a small quantity of ammonia or fresh air. Liquid bromine is strong caustic, which can damage skin and cause inflammation. Fetching liquid bromine, you should wear rubber gloves. Bromine water is less inflammation than liquid bromine, but when using it you shouldn't get it from bottles directly but use it with dropping pipette. If it is splashed on hand carelessly, wash the hand with water quickly and then deal with bandage which has been soaked fully in sodium hyposulfite.

3. Hydrogen fluoride is rank poisonous and caustic. It mainly harms skeleton, hemopoiesis system, nervous system, teeth, skin mucosa. Breathing in hydrogen fluoride will cause toxicosis. It also can burn skin. Performing experiments with it, you should carry out in fume hood and wear gloves and using plastic tube when fetching it.

4. Potassium chlorate is a strong oxidizing agent, improper preservation may cause explosion. The mixture of potassium chlorate, sulfur and phosphor is dynamite. So you should never mix them together. $KClO_3$ decomposes easily, so don't grind emphatically or oven dry. Carrying out the experiments with $KClO_3$, as you do experiment with other reagents of strong oxidizability, surplus reagent should be poured into recovery bottle, no pouring to acid jar.

# 实验十七　氢、氧、过氧化氢

## 目的要求

1. 掌握实验室制备氢气、氧气的方法。
2. 掌握过氧化氢的化学性质。
3. 掌握氢气使用的安全知识。

## 预习与思考

1. 预习"无机化学实验基本操作"中有关气体的发生、净化、收集等有关内容。
2. 思考下列问题：
(1) 点燃氢气时要注意的安全事项包括什么内容？
(2) 在实验室如何制备 $H_2O_2$ 和 $Na_2O_2 \cdot 8H_2O$？反应条件如何？

## 仪器、药品和材料

玻璃缸，燃烧匙，玻璃片，胶塞，广口瓶。

锌粒，铜粉，铁粉，硫磺粉，红磷，细铁丝，$CuO(s)$，$KClO_3(s)$，$MnO_2(s)$，$BaO_2 \cdot 2H_2O(s)$，$Na_2O_2(s)$，$KMnO_4(0.01\ mol \cdot L^{-1})$，$K_2CrO_4(0.1\ mol \cdot L^{-1})$，$K_2Cr_2O_7(0.1\ mol \cdot L^{-1})$，$H_2O_2\ 3\%(m)$，$NaOH\ 40\%(m)$，$H_2S(饱和水溶液)$，乙醚，无水乙醇，$HCl(6\ mol \cdot L^{-1})$，$H_2SO_4(1\ mol \cdot L^{-1})$，$KI(0.1\ mol \cdot L^{-1})$，$Pb(NO_3)_2(0.1\ mol \cdot L^{-1})$，$AgNO_3(0.1\ mol \cdot L^{-1})$，$MnSO_4(0.1\ mol \cdot L^{-1})$。

pH 试纸，碘化钾-淀粉试纸。

## 实验内容

### 1. 氢气的制备与性质

(1) 氢气的制备

装置如图 17-1 所示，大试管内加入 10 mL 6 mol·L⁻¹ HCl 溶液和 4～5 粒锌粒。用一个带尖嘴的玻璃管的胶塞塞住管口，观察氢气的生成。收集产生的氢气并试验其纯度。

(2) 氢气的点燃

在证实氢气的纯度后，点燃氢气，观察氢气在空气中的燃烧。把一个底部干燥的盛有冷水的小烧杯放在氢气火焰的顶部，观察现象，写出反应式。

(3) 氢气的还原性

装置如图 17-2 所示，在一支干燥的试管中装入少量干燥的氧化铜粉末及少许作催化剂用的铜粉，用铁夹固定好并使试管底部略高于管口（为什么？），用另一支大试管制备氢气。当确证氢气的纯度后，将氢气导入装有氧化铜粉末的试管中，导管要插到粉末上方。加热氧化铜，观察现象，写出反应式。结束实验时，必须先停止加热并继续通入氢气，直至试管冷却到室温为止（为什么？）。

### 2. 氧气的制备与性质

（1）氧气的制备与收集

装置如图 17-3 所示。7 g 干燥的 $KClO_3$ 粉末和 1 g 经灼烧过的 $MnO_2$ 粉末放在蒸发皿中混合均匀后再装入一支干燥的大试管中。加热试管，使气体均匀逸出，用广口瓶收集三瓶氧气。进行下列实验。

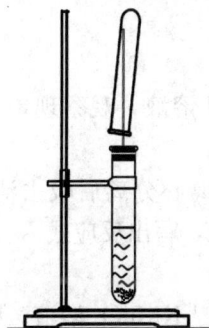

图 17-1 制备和收集氢气　　图 17-2 氢气还原氧化铜装置　　图 17-3 制备和收集氧气装置

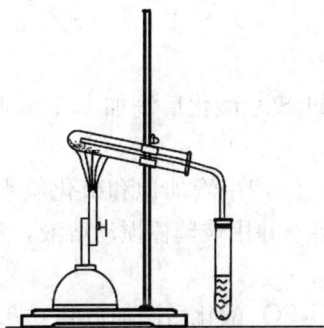

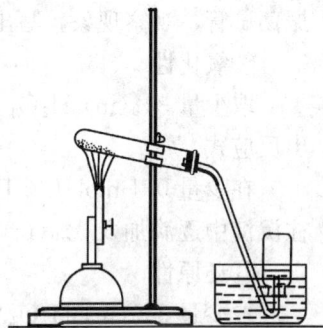

（2）氧气的性质

①硫在氧气中的燃烧（在通风橱内进行）

在燃烧匙中放少量硫粉，于煤气灯上加热硫粉熔化至近沸，立即将燃烧匙放入盛有氧气的广口瓶中，观察现象，写出反应式。用水吸收瓶内气体并试验溶液的酸碱性。

②磷在氧气中的燃烧（在通风橱内进行）

燃烧匙中放入少量红磷，于煤气灯上加热至红热后重复 2.（2）.①的操作，观察现象，写出反应式。

③铁丝在氧气中燃烧

用坩埚夹取一段卷成螺旋状的细铁丝，铁丝末端系上一根火柴梗，点燃火柴后即将铁丝放入盛有氧气的广口瓶中，为防止反应过程中生成的赤热物质落到瓶底而使玻璃瓶破裂，可在瓶内先加入少量水或加入一层细砂。观察现象，写出反应式。

### 3. 臭氧的制备与性质

取少量固体 $BaO_2 \cdot 2H_2O$ 于小试管中，慢慢地加入约 2 mL 浓硫酸，将反应物放在冰水中冷却，用碘化钾−淀粉试纸检验反应逸出的气体。反应式为：

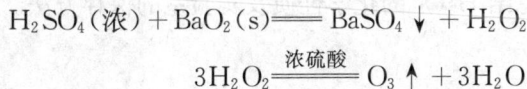

$$H_2SO_4(浓) + BaO_2(s) = BaSO_4 \downarrow + H_2O$$

$$3H_2O_2 \xrightarrow{浓硫酸} O_3 \uparrow + 3H_2O$$

### 4. 过氧化氢的制备与性质

（1）过氧化氢的制备

取少量 $Na_2O_2(s)$ 于小试管中，加入少量蒸馏水溶解后放在冰水中冷却，并加以搅拌。用试纸检验溶液的酸碱性，再往试管中滴加已用冰水冷却过的 $1\ mol \cdot L^{-1}\ H_2SO_4$ 溶液至酸性为止（目的是什么？），写出反应式。

（2）过氧化氢的鉴定

取以上制得的 $H_2O_2$ 溶液，加入约 0.5 mL 乙醚，并加入少量 1 mol·L$^{-1}$ $H_2SO_4$ 酸化溶液，再加入 2~3 滴 0.1 mol·L$^{-1}$ $K_2CrO_4$ 溶液，振荡试管，观察水层和乙醚层颜色的变化，写出反应式。

（3）过氧化氢的性质

①酸性

在小试管中加入少量 40%（m）NaOH 溶液，约 1 mL $H_2O_2$ 溶液及约 1 mL 无水乙醇，振荡试管，观察现象，写出反应式。

②氧化性

取少量 3%（m）$H_2O_2$ 溶液以 $H_2SO_4$ 酸化后滴加 0.1 mol·L$^{-1}$ KI 溶液，观察现象，写出反应式。

在少量 0.1 mol·L$^{-1}$ $Pb(NO_3)_2$ 溶液中滴加饱和硫化氢水溶液，离心分离后吸去清液，往沉淀中逐滴加入 3%（m）$H_2O_2$ 溶液并用玻璃棒搅动溶液，观察现象，写出反应式。

③还原性

取少量 3%（m）$H_2O_2$ 溶液用 $H_2SO_4$ 酸化后滴加数滴 0.01 mol·L$^{-1}$ KMnO$_4$ 溶液，观察现象。用火柴余烬检验反应生成的气体，写出反应式。

在少量 0.1 mol·L$^{-1}$ AgNO$_3$ 溶液中滴加 40%（m）NaOH 溶液至棕色沉淀生成，再加入少量 3%（m）$H_2O_2$ 溶液，观察现象。用火柴余烬检验反应生成的气体，写出反应式。另取少量 0.1 mol·L$^{-1}$ AgNO$_3$ 溶液，加入少量 3%（m）$H_2O_2$ 溶液，现象又有何不同？试解释之。

④介质酸碱性对 $H_2O_2$ 氧化还原性质的影响

在少量 3%（m）$H_2O_2$ 溶液中加入 40%（m）NaOH 溶液数滴，再加入 0.1 mol·L$^{-1}$ MnSO$_4$ 溶液数滴，观察现象，写出反应式。溶液经静置后倾去清液，往沉淀中加入少量 $H_2SO_4$ 溶液后滴加 3%（m）$H_2O_2$ 溶液，观察又有什么变化？写出反应式并给予解释。

⑤过氧化氢的分解

加热约 2 mL 3%（m）$H_2O_2$ 溶液，有什么现象发生？用火柴余烬检验产生的气体，写出反应式。

在少量 3%（m）$H_2O_2$ 溶液中加入少量 MnO$_2$ 固体，观察现象，用火柴余烬检验反应产生的气体，写出反应式。

在少量 3%（m）$H_2O_2$ 溶液中加入少量铁粉，观察现象，用火柴余烬检验反应产生的气体，写出反应式。

通过以上实验简单总结 $H_2O_2$ 的化学性质及实验室的保存方法。

## 思考题

1. 过氧化氢是否既可作氧化剂又可作还原剂？什么条件下过氧化氢可将 Mn$^{2+}$ 氧化为 MnO$_2$？什么条件下 MnO$_2$ 又可将过氧化氢氧化而产生氧气？它们相互矛盾吗？为什么？

2. 在制备氧气的试验中，为什么要在 KClO$_3$ 中加入 MnO$_2$？MnO$_2$ 又为什么必须预先经过灼烧？

3. 过氧化钡与浓硫酸、稀硫酸作用时产物分别是什么？如何证实？

## 安全注意事项

　　氢气与氧气以 2:1（体积比）混合后一经点燃即发生非常剧烈的爆炸，氢气与空气中的氧混合时也可以形成爆炸性的气体。因此在点燃或使用氢气前都必须注意检查氢气的纯度。检查氢气纯度的方法可以用一支小试管罩在氢气的出口管上，收集约半分钟后慢慢地取出试管，并用拇指紧堵管口（管口朝下）将管口移至火焰的上方，松开拇指如无尖锐的爆鸣声，则表明氢气基本上已经纯净，如有尖锐爆鸣声，则表示管内的氢气中还有空气，此时应另取一支试管重做上述试验，（如果仍用同一支试管重复试验时，则必须保证在试管内没有残余的氢气火焰）直至确证制备的氢气基本纯净为止。

# Exp 17   Hydrogen, Oxygen and Hydrogen Peroxide

## Objectives

1. To grasp the methods of producing hydrogen, oxygen in laboratory.
2. To grasp the chemical properties of peroxide.
3. To grasp the method of using hydrogen safely.

## Previewing and Thinking

1. Preview the content of generation, purification and collection of gas in the chapter 'basic operation of inorganic chemistry experiment'.

2. Think the following questions:

(1) What items should you know before igniting hydrogen?

(2) How to prepare $H_2O_2$ and $Na_2O_2 \cdot 8H_2O$ in the laboratory? What is the reaction condition?

## Instruments, Reagents and Materials

glass trough, combustion spoon, glass plate, rubber stopple, wide-mouthed bottle.

zinc grain, copper powder, iron powder, sulfur powder, red phosphorus, thin iron wire, $CuO(s)$, $KClO_3(s)$, $MnO_2$, $BaO_2 \cdot 2H_2O(s)$, $Na_2O_2(s)$, $KMnO_4(0.01mol \cdot L^{-1})$, $K_2CrO_4(0.1 mol \cdot L^{-1})$, $K_2Cr_2O_7(0.1 mol \cdot L^{-1})$, $H_2O_2 3\%(m)$, $NaOH40\%(m)$, $H_2S$ (saturated), ether, anhydrous alcohol, $HCl(6 mol \cdot L^{-1})$, $H_2SO_4(1 mol \cdot L^{-1})$, $KI(0.1 mol \cdot L^{-1})$, $Pb(NO_3)_2(0.1 mol \cdot L^{-1})$, $AgNO_3(0.1 mol \cdot L^{-1})$, $MnSO_4(0.1 mol \cdot L^{-1})$.

pH test paper, KI-starch test paper.

## Experimental Procedures

### 1. Preparation and properties of hydrogen

(1) Preparation

The instruments are as following Fig. 17-1, adding 10 mL 6 mol·L⁻¹ HCl and 4~5 zinc grains to a test tube. Then use glue of a glass tube with acute beak to plug up the exit, observe the generation of hydrogen. Collect the gas produced, and check its purity.

(2) Igniting hydrogen

After checking the purity of hydrogen, ignite it, observe the burning phenomena of hydrogen in the air. Put a small dry beaker to the anodic position of the flame, observe the phenomena then write down the reaction equation.

(3) Reducibility of hydrogen

Instrument as Fig. 17-2, add a small quantity of dry copper oxide and copper powder as catalyst to a dry tube, use iron clip to fix it and ensure the basefond higher than the opening. (why?). Use another large tube to prepare hydrogen. After ensuring the purity of hydrogen, ventilate it to the tube charged with copper oxide powder. Ensure vessel be inserted at the bottom of the tube. When heating the copper oxide powder, observe the phenomena, and then write down the reaction equation. At the end of the experiment, you must stop heating and go on fluxing hydrogen, upon cooling to room temperature, stop ventilating hydrogen (Why?).

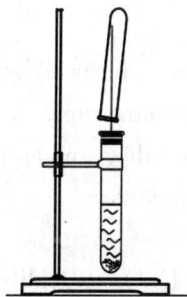

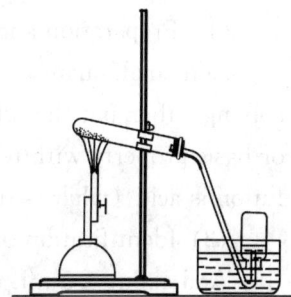

Fig. 17-1   Instrument for prepara-        Fig. 17-2   Instrument for CuO        Fig. 17-3   Instrument for prepara-
        tion and collecting $H_2$        being reduced by $H_2$                tion and collecting $O_2$

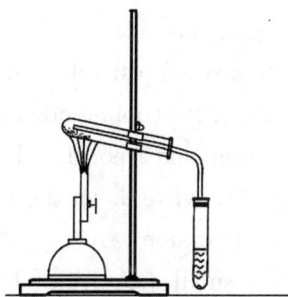

## 2.   Preparation and properties of oxygen

(1) Preparation and collecting oxygen

Instrument as Fig. 17-3. Mix 7g dry $KClO_3$ and 1g calcined $MnO_2$ powder evenly in e-vaporating dish, then loading them to a dry test tube. Heat the tube, the gas will flow outward. Then use a wide-mouthed bottle to collect the gas for the following experiments.

(2) The properties of oxygen

①Burning of sulfur in oxygen (carrying out in fuming hood)

Add some sulfur powder in burning spoon, heat it on gas lamp. Stop heating when sulfur nearly boils. Put burning spoon to wide-mouthed bottle, which is charged with oxygen quickly, observe the phenomena and write down the reaction equation. Use water to absorb the gas and check its acidity or basicity.

②Burning of phosphorus in oxygen (carrying out in fuming hood)

Add some phosphorus powder in combustion spoon, heat it on gas lamp then stop heating when the powder turns red, and repeat the operation as 2. (2). ①, observe the phenomena and write down the reaction equation.

③Burning of iron wire in oxygen

Clip helical iron wire with crucible tongs, bundle a match at the end. After igniting the match, put the iron wire to wide-mouthed bottle quickly which is charged by oxygen. In order to avoid disruption of glass bottle, you should place some water or fine sand at the bottom. Observe the phenomena and write down the reaction equation.

## 3.   Preparation and properties of oxone

Add small solid of $BaO_2 \cdot 2H_2O$ to a small test tube, then about 2 mL concentrated sulfuric acid is added slowly. Use KI-starch test paper to test the gas produced.

The reaction equations are:

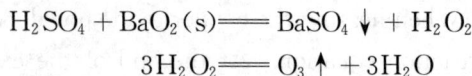

$$H_2SO_4 + BaO_2(s) \!\!=\!\!= BaSO_4 \downarrow + H_2O_2$$
$$3H_2O_2 \!\!=\!\!= O_3 \uparrow + 3H_2O$$

### 4. Preparation and properties of hydrogen peroxide

(1) Preparation and of hydrogen peroxide

Add small quantity of $H_2O_2$ to a small test tube, some distilled water is added after dissolving, then put the tube in ice water for cooling down, stir at the same time. Test its acid or basic property with test paper, then add $1mol \cdot L^{-1}$ $H_2SO_4$ to the tube, and assure the solution is acid (what is the purpose?), write down the reaction equations.

(2) Identification of hydrogen peroxide

Add about 0.5 mL ether and a small quantity of $1\ mol \cdot L^{-1}$ $H_2SO_4$ to acidify the solution, then 2~3 drops of 0.1 $mol \cdot L^{-1}$ $K_2CrO_4$ are added, shake the test tube, observe the color change between water layer and ether layer, write down the reaction equation.

(3) The properties of hydrogen peroxide

①The acidity of hydrogen peroxide

Add a small quantity of 40%(m) NaOH solution to the test tube, about 1 mL $H_2O_2$ and 1 mL anhydrous alcohol, shake the tube, observe the phenomena, write down the reaction equation.

②Oxidizability of $H_2O_2$

Add $H_2SO_4$ to acidify 3% $H_2O_2$, then 0.1 $mol \cdot L^{-1}$KI is added, observe the phenomena, and write down the reaction equation.

Add saturated $H_2S$ solution to a test tube containing a small quantity of $Pb(NO_3)_2$ solution. After precipitate forms, centrifugalize, separate, and decant the above opaque solution, add 3%(m) $H_2O_2$ to the precipitate dropwise, stir the solution at the same time. Centrifugalize and decant the above opaque solution. Observe the phenomena, and write down the reaction equation.

③Reducibility of $H_2O_2$

Acidify 3%(m) $H_2O_2$ with $H_2SO_4$, then add several drops of 0.1 $mol \cdot L^{-1}$KMnO_4, observe the phenomena. Test the gas produced with match ember.

Add 2 $mol \cdot L^{-1}$NaOH solution to 0.1 $mol \cdot L^{-1}$ $AgNO_3$ solution when brown precipitate is produced, then add a small quantity of 3%(m) $H_2O_2$ solution, observe the phenomena. Use match ember to test the gas produced, write down the reaction equation. Get a small quantity of 0.1 $mol \cdot L^{-1}$ $AgNO_3$ solution, add a small quantity of 3% (m) $H_2O_2$ solution, is there any difference? Please explain above phenomena.

④Acidity or alkalinity of medium's influence to oxidizability and reducibility of $H_2O_2$

Add several drops of 40%(m) NaOH solution to a small quantity of 3%(m) $H_2O_2$ solution, then add several drops of 0.1 $mol \cdot L^{-1}$ $MnSO_4$ solution, observe the phenomena, write down the reaction equation. Keep the solution standing, decant the clear solution, and add a small quantity of $H_2SO_4$ solution to the precipitate. Then $3\%(m)$ $H_2O_2$ solution is added, what will you observe? Write down the reaction equation and explain.

⑤Decomposition of hydrogen peroxide

Heat 2 mL solution of $3\%(m)$ hydrogen peroxide, what will happen? Test the gas produced by match ember, and write down the reaction equation.

Add a small quantity of $MnO_2$ to $3\%$ $(m)$ $H_2O_2$ solution, observe the phenomena, test the gas produced by match ember, and write down the reaction equation.

Add a small quantity of iron powder to $3\%$ $(m)$ $H_2O_2$ solution, observe the phenomena. Test the gas produced by match ember, write down the reaction equation.

Summarize the chemical properties of $H_2O_2$ and preserving methods after the above experiments being accomplished.

## Questions

1. Can $H_2O_2$ be used as oxidant and reductant? What is the medium needed if $H_2O_2$ oxidizes $Mn^{2+}$ to $MnO_2$? What is the medium needed if $MnO_2$ oxidizes $H_2O_2$ to produce $O_2$? Are they contrary to each other? Why?

2. In the experiment of preparation of oxygen, why do you add $KClO_3$ to $MnO_2$? And why the solid $MnO_2$ is calcined in advance?

3. What products are produced when barium peroxide reacts with concentrated sulfuric acid or dilute sulfuric acid respectively? How to identify it?

## Safety Details should be Known

When we mix hydrogen and oxygen (volume proportion: 2:1) together, they can explode as rapidly as we ignite the mixture. When hydrogen and air are mixed together, they can also form explosive gas. So we must check the purity of hydrogen before we use it. The checking method is: use a test tube, and cover the outlet of hydrogen, collect it for 30 s or so. Then take out of it slowly, use you thumb to block up the exit of the tube, put it to the anodic position of the flame. If there is no acute cracking, this phenomena manifest that the hydrogen is pure on the whole, on the contrary, you should change another tube to repeat the above operation, ensure its purity is good.

# 实验十八　硫及其化合物

## 目的要求

1. 掌握硫化氢、硫代硫酸盐的还原性，二氧化硫的氧化还原性及过硫酸盐的强氧化性。
2. 掌握硫的含氧酸及其盐的性质。

## 预习与思考

1. 复习有关 $SO_3^{2-}$，$SO_4^{2-}$，$S_2O_3^{2-}$，$S_2O_8^{2-}$ 和 $S^{2-}$ 离子的性质及有关反应。
2. 在有硫化氢或二氧化硫参与或产生的实验中，应注意哪些安全措施？

## 仪器和药品

蒸馏烧瓶，分液漏斗，蒸发皿。

硫磺粉，锌粉，汞，活性炭，品红溶液，$H_2S$（饱和水溶液），氯水，碘水，浓硫酸，$Na_2SO_3(s)$，$K_2S_2O_8(s)$，$CdCO_3(s)$，$MnSO_4(0.002\ mol \cdot L^{-1})$，$KI(0.1\ mol \cdot L^{-1})$，$K_2Cr_2O_7(0.1\ mol \cdot L^{-1})$，$KMnO_4(0.1\ mol \cdot L^{-1}$，$0.01\ mol \cdot L^{-1})$，$H_2SO_4(1\ mol \cdot L^{-1})$，$HCl(2\ mol \cdot L^{-1})$，$AgNO_3(0.1\ mol \cdot L^{-1})$。

待鉴别溶液：$Na_2S$，$Na_2SO_3$，$Na_2SO_4$，$Na_2S_2O_3$ 和 $K_2S_2O_8$ 均为 $0.1\ mol \cdot L^{-1}$

## 实验内容

### 1. 单质硫的性质

（1）硫的熔化和弹性硫的生成

约 3 g 硫粉加入试管中缓慢加热，观察硫磺色态的变化。待硫粉熔化至沸后迅速倾入一盛有冷水的烧杯中，观察色态变化并试验其弹性。弹性硫放置一段时间后又有什么变化？试给予解释。

（2）硫的化学性质

① 硫与汞的反应

在一瓷坩埚中加入一小滴汞，然后加入少量硫磺粉，用玻璃棒搅动使之混合。观察现象，写出反应式。产物最后集中回收。

②硫与浓硝酸的反应（在通风橱内进行）

少量硫粉在试管内与浓硝酸加热反应数分钟，观察现象，写出反应式。自行设计方案验证反应产物。

③硫的氧化性质

在蒸发皿内混合好约 1 g 锌粉及 2 g 硫粉。用烧红了的玻璃棒接触混合物，观察现象，写出反应式。设计方案验证反应产物。

### 2. 硫的氧化物——二氧化硫

（1）二氧化硫的制备（在通风橱内进行）

装置如图 18-1 所示，蒸馏瓶内放入 5 g $Na_2SO_3$ 固体，分液漏斗内装浓硫酸。缓慢向蒸馏瓶滴加浓 $H_2SO_4$，观察现象，写出反应式。

（2）二氧化硫的性质

①还原性　取 1 mL 0.01 mol·$L^{-1}$ $KMnO_4$ 溶液，用 $H_2SO_4$ 酸化后通入 $SO_2$ 气体。观察现象，写出反应式。

②氧化性　向饱和硫化氢水溶液中通入 $SO_2$ 气体，观察现象，写出反应式。

③漂白作用　品红溶液中通入 $SO_2$ 气体，观察现象。

（3）$SO_3^{2-}$ 的检出

由于含 $SO_3^{2-}$ 的溶液中往往还含有少量 $SO_4^{2-}$，会干扰 $SO_3^{2-}$ 的检出，因此需将 $SO_4^{2-}$ 预先除去。请自行设计分离步骤并验证某试样中含有 $SO_3^{2-}$，写出分离过程示意图及有关反应方程式。

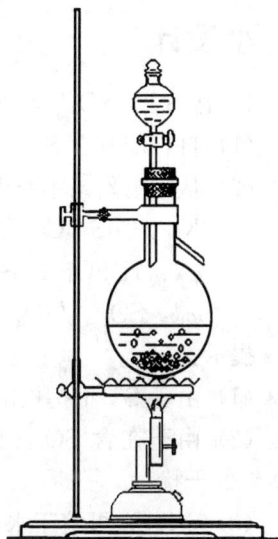

图 18-1　实验室制 $SO_2$ 装置

### 3. 硫代硫酸盐的制备与性质

（1）制备

烧杯中加入约 8 g $Na_2SO_3$(s)，3 g 已研细了的硫磺粉及 50 mL 水。在不断搅拌下煮沸 5 min。待反应完毕后加入少量活性炭粉作脱色剂。过滤并弃去残渣，滤液转移至蒸发皿中水浴加热浓缩至液体表面出现结晶为止。自然冷却，晶体析出后抽滤。写出反应式。产物留作下面实验用。

（2）性质

取少量自制的 $Na_2S_2O_3·5H_2O$ 晶体溶于约 5 mL 水中，进行以下实验。

①向溶液中滴加 2 mol·$L^{-1}$ HCl 溶液，观察现象，写出反应式。该现象说明 $Na_2S_2O_3$ 什么性质？

②溶液中滴加碘水，观察现象，写出反应式。该实验说明 $Na_2S_2O_3$ 什么性质？

③溶液中滴加氯水，设法证实反应后溶液中有 $SO_4^{2-}$ 存在。写出反应式。

④往有 4 滴 0.1 mol·$L^{-1}$ $AgNO_3$ 的溶液中滴加 $Na_2S_2O_3$ 溶液，仔细观察反应现象，写出反应式。该实验说明 $Na_2S_2O_3$ 什么性质？

### 4. 过二硫酸钾的氧化性

（1）往有 2 滴 0.002 mol·$L^{-1}$ $MnSO_4$ 溶液的试管中加入约 5 滴 1 mol·$L^{-1}$ $H_2SO_4$，2 滴 $AgNO_3$ 溶液，再加入少量 $K_2S_2O_8$ 固体，水浴加热，溶液的颜色有什么变化？

另取一支试管，不加入 $AgNO_3$ 溶液，进行同样实验。

比较上述两个实验的现象有什么不同，为什么？写出反应式。

（2）取少量 0.1 mol·$L^{-1}$ KI 溶液用硫酸酸化后再加入少量 $K_2S_2O_8$ 固体。观察现象，写出反应式。

### 5. 硫化氢的还原性

（1）取几滴 0.1 mol·$L^{-1}$ $KMnO_4$ 溶液用硫酸酸化后通入硫化氢气体，观察现象，写出反应式。

(2) 取几滴 $0.1\ mol\cdot L^{-1}\ K_2Cr_2O_7$ 溶液用硫酸酸化后通入硫化氢气体，观察现象，写出反应式。

## 小设计

1. 含 $S^{2-}$，$SO_3^{2-}$ 与 $S_2O_3^{2-}$ 混合液的分离检出，要求：

(1) 自行配制含 $S^{2-}$，$SO_3^{2-}$，$S_2O_3^{2-}$ 的混合溶液。

(2) 从本书附录中查出以下数据，自行设计分离步骤，分别检出 $S^{2-}$，$SO_3^{2-}$，$S_2O_3^{2-}$。

$K_{sp}$：$CdS$，$CaCO_3$，$SrCO_3$，$SrSO_4$，$BaSO_4$，$BaS_2O_3$，$SrSO_3$，$BaSO_3$

$\varphi_A^\ominus$：$H_2SO_4/H_2SO_3$，$H_2SO_3/S$，$S/H_2S$，$H_2O_2/H_2O$，$Br_2/Br^-$

$\varphi_B^\ominus$：$SO_4^{2-}/SO_3^{2-}$，$SO_3^{2-}/S$，$S/S^{2-}$

提示：

(1) 由于 $S^{2-}$ 干扰检出，因此可用 $CdCO_3$ 首先把它从溶液中分离出去。

(2) 由于在含 $SO_3^{2-}$ 的溶液中往往含有 $SO_4^{2-}$，故 $SO_3^{2-}$ 的检出必须考虑分离除去 $SO_4^{2-}$ 的干扰 。

(3) $SrS_2O_3$ 可溶于水。

思考以下问题：

(1) 用 $CdCO_3(s)$ 分离离子彻底吗？为什么？体系中加入了 $CdCO_3$ 后将引入什么离子？如何除去？

(2) 如何证实混合液中有 $S_2O_3^{2-}$？

(3) 为什么 $S_2O_3^{2-}$ 与 $SO_3^{2-}$ 的分离用 $Sr^{2+}$ 而不用 $Ba^{2+}$？

(4) 为什么在含 $SO_3^{2-}$ 的试液中加入 $BaCl_2$ 溶液后生成白色沉淀还不能证实是 $SO_3^{2-}$？

2. 现有五种已失落标签的试剂，分别是 $Na_2S$，$Na_2SO_3$，$Na_2S_2O_3$，$Na_2SO_4$ 和 $K_2S_2O_8$，试设法用实验方法加以鉴别。

## 思考题

1. 如何证实亚硫酸盐中存在 $SO_4^{2-}$？为什么亚硫酸盐中常常有硫酸盐，而硫酸盐中却很少有亚硫酸盐？怎样检验 $SO_4^{2-}$ 盐中的 $SO_3^{2-}$？

2. 比较 $S_2O_8^{2-}$ 与 $MnO_4^-$ 氧化性的强弱，$S_2O_3^{2-}$ 与 $I^-$ 还原性的强弱。为什么 $K_2S_2O_8$ 与 $Mn^{2+}$ 的反应要在酸性介质中进行？$Na_2S_2O_3$ 与 $I_2$ 的反应能否在酸性介质中进行？为什么？

## 安全知识

1. 二氧化硫具有刺激性气味，对人体及环境带来毒害与污染。主要对人体造成黏膜及呼吸道损害，引起流泪、流涕、咽干、咽痛等症状及呼吸系统炎症。大量吸入会导致窒息死亡。因此，凡涉及产生二氧化硫的反应都要采取相应措施，减少二氧化硫逸出并在通风橱内进行。

2. 硫化氢具有强烈的臭鸡蛋气味，是毒性较大的气体。主要引起中枢神经系统中毒，与呼吸酶中的铁质结合使酶活性减弱，造成黏膜损害及呼吸系统损害。轻度时产生头晕、头痛、呕吐，严重时可引起昏迷，意识丧失，窒息而致死亡。因此，凡涉及硫化氢参与的反应都应在通风橱内进行。

# Exp 18    Sulfur and Its Compounds

## Objectives

1. To grasp the reducibility of hydrogen sulfide and thiosulfate, oxidizability and reducibility properties of sulfur dioxide, and strong oxidizability of persulfate.

2. To grasp the properties of oxyacids and salts containing sulfur.

## Previewing and Thinking

1. Review the properties of $SO_3^{2-}$, $SO_4^{2-}$, $S_2O_3^{2-}$, $S_2O_8^{2-}$, $S^{2-}$ and related reactions.

2. In the reactions which $H_2S$ and $SO_2$ are participating, what safety precaution should be taken?

## Instruments and Reagents

distilled flask, separating funnel, evaporating dish.

sulfur powder, zinc powder, mercury, activated carbon, fuchsine solution, $H_2S$ (saturated water solution), chlorine water, iodine water, $Na_2SO_3$(s), $K_2S_2O_8$(s), $CdCO_3$(s), $MnSO_4$(0.002 mol·$L^{-1}$), KI(0.1 mol·$L^{-1}$), $K_2Cr_2O_7$(0.1 mol·$L^{-1}$), $KMnO_4$ (0.1 mol·$L^{-1}$, 0.01 mol·$L^{-1}$), $H_2SO_4$(1 mol·$L^{-1}$), HCl(2 mol·$L^{-1}$), $AgNO_3$(0.1 mol· $L^{-1}$).

To be identified solutions: $Na_2S$, $Na_2SO_4$, $Na_2S_2O_3$ and $K_2S_2O_8$(all of the concentrations are 0.1 mol·$L^{-1}$)

## Experimental Procedures

### 1.  Properties of sulfur

(1) Melting of sulfur and producing of elastic sulfur

About 3 g of sulfur is added to a test tube and heat slowly, observe the color and state change. As soon as the sulfur melts and boils, pour it to a beaker charged with cold water quickly, observe the color and state change and test its elasticity. What changes does it happen after placing for some time? Try to explain.

(2) Chemical properties of sulfur

①Sulfur reacts with mercury

Add a drop of mercury to a crucible, and then some sulfur powder is added, agitate it with glass rod. Observe the phenomenon, and write down the reaction equation. Products are recycled at the end of the experiment.

②Sulfur reacts with concentrated sulfuric acid (carry out in the fume hood)

Some sulfur powder and concentrated sulfuric acid are heated for several minutes in a test tube. Observe the phenomenon, and write down the reaction equation. Design a scheme by yourself to identify the product.

③Oxidizability of sulfur

Mix about 1 g zinc powder and 2 g of sulfur powder in the evaporating dish. Using glass rod burned to get in touch with the mixture, observe the phenomenon, write down the reaction equations. Design a scheme by yourself to identify the product.

## 2. Oxide of sulfur: sulfur dioxide

(1) Preparation of sulfur dioxide (carrying out in fume hood)

The equipment is as in Fig. 18-1, add 5g solid $Na_2SO_3$ in evaporating dish, the separating funnel equipped with concentrated sulfuric acid. Drop concentrated sulfuric acid slowly to the distilled flask, observe the phenomenon, write down the reaction equation.

(2) Properties of sulfur dioxide

①Reducibility of sulfur dioxide

Add 1 mL 0.01mol · $L^{-1}$ $KMnO_4$, acidify with $H_2SO_4$, then ventilate sulfur dioxide. Observe the phenomenon, and write down the reaction equation.

②Oxidizability of sulfur dioxide

$SO_2$ is bubbled to saturated hydrogen sulfide solution, observe the phenomenon, and write down the reaction equations.

③Bleaching activity of sulfur dioxide

$SO_2$ is bubbled to fuchsine solution, observe the phenomenon.

(3) Detection of $SO_3^{2-}$

Solution with $SO_3^{2-}$ also contains $SO_4^{2-}$, and it will interfere the detection of $SO_3^{2-}$. Therefore, when we detect the $SO_3^{2-}$, $SO_4^{2-}$ should be removed. Please design a separating scheme and analyze a sample contains $SO_3^{2-}$, write down the separating procedures and related reaction equations.

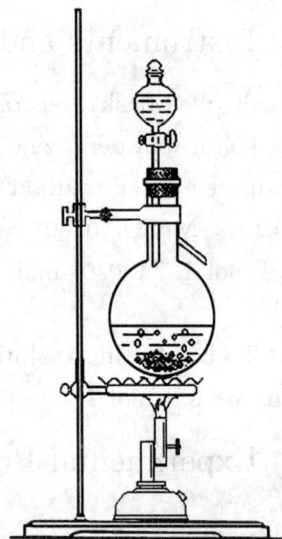

Fig. 18-1 The equipment for preparation of $SO_2$

## 3. Preparation and properties of thiosulfate

(1) Preparation

Add about 8 g $Na_2SO_3$(s), 3 g comminuted sulfur powder and 50 mL water. Stir and boil for 5 min. At the end of the reaction, add some activated carbon as depigmenting agent. Filter, transfer the filtrate to evaporating dish, then heat in water-bath until crystal turns out. Cool down at room temperature, when crystal comes out, filter under reduced pressure. Write down the reaction equation. Conserve the product for following experiment.

(2) Properties

Add some crystal of $Na_2S_2O_3 \cdot 5H_2O$ to about 5 mL water, carry out the following experiments.

①Drop 2 mol·$L^{-1}$ HCl to the above solution, observe the phenomenon, write down the reaction equation. What properties does it illustrate from this phenomenon?

②Add iodine water to the solution, observe the phenomenon, write down the reaction equation. What properties does it illustrate from this phenomenon?

③Add chlorine water to the solution, try to identify the existence of $SO_4^{2-}$, write down the reaction equation.

④Add 4 drops of 0.1 mol·$L^{-1}$ $AgNO_3$ to $Na_2S_2O_3$, observe the phenomenon carefully, write down the reaction equation. What properties does it illustrate from this phenomenon?

**4. Properties of potassium persulfate**

(1) Add about 5 drops of 1 mol·$L^{-1}$ $H_2SO_4$, 2 drops of $AgNO_3$ and some $K_2S_2O_8$ to 2 drops of 0.002 mol·$L^{-1}$ $MnSO_4$, heat in water bath, what will we observe on the color of the solution?

Take another test tube, add all of the reagents in above experiment except $AgNO_3$, observe the phenomena, explain it.

Compare the difference between the above two experiments, why? Write down the reaction equation.

(2) Get a small quantity of 0.1 mol·$L^{-1}$ KI, acidifying with sulfuric acid, then a small quantity of $K_2S_2O_8$ is added. Observe the phenomenon, write down the reaction equation.

**5. Reducibility of hydrogen sulfide**

(1) In several drops of 0.1 mol·$L^{-1}$ $KMnO_4$, acidfying with $H_2SO_4$, gaseous $H_2S$ is bubbled to it. Observe the phenomenon, write down the reaction equation.

(2) In several drops of 0.1 mol·$L^{-1}$ $K_2Cr_2O_7$, acidfying with $H_2SO_4$, gaseous $H_2S$ is bubbled to it. Observe the phenomenon, write down the reaction equation.

## Designing experiments

1. Separation and detection of mixture containing $S^{2-}$, $SO_3^{2-}$ and $S_2O_3^{2-}$

Requirement:

(1) Prepare mixed solutions containing $S^{2-}$, $SO_3^{2-}$, $S_2O_3^{2-}$ by yourself.

(2) Design separating schemes by yourself, detect $S^{2-}$, $SO_3^{2-}$, $S_2O_3^{2-}$ by turn.

Look up the following data from your textbook:

$K_{sp}$: CdS, $CaCO_3$, $SrCO_3$, $SrSO_4$, $BaSO_4$, $BaS_2O_3$, $SrSO_3$, $BaSO_3$

$\varphi_A^{\ominus}$: $H_2SO_4/H_2SO_3$, $H_2SO_3/S$, $S/H_2S$, $H_2O_2/H_2O$, $Br_2/Br^-$

$\varphi_B^{\ominus}$: $SO_4^{2-}/SO_3^{2-}$, $SO_3^{2-}/S$, $S/S^{2-}$

Notes:

(1) Because of the interference of $S^{2-}$, you should use $CdCO_3$ to remove it from the solutions.

(2) In the solution of $SO_3^{2-}$, there also contains $SO_4^{2-}$, so $SO_4^{2-}$ should be separated first before you detect $SO_3^{2-}$.

(3) $SrS_2O_3$ can dissolve in water.

Thinking the following questions:

(1) Using $CdCO_3$ to separate ions, can it be separated thoroughly? Why? What ions will be led in if you add $CdCO_3$ to the system? How to remove it?

(2) How to identify $S_2O_3^{2-}$ in the solution?

(3) Why do we use $Sr^{2+}$ when separating $S_2O_3^{2-}$ from $SO_3^{2-}$?

(4) Why the existence of $SO_3^{2-}$ can't be confirmed if $BaCl_2$ is added then precipitate produced?

2. Now, there are five reagents missing labels: $Na_2S$, $Na_2SO_3$, $Na_2S_2O_3$, $Na_2SO_4$, $K_2S_2O_8$, try to identify them.

## Questions

1. How to confirm the existence of $SO_4^{2-}$ in sulfite? There always contains sulfite in sulfate, but sulfate always doesn't contain sulfite, why? How to detect $SO_3^{2-}$ in sulfate?

2. Compare the oxidizability of $S_2O_8^{2-}$ and $MnO_4^-$, reducibility of $S_2O_3^{2-}$ and $I^-$. Why the reaction that $Mn^{2+}$ and $K_2S_2O_8$ are in should be carried out in acid medium?

## Safety details

1. $SO_2$ is a kind of gas with pungent odor, it is poisonous and will bring pollution to people and environment. It mainly damages our mucous membrane and respiratory tract, causing tearing, snotting, throat aridity and soreness, inflammation of respiratory passage. Breathing in too much will cause asphyxia to die. So all reactions related to $SO_2$ should be taken actions correspondingly, reduce its leaking and carrying out in a fume hood.

2. $H_2S$ is a kind of gas with notorious egg odor, it is poisonous gas. It mainly causes intoxication of central nerve system, combines with iron in respiratory enzyme, results in contamination of mucous membrane and respiratory apparatus system. Small amounts of it causes dizziness, headache, disgorging. Severe conditions are cataphora, consciousness lost, asphyxia to die. So all reactions related to $H_2S$ should be carried out in fume hood.

# 实验十九　氮 和 磷

## 目的要求

1. 掌握氨和铵盐、硝酸和硝酸盐的主要性质。
2. 掌握磷酸盐的主要性质。
3. 掌握亚硝酸及其盐的性质。

## 预习与思考

1. 复习氮磷及其主要化合物的性质以及进行氮磷实验时应注意的安全措施。
2. 思考下列问题：

（1）使用浓硝酸和硝酸盐时应注意哪些安全问题？

（2）浓硝酸和稀硝酸与金属、非金属及一些还原性化合物反应时，$N(V)$ 的主要还原产物是什么？

（3）为什么在一般情况下不使用 $HNO_3$ 作为酸性反应介质？

## 实验仪器和药品

温度计，水槽。

钢片，锌片，硫磺粉，铝屑。

$NH_4NO_3(s)$，$NH_4Cl(s)$，$Ca(OH)_2(s)$，$KNO_3(s)$，$Cu(NO_3)_2(s)$，$AgNO_3$，$FeSO_4 \cdot 7H_2O(s)$，$Na_2HPO_4(s)$，$(NH_4)_2SO_4(s)$，$PCl_5(s)$，$NaNO_3$（饱和，$0.1\ mol \cdot L^{-1}$，$0.5\ mol \cdot L^{-1}$），$Na_4P_2O_7(0.1\ mol \cdot L^{-1})$，$NaPO_3(0.1\ mol \cdot L^{-1})$，$Na_2HPO_4(0.1\ mol \cdot L^{-1})$，$Na_3PO_4(0.1\ mol \cdot L^{-1})$，$NaH_2PO_4(0.1\ mol \cdot L^{-1})$，$NaNO_3(0.5\ mol \cdot L^{-1})$，$NH_3 \cdot H_2O$（浓，$2\ mol \cdot L^{-1}$），$H_2S$（饱和水溶液），$NaOH(2\ mol \cdot L^{-1}$，$6\ mol \cdot L^{-1})$，$HAc(2\ mol \cdot L^{-1}$，$6\ mol \cdot L^{-1})$，$KI(0.1\ mol \cdot L^{-1})$，$H_2SO_4(3\ mol \cdot L^{-1})$，$HNO_3$（concentrated，$2\ mol \cdot L^{-1}$），$AgNO_3(0.1\ mol \cdot L^{-1})$，$NaOH(40\%$，m$)$，$CaCl_2(0.1\ mol \cdot L^{-1})$，$HCl\ (2\ mol \cdot L^{-1})$。

奈斯勒试剂（$K_2[HgI_4]+KOH$），对氨基苯磺酸，α-萘胺，四氯化碳，蛋白溶液，酚酞溶液，石蕊试纸，pH 试纸。

## 实验内容

### 1. 氨和铵盐的性质

（1）氨的实验室制备及其性质

①制备

3 g $NH_4Cl(s)$ 及 3g $Ca(OH)_2(s)$ 混合均匀后装入一支干燥的大试管中，按图 19-1 装置制备和收集氨气（制备过程应注意什么问题？）。用塞子塞紧氨气收集管管口，留作下列实验使用。

②性质

　　a. 在水中的溶解

　　把盛有氨气的试管倒置在盛有水的大烧杯或水槽中，水槽中加入 2~3 滴酚酞，在水下打开塞子，轻轻摇动试管，观察有何现象发生？当水柱停止上升后，用手指堵住管口并将试管自水中取出。

　　b. 氨水的酸碱性

　　试验上述试管内溶液的酸碱性。

　　c. 氨的加合作用

　　在一小坩埚内滴入几滴浓氨水，再把一个内壁用浓盐酸湿润过的烧杯罩在坩埚上，观察现象，写出反应式。

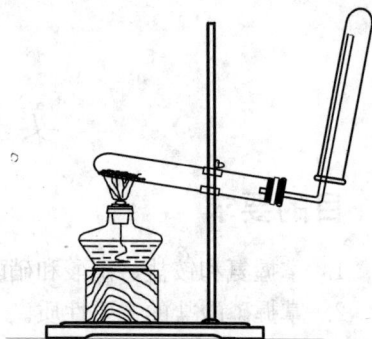

图 19-1　制备氨的装置

　　(2) 铵盐的性质及检出

　　①铵盐在水中溶解的热效应

　　试管中加入 2 mL 水，测量水温后再加入 2 g $NH_4NO_3(s)$，用小玻璃棒轻轻搅动溶液，再次测量溶液温度，记录温度变化，并作理论解释。$(NH_4)_2SO_4(s)$，$NH_4Cl(s)$ 等铵盐溶于水时将是吸热还是放热？为什么？

　　②铵盐的热分解

　　分别在三支已干燥的小试管中加入约 0.5 g $NH_4Cl(s)$，$NH_4NO_3(s)$ 和 $(NH_4)_2SO_4$(s)，用试管夹夹好，管口贴上一条已湿润的石蕊试纸，均匀加热试管底部。观察这三种铵盐的热分解的异同，分别写出反应式。在 $NH_4Cl$ 试管中较冷的试管壁上附着的白色霜状物质是什么？如何证实？

　　③铵盐的检出反应

　　a. 气室法检出

　　取几滴铵盐溶液置于一表面皿中心，另一表面皿中心贴附有一小条湿润的 pH 试纸，然后在铵盐溶液中滴加 6 mol·$L^{-1}$ NaOH 溶液至呈碱性，将贴有 pH 试纸的表面皿盖在铵盐的表面皿上形成"气室"，将气室置于水浴上微热，观察 pH 试纸颜色的变化。

　　b. 取几滴铵盐溶液，加入 2 滴 2 mol·$L^{-1}$ NaOH 溶液，然后再加入 2 滴奈斯勒试剂（$K_2[HgI_4]$+KOH），观察红棕色沉淀的生成。反应式为：

$$NH_4Cl+2K_2[HgI_4]+4KOH =\!\!= \left[ O\!\!\begin{array}{c} Hg \\ \\ Hg \end{array}\!\!NH_2 \right] I\!\downarrow + KCl+7KI+2H_2O$$

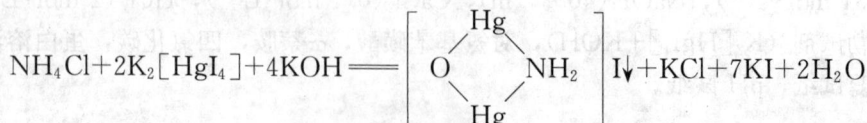

## 2. 亚硝酸及其盐的性质（注意亚硝酸及其盐有毒，切勿入口！）

　　(1) 亚硝酸的生成与分解

　　把已用冰水冷冻过的约 1 mL 饱和 $NaNO_2$ 溶液与约 1 mL 3 mol·$L^{-1}$ $H_2SO_4$ 混合均匀。观察现象，溶液放置一段时间后又有什么变化？为什么？反应式为：

$$NaNO_2 + H_2SO_4 =\!\!= 2HNO_2 + Na_2SO_4$$
$$2HNO_2 =\!\!= N_2O_3（蓝色） + H_2O$$
$$N_2O_3 =\!\!= NO + NO_2（棕色）$$

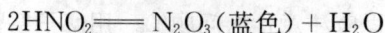

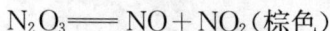

　　(2) 亚硝酸的氧化性

取少量 $0.1\ mol \cdot L^{-1}$ KI 溶液用 $H_2SO_4$ 酸化，再加入几滴 $NaNO_2$ 溶液，观察反应物及产物的色态，微热试管，又有什么变化？写出反应式。

（3）亚硝酸的还原性

将几滴 $KMnO_4$ 溶液用硫酸酸化后再滴加 $0.1\ mol \cdot L^{-1}$ $NaNO_2$ 溶液，观察现象，写出反应式。

（4）亚硝酸根的检出

①取 $1\sim2$ 滴 $0.01\ mol \cdot L^{-1}$ $NaNO_2$ 溶液，加入几滴 $6\ mol \cdot L^{-1}$ HAc 酸化后再加入一滴对氨基苯磺酸和一滴萘胺溶液，溶液应显红色，表明溶液中含有 $NO_2^-$。应注意 $NO_2^-$ 的浓度不宜太大，否则红紫色将很快褪去，生成褐色沉淀及黄色溶液。

②在少量 $NaNO_2$ 溶液中加入 $0.1\ mol \cdot L^{-1}$ KI 溶液 $1\sim2$ 滴，用 $H_2SO_4$ 酸化后再加入几滴四氯化碳，振荡试管，观察现象。四氯化碳层显紫色，表明 $NO_2^-$ 的存在。

### 3. 硝酸及其盐的性质

（1）硝酸的氧化性

分别实验浓硝酸与硫（见硫族实验1.（2）.②）；浓硝酸与硫化氢；浓硝酸与金属铜；稀硝酸与金属铜；稀硝酸与活泼金属（锌）的反应，产物各是什么？写出它们的反应式。总结稀硝酸与浓硝酸被还原的规律，并检验稀硝酸与 Zn 反应产物中 $NH_3$ 或 $NH_4^+$ 的存在。

（2）硝酸盐的热分解

分别试验 $KNO_3(s)$，$Cu(NO_3)_2(s)$，$AgNO_3(s)$ 的热分解，用火柴余烬检验反应生成的气体，说明它们热分解反应的异同。写出反应式并作出解释。

（3）硝酸盐的检验

①试液加入 $40\%(m)$NaOH 溶液呈强碱性，再加入少量铝屑，用 pH 试纸检验反应产生的气体，证实 $NO_3^-$ 的存在。写出反应式。

②置少量固体 $FeSO_4 \cdot 7H_2O$ 于试管中，滴加一滴 $0.5\ mol \cdot L^{-1}$ $NaNO_3$ 溶液及一滴浓硫酸，观察现象。反应式为：

$$3Fe^{2+} + NO_3^- + 4H^+ == 3Fe^{3+} + NO + H_2O$$

$$Fe^{2+} + NO + SO_4^{2-} == Fe(NO)SO_4(棕色)$$

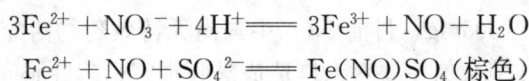

### 4. 磷酸盐的性质

（1）磷酸盐的酸碱性

①分别检验正磷酸盐、焦磷酸盐、偏磷酸盐水溶液的 pH。

②分别检验 $Na_3PO_4$，$Na_2HPO_4$，$NaH_2PO_4$ 水溶液的 pH。以等量的 $0.1\ mol \cdot L^{-1}$ $AgNO_3$ 的溶液分别加入到这些溶液中产生沉淀后溶液的 pH 又有什么变化？请给予解释。

（2）磷酸钙盐的生成与性质

分别向 $0.1\ mol \cdot L^{-1}$ $Na_3PO_4$，$Na_2HPO_4$ 和 $NaH_2PO_4$ 溶液中加入 $CaCl_2$ 溶液，观察有无沉淀生成？再加入 $2\ mol \cdot L^{-1}$ $NH_3 \cdot H_2O$ 后又有何变化？继续加入 $2\ mol \cdot L^{-1}$ HCl 后又有什么变化？请给予解释并写出反应式。

（3）磷酸根、焦磷酸根、偏磷酸根的鉴别

①分别向 $0.1\ mol \cdot L^{-1}$ $Na_2HPO_4$，$Na_4P_2O_7$ 和 $NaPO_3$ 水溶液中滴加 $0.1\ mol \cdot L^{-1}$ $AgNO_3$ 溶液，各有什么现象发生？生成的沉淀溶于 $2\ mol \cdot L^{-1}$ $HNO_3$ 吗？

②以 $2\ mol\cdot L^{-1}$ HAc 溶液酸化磷酸盐溶液、焦磷酸盐溶液后分别加入蛋白溶液，各有什么现象发生？

把以上实验结果填在表 19-1 中，并说明磷酸根、焦磷酸根、偏磷酸根的鉴别方法。

表 19-1 磷的含氧酸盐的性质

|  | $PO_4^{3-}$ | $P_2O_7^{4-}$ | $PO_3^{-}$ |
|---|---|---|---|
| 滴加 $AgNO_3$ |  |  |  |
| 沉淀在 $2\ mol\cdot L^{-1} HNO_3$ 中 |  |  |  |
| HAc 酸化后加入蛋白溶液 |  |  |  |

（4）磷酸盐的转化

在坩埚中放入少许研细了的 $Na_2HPO_4$ 粉末，小火加热，待水份完全逃逸后大火灼烧 15 min，冷却、检验产物中磷酸根的存在形式，写出反应式。（注：用 $AgNO_3$ 鉴定产物时，加 HAc 溶液可以消除少量 $PO_4^{3-}$ 对其他离子的干扰。）

## 小设计

取少量 $PCl_5$(s) 溶于水中，令其水解彻底。请自行设计方案检验 $PCl_5$ 的水解产物。

### 思考题

1. 实验室中用什么方法制备氨气？直接加热 $NH_4NO_2$ 的方法可以吗？为什么？

2. 如何分别检出 $NaNO_2$，$Na_2S_2O_3$ 和 KI 溶液？

3. $PCl_5$ 水解后加入 $AgNO_3$ 时为什么只有 AgCl 沉淀出来而 $Ag_3PO_4$ 却不沉淀？如何使 $Ag_3PO_4$ 沉淀？有关 $K_{sp}$ 数据请自行查阅相关书籍。

### 安全知识

1. 除 $N_2O$ 外，所有氮的氧化物都有毒，其中尤以 $NO_2$ 为甚。在大气中 $NO_2$ 的允许含量为每升空气不得超过 0.005mg，目前 $NO_2$ 中毒尚无特效药物治疗，一般只能输入氧气以帮助呼吸和血液循环。二氧化氮主要对人体造成黏膜损害引起肿胀充血；呼吸系统损害引起各种炎症；神经系统损害引起眩晕、无力、痉挛、面部发绀等；造血系统损害破坏血红素等等。吸入高浓度的氮氧化物将迅速出现窒息以至死亡。因此，凡涉及氮氧化物生成的反应均应在通风橱内进行。

2. 实验室常见的磷有白磷及红磷。红磷毒性较小，白磷为蜡状结晶体，燃点为 318 K，在空气中易氧化，毒性很大，常保存于水中或油中。磷化氢是无色恶臭剧毒的气体。$PCl_3$ (l)和 $PCl_5$(s)都有腐蚀性，使用时应注意。

# Exp 19   Nitrogen and Phosphor

## Objectives

1. To master the main properties of ammonia and ammonium salt, nitric acid and nitrate.

2. To master the main properties of phosphate.

3. To master the properties of nitrous acid and nitrites.

## Previewing and Thinking

1. Go over the main properties of nitrogen, phosphor and their major compounds and the safety measures when experimenting with nitrogen and phosphor.

2. Think the following questions:

(1) What security regularities must be paid attention to when using concentrated nitric acid and nitrate?

(2) What is the major reduction product of N(V) when the concentrated or dilute nitric acids reacting with metals and non-metals as well as some reducible compounds.

(3) Why nitric acid is not chosen as making a normal acid medium?

## Instruments and Reagents

thermometer, gutter.

steel, zinc, sulfur powder, aluminium pouder.

$NH_4NO_3(s)$, $NH_4Cl(s)$, $Ca(OH)_2(s)$, $KNO_3(s)$, $Cu(NO_3)_2(s)$, $AgNO_3$, $FeSO_4 \cdot 7H_2O(s)$, $Na_2HPO_4(s)$, $(NH_4)_2SO_4(s)$, $PCl_5(s)$, $NaNO_3$(saturated, 0.1 mol·$L^{-1}$, 0.5 mol·$L^{-1}$), $Na_4P_2O_7$(0.1 mol·$L^{-1}$), $NaPO_3$(0.1 mol·$L^{-1}$), $Na_2HPO_4$(0.1 mol·$L^{-1}$), $Na_3PO_4$(0.1 mol·$L^{-1}$), $NaH_2PO_4$(0.1 mol·$L^{-1}$), $NaNO_3$(0.5 mol·$L^{-1}$), $NH_3 \cdot H_2O$(2 mol·$L^{-1}$, concentrated), $H_2S$(saturated solution), $NaOH$ (2 mol·$L^{-1}$, 6 mol·$L^{-1}$), $HAc$(2 mol·$L^{-1}$, 6 mol·$L^{-1}$), $KI$(0.1 mol·$L^{-1}$), $H_2SO_4$(3 mol·$L^{-1}$), $HNO_3$(concentrated, 2 mol·$L^{-1}$), $AgNO_3$(0.1 mol·$L^{-1}$), $NaOH$(40%, m), $CaCl_2$(0.1 mol·$L^{-1}$), $HCl$ (2 mol·$L^{-1}$).

Nessler reagent ($K_2[HgI_4]$+KOH), $p$-aminobenzene sulfonic acid, $\alpha$-naphthylamine, $CCl_4$, albumen solution, phenolphthalein solution, litmus paper, pH test paper.

## Experimental Procedures

### 1.  The properties of ammonia and ammonium salt

(1) The preparation of ammonia and its properties

①Preparation

Mix 3 g $NH_4Cl(s)$ and 3 g $Ca(OH)_2(s)$ in a big dry test tube and install the instrument and collect the alkaline gas just as Fig. 19-1. (What would be noticed in the procedures?) Plug the test tube containing alkaline gas, conserve it for following experiments.

②Properties

a. Solubility in water

Place the test tube with alkaline gas up side down in a vast flask or in a big flask or in the gutter with water, a few drops of phenolphthaleins are dropped,

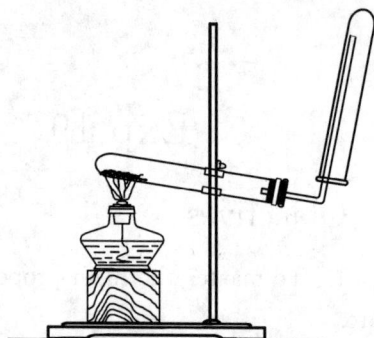

Fig. 19-1 Instrument of $NH_3$ generation

and then open the plug meanwhile swinging the test tube slightly. Observe the phenomena. When the water stops rising, close the outlet of the test tube with one of your fingers before dislodging the test tube from water.

b. The acidity and basiuty of ammonia water

Test the acidity or alkalinity of the solution in the test tube in the above experiment.

c. Conjugation of ammonia

Add several drops of concentrated ammonia water to a small crucible, and then cover it with a beaker whose inside wall has been moisten by concentrated hydrochloric acid. Observe the phenomena and write down the reaction equation.

(2) The properties of ammonium salts and their identification method

①The caloric effect of ammonium salts dissolving in the water

Add 2 mL water in a test tube, measure the temperature before adding 2 g $NH_4NO_3$ (s). Swing the solution slightly with a small glass rod. Measure the temperature again, record the temperature change and explain it theoretically. Does it absorb heat or heat emission when some ammonium salts such as $(NH_4)_2SO_4(s)$, $NH_4Cl(s)$ dissolving in water? Why ?

②Thermal decomposition of ammonium salts

Add about 0. 5 g $NH_4Cl$ (s) and $NH_4NO_3(s)$, $(NH_4)_2SO_4(s)$ to three small test tubes respectively and fix them with test tube clamps. Paste a moisten litmus test paper at the opening orifice, and heat them at the bottom slightly. Observe the differences of the three salts when they are heated, and write down each reaction equation. You may find some white frost-like substance at the wall of the test tube containing $NH_4Cl$, what is it? How to confirm your judgment?

③Identification of $NH_4^+$

a. Air chamber way

Add several drops of ammonium salts solution to the center of the watch glass and another watch glass is attached a piece of moisten pH test paper in the center. Add 6 $mol \cdot L^{-1}$ NaOH solution to the ammonium salt solution until the solution is alkaline. At this moment, cover the ammonium salt with the watch glass attached with pH test paper to form one air chamber. Heat the air chamber on the water bath slightly, observe the change of the pH test paper.

　　b.　Add 2 drops of 2 mol·L$^{-1}$ NaOH solution to the some ammonium salts solution, and then add 2 drops of Nessler reagent ($K_2[HgI_4]+KOH$), observe the marron precipitate. The reaction equation is:

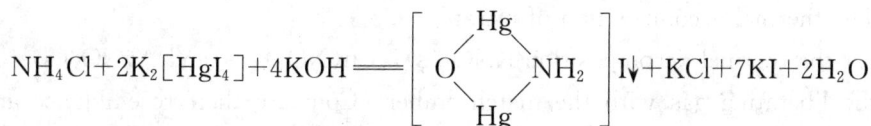

$$NH_4Cl+2K_2[HgI_4]+4KOH = \left[ \begin{array}{c} Hg \\ O \diagup \hspace{0.5cm} \diagdown NH_2 \\ \diagdown \hspace{0.8cm} \diagup \\ Hg \end{array} \right] I\downarrow +KCl+7KI+2H_2O$$

### 2.　The properties of nitrous acids and nitrites （Notice: they are toxic, never eating!）

　　(1)　The generation and decomposing of nitrous acids

　　Mix 1 mL saturated NaNO$_2$ solution which has been cooled down by ice-water and about 1 mL 3 mol·L$^{-1}$ H$_2$SO$_4$.　Observe the phenomena.　What will it be when the solution is laid up for a while.　Why?　The reaction equations are:

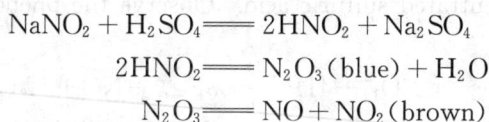

$$NaNO_2 + H_2SO_4 = 2HNO_2 + Na_2SO_4$$
$$2HNO_2 = N_2O_3 (blue) + H_2O$$
$$N_2O_3 = NO + NO_2 (brown)$$

　　(2)　Oxidizability of nitrous acid

　　Acidify 0.1 mol·L$^{-1}$ KI solution with H$_2$SO$_4$, and then add several drops of NaNO$_2$ solution.　Observe the color and state of the reactant and product.　Heat the test tube slightly, what change would take place? Write down the reaction equation.

　　(3)　Reducibility of nitrous acid

　　Acidify the KMnO$_4$ solution with concentrated H$_2$SO$_4$, and add several drops of 0.1 mol·L$^{-1}$ NaNO$_2$ solution.　Observe the phenomena and write down the reaction equations.

　　(4)　Identification of NO$_2^-$

　　①In one or two drops of 0.01 mol·L$^{-1}$ NaNO$_2$ solution acidified by several drops of 6 mol·L$^{-1}$ HAc , one drop of $p$-aminobenzene sulfonic acid and naphthylamine solution are added, at this moment the mixture should turn red, and this phenomenon shows that the solution contains NO$_2^-$.　（Notice: the concentration of NO$_2^-$ should not be too much high, if not, the magenta would fade fast, a kind of brown precipitate will form and the solution will become yellow.）

　　②Add one or two drops of 0.1 mol·L$^{-1}$ KI solution to small quantity of NaNO$_2$ solution, and then add several drops of CCl$_4$ after being acidified by H$_2$SO$_4$.　Vibrate the test tube and observe the phenomena.　If the CCl$_4$ layer turns out to be purple, it shows existence of NO$_2^-$.

### 3.　The properties of nitric acid and the salts

　　(1)　Oxidizability of nitric acid

　　In the experiments of concentrated nitric acid with sulfur （see the experiment of sulfur, chapter 1. (2). ②) and concentrated nitric acid with hydrogen sulfide and concentrated nitric acid with copper and dilute nitric acid with active metals such as zinc respectively, what are the products? Write down the reaction equations. Summarize the principles of dilute and concentrated nitric acids being reduced and test whether the $NH_3$ or $NH_4^+$ exist when the dilute nitric acid reacts with Zn.

(2) The thermal decomposition of nitrate

To test the thermal decomposition $KNO_3(s)$, $Cu(NO_3)_2(s)$ and $AgNO_3(s)$ respectively, test the liberating gas with the match ember. Compare their resemblance and differences. Write down the reaction equations and explain them.

(3) Identification of nitrate

①Add 40%(m) NaOH solution to alkalify the test solution, and then add a little aluminium, later test the generated gas with pH test paper to justify the existence of $NO_3^-$. Write down the reaction equation.

②Place some $FeSO_4 \cdot 7H_2O(s)$ in a test tube, and then add a drop of 0.5 mol·L$^{-1}$ $NaNO_3$ solution and concentrated sulfuric acid. Observe the phenomena and its reaction equations are as follows:

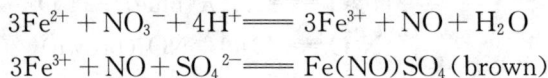

$$3Fe^{2+} + NO_3^- + 4H^+ = 3Fe^{3+} + NO + H_2O$$
$$3Fe^{3+} + NO + SO_4^{2-} = Fe(NO)SO_4 (brown)$$

### 4. Properties of phosphates

(1) The acidity or alkalinity of phosphates

①To test the pH value of the aqueous solution of orthophosphate, pyrophosphate and metaphophate.

②Test the pH value of the aqueous solutions of $Na_3PO_4$, $Na_2HPO_4$ and $NaH_2PO_4$. Add equal amount of $AgNO_3$ solution to these solutions, test the change of pH value. Explain it.

(2) The generation and properties of calcium phosphate

Add 0.1 mol·L$^{-1}$ $CaCl_2$ solution into 0.1 mol·L$^{-1}$ $Na_3PO_4$, 0.1 mol·L$^{-1}$ $Na_2HPO_4$ and 0.1 mol·L$^{-1}$ $NaH_2PO_4$ solutions, observe whether there will be some precipitate. And then add 2 mol·L$^{-1}$ $NH_3 \cdot H_2O$, what will be observed? What will be observed when adding 2 mol·L$^{-1}$ HCl. Explain the phenomena and write down the reaction equations.

(3) Identification of $PO_4^{3-}$, $P_2O_7^{4-}$ and $PO_3^-$

①Drop 0.1 mol·L$^{-1}$ $AgNO_3$ solution respectively into the 0.1 mol·L$^{-1}$ $Na_2HPO_4$, 0.1 mol·L$^{-1}$ $Na_4P_2O_7$ and 0.1 mol·L$^{-1}$ $NaPO_3$, what will happen? Could the generated precipitate be dissolved by 2 mol·L$^{-1}$ $HNO_3$?

②Acidify the solutions of phosphate and pyrophosphate with 2 mol·L$^{-1}$ HAc, and then albumen solution is added, what will take place in each test tube?

Fill the following Table 19-1 with the results from the experiments above and illustrate the method to identify $PO_4^{3-}$, $P_2O_7^{4-}$ and $PO_3^-$.

**Table 19-1   The properties of phosphorous oxysalt**

|  | $PO_4^{3-}$ | $P_2O_7^{4-}$ | $PO_3^-$ |
|---|---|---|---|
| dropping $AgNO_3$ |  |  |  |
| adding 2 mol·L$^{-1}$ $HNO_3$ to precipitate |  |  |  |
| acidifying with HAc then adding albumen solution |  |  |  |

(4)  The transformation of phosphate

Place small quantity of $Na_2HPO_4$ has powder in a crucible, heat slightly, and then burn it about 15 min after moisture escaped. Cool down and test the state of $PO_4^{3-}$ in the product. Write down the reaction equations. (Notice: unknown substance is identified by $AgNO_3$, the HAc solution would eliminate the other ions interference with $PO_4^{3-}$).

## A design

Place small quantity of $PCl_5$ (s) in the water and hydrolyze it completely. Design a scheme to test the hydrolysis product of $PCl_5$ by yourself.

## Questions

1.  How to produce nitrogen in the laboratory? Does it work if heating $NH_4NO_2$ directly? Why?

2.  How to identify $NaNO_2$, $Na_2S_2O_3$ and KI solution?

3.  Why does only the AgCl precipitate but no $Ag_3PO_4$ came out when adding $AgNO_3$ solution to the solution in which $PCl_5$ is hydrolyzed? How to deposit $Ag_3PO_4$ precipitate? The data of $K_{sp}$ can be found in related book by yourself.

## Knowledge of security

1.  All the nitrogen oxides are toxic except $N_2O$, among which $NO_2$ is especially rank poisonous. The permitted density of $NO_2$ is 0.005mg per liter in the atmosphere. So far there is no such specific drug to treat the $NO_2$ intoxication. Generally the therapy method is to input oxygen to assist breathing and blood circulation. Nitrogen dioxide mainly causes membrane damage and causes swelling epidemic; damage respiratory inflammation; nervous system is damaged to cause dizziness, weakness, cramps, facial membranes; blood-making system to undermine heme, *etc*. You would feel suffocated or even be pushed to death when sucking in nitrogen oxides in high density. So, all the reactions must be done in the fume hood relating to nitrogen oxides.

2.  The phosphorus such as white phosphorus and red phosphorus are common in the laboratory. The red phosphorus is less harmful and the white phosphorus is a kind of wax-like crystal whose ignition point is 318 K, which would be oxidized in the atmosphere and the toxicity is very high. It is often preserved in water or oil. $PH_3$ is a kind of gas which is achromasy and hypertoxic with evil smelling. Both $PCl_3$ (l) and $PCl_5$ (s) are caustic so that you must be careful.

# 实验二十　砷、锑、铋

## 实验目的

1. 通过试验＋Ⅲ氧化态的砷、锑、铋氧化物、氢氧化物的酸碱性以及＋Ⅲ氧化态砷、铋盐的还原性和＋Ⅴ氧化态砷、锑、铋盐的氧化性，总结出它们的变化规律。

2. 掌握砷、锑、铋硫代酸盐及硫化物的生成和性质。

## 仪器、药品及材料

试管、离心试管、离心机、烧杯、酒精灯。

$As_2O_3(s)$，$SbCl_3(s)$，$Bi(NO_3)_3(s)$，$NaBiO_3(s)$，$HCl(2\ mol \cdot L^{-1}$，$6\ mol \cdot L^{-1}$，浓)，$HNO_3(2\ mol \cdot L^{-1}$，$6\ mol \cdot L^{-1})$，$NaOH\ (2\ mol \cdot L^{-1}$，$6\ mol \cdot L^{-1})$，$SbCl_3(0.1\ mol \cdot L^{-1})$，$Bi(NO_3)_3(0.1\ mol \cdot L^{-1})$，$Na_3AsO_4(0.1\ mol \cdot L^{-1})$，$Na_2S(0.5\ mol \cdot L^{-1})$，$MnSO_4(0.02\ mol \cdot L^{-1})$，硫代乙酰胺(5%)，碘水，氯水，$Na_3AsO_3(0.1\ mol \cdot L^{-1})$，$CCl_4$，硫化氢气体。

pH试纸，淀粉碘化钾试纸，醋酸铅试纸。

## 实验内容

### 1. ＋Ⅲ氧化态砷、锑、铋氧化物和氢氧化物的酸碱性

(1) $As_2O_3$ 的性质

①取绿豆大小的 $As_2O_3$（极毒）固体，溶于 1 mL 水（微热）中，用 pH 试纸检验水溶液的酸碱性。

②取两份绿豆大小的 $As_2O_3$ 固体，分别注入约 10 滴 $6\ mol \cdot L^{-1}$ HCl 溶液和浓 HCl 中，微热后，观察溶解情况。写出反应方程式。

③取绿豆大小的 $As_2O_3$ 固体放入试管中，注入约 10 滴 $2\ mol \cdot L^{-1}$ NaOH 溶液，振荡试管，观察溶解情况，写出反应方程式，保留溶液供下面的实验用。

(2) 锑、铋（Ⅲ）氢氧化物的酸碱性

分别在两支试管中各滴入 10 滴 $0.1\ mol \cdot L^{-1}$ $SbCl_3$ 和 $Bi(NO_3)_3$ 溶液，再各滴入 5 滴 $2\ mol \cdot L^{-1}$ NaOH 溶液，观察反应产物的颜色和状态。将每种沉淀分成两份，在其中一份加入 $6\ mol \cdot L^{-1}$ NaOH 溶液，另一份中加入 $6\ mol \cdot L^{-1}$ 盐酸溶液，观察每份中的反应情况，写出反应方程式。

根据以上的实验结果，比较砷(Ⅲ)、锑(Ⅲ)、铋(Ⅲ)氧化物和氢氧化物的酸碱性，并指出它们的变化规律。

### 2. 锑（Ⅲ）盐的水解作用

在干燥的试管内放入绿豆大小 $SbCl_3$ 固体，加水溶解。向试管滴加 $6\ mol \cdot L^{-1}$ 盐酸至沉淀恰好溶解，再加水稀释，观察并解释上述现象，写出反应方程式。

### 3. 砷（Ⅲ）、铋（Ⅲ）的还原性和砷（V）、铋（V）的氧化性

（1）取 5 滴本实验1.（3）.③制得的亚砷酸钠溶液，用 2 mol·L$^{-1}$盐酸调 pH 8~9，滴 1~2 滴碘水，观察有何现象发生。然后将溶液用浓盐酸酸化，再加 5 滴四氯化碳，观察四氯化碳层溶液的颜色。写出反应方程式并加以解释。

（2）在试管中注入 10 滴 Bi(NO$_3$)$_3$ 溶液，再注入 6 mol·L$^{-1}$ NaOH 溶液和氯水，加热，并观察现象。倾去溶液，洗涤沉淀，再加浓盐酸于沉淀物中，有何现象发生？试鉴定气体产物，写出反应方程式。

（3）在 1 滴 0.02 mol·L$^{-1}$ MnSO$_4$ 溶液中，加入 2 mL 2 mol·L$^{-1}$硝酸，再加入绿豆大小的铋酸钠固体，微热，加入 1~2 mL 水稀释，观察溶液颜色的变化，写出反应方程式。

由以上实验，总结出砷、锑、铋高低价态氧化还原性的变化规律。

### 4. 砷、锑、铋硫化物和硫代酸盐

（1）三硫化二砷、硫代亚砷酸钠的生成和性质

分别向三支小试管中滴加 10 滴亚砷酸钠溶液和 10 滴 6 mol·L$^{-1}$盐酸溶液，再分别向三支小试管滴入 10 滴 5‰硫代乙酰胺溶液，微热，观察反应产物的颜色和状态。将沉淀离心分离后，弃去清液，分别在沉淀上注入浓盐酸、2 mol·L$^{-1}$ NaOH 和 0.5 mol·L$^{-1}$ Na$_2$S 溶液，观察沉淀是否溶解并写出反应方程式。

（2）Sb$_2$S$_3$，硫代亚锑酸盐的生成和性质

将三氯化锑溶液代替亚砷酸钠溶液，按本实验4.（1）步骤进行类似实验。

（3）Bi$_2$S$_3$ 的生成和性质

将硝酸铋溶液代替亚砷酸钠溶液，按本实验4.（1）步骤进行类似实验。

（4）五硫化二砷与硫代砷酸盐的生成和性质

将盛有 1 mL 0.1 mol·L$^{-1}$砷酸钠溶液的离心试管和盛有 1 mL 浓盐酸的离心试管同时放在冰水中冷却 5~10 min 后混和，并通入硫化氢气体（通风橱中进行）。观察反应产物的颜色和状态。离心分离，弃去清液，将沉淀分成三份，按本实验4.（1）步骤一样进行实验，观察现象，写出反应方程式。

将上面实验结果归纳在表 20-1 中，并比较砷、锑、铋硫化物的性质。

**表 20-1　As，Sb，Bi 的硫化物性质**

| 硫化物<br>颜色和试剂 | As$_2$S$_3$ | Sb$_2$S$_3$ | Bi$_2$S$_3$ | As$_2$S$_5$ |
|---|---|---|---|---|
| 颜色 | | | | |
| 浓 HCl | | | | |
| 2 mol·L$^{-1}$ NaOH | | | | |
| 0.5 mol·L$^{-1}$ Na$_2$S | | | | |

## 思考题

1. 在本实验中，哪一个实验能够说明溶液的酸碱性会影响氧化还原反应方向。

2. 实验室中如何配制硝酸铋溶液？

3. 用标准电极电势判断下列反应能否发生？如能发生，请写出反应的产物和现象。

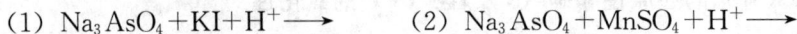

(1) $Na_3AsO_4 + KI + H^+ \longrightarrow$　　　(2) $Na_3AsO_4 + MnSO_4 + H^+ \longrightarrow$

(3) $NaBiO_3 + MnSO_4 + H^+ \longrightarrow$　　　(4) $NaSb(OH)_6 + KI + H^+ \longrightarrow$

## 安全知识

1. 砷、锑、铋及其化合物都是有毒物质。特别是三氧化二砷（俗称砒霜）和胂（$AsH_3$）及其他可溶性的砷化物都是剧毒物质。要在教师指导下使用。取用量要少，切勿进入口内或与有伤口的地方接触。实验后一定要洗手，若万一失误，应立即去医疗室。也可用乙二硫醇（$HSCH_2CH_2SH$）解毒，其反应式是：

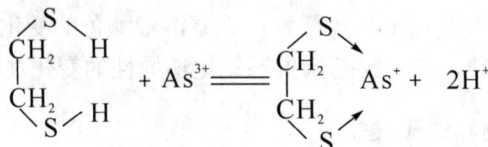

$$
\begin{array}{c}
\begin{matrix} CH_2 \\ | \\ CH_2 \end{matrix}
\begin{matrix} \diagdown S \diagup H \\ \\ \diagup S \diagdown H \end{matrix}
+ As^{3+} ===
\begin{matrix} CH_2 \\ | \\ CH_2 \end{matrix}
\begin{matrix} \diagdown S \diagup \\ \\ \diagup S \diagdown \end{matrix}
As^+ + 2H^+
\end{array}
$$

2. 马氏验砷法

砷化氢（$AsH_3$）是最剧毒的无机毒物之一。因此，必须在通风橱中进行实验，而且要特别谨慎小心。可用下述方法检验微量的砷，装置如图 20-1 所示。

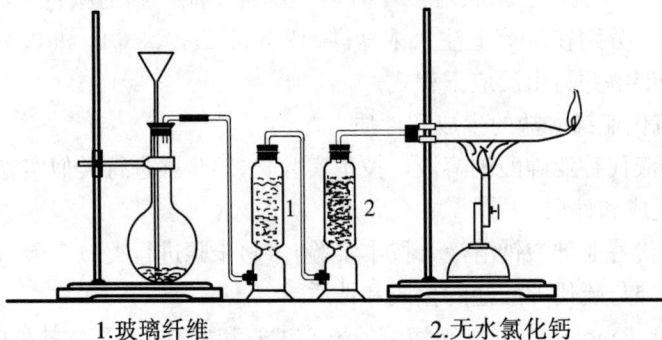

1.玻璃纤维　　　　　　　　2.无水氯化钙

图 20-1　马氏验砷法

首先由长颈漏斗将 $20\%H_2SO_4$ 注入盛有不含砷的锌的烧瓶中，使漏斗颈的下端恰好被酸淹没。产生的氢气经过净化干燥和纯度检验后点火燃烧，然后从漏斗中加入 3 mL 任何砷化合物溶液，则有 $AsH_3$ 生成。

$$2Na_3AsO_3 + 9H_2SO_4 + 6Zn === 3Na_2SO_4 + 6H_2O + 2AsH_3 + 6ZnSO_4$$

由于这个反应的结果，氢焰改变它的形状和颜色。火焰伸长，变成淡青色，并生成 $As_2O_3$ 白烟。在试管中间部位加热，砷化氢（胂）即分解，在试管冷却处有灰黑色斑点（砷镜）生成。

$$2AsH_3 \xrightarrow{\triangle} 2As + 3H_2$$

实验完毕后，先用水灌满烧瓶，然后才可将仪器拆开，用等体积的稀 $NaOH$ 溶液和 $H_2O_2$ 洗掉试管中砷的斑点。

# Exp 20   Arsenic, Antimony and Bismuth

## Objectives

1. To test the acidity and alkalinity of the oxides and hydroxides of As, Sb and Bi in the quantivalence of $+\text{III}$, and the reducibility of As and Bi in the quantivalence of $+\text{III}$ and the oxidizability of them in the quantivalence of $+\text{V}$. To summarize the principles.

2. To master the preparation methods and properties of the sulfurets and sulfoacid compounds of As, Bi and Sb.

## Instruments, Reagents and Materials

test tube, centrifugal test tubes, centrifugal machine, beakers and alcohol burner.

$As_2O_3(s)$, $SbCl_3(s)$, $Bi(NO_3)_3(s)$, $NaBiO_3(s)$, HCl (2 mol·$L^{-1}$, 6 mol·$L^{-1}$, concentrated), $HNO_3$ (2 mol·$L^{-1}$, 6 mol·$L^{-1}$), NaOH (2 mol·$L^{-1}$, 6 mol·$L^{-1}$), $SbCl_3$ (0.1 mol·$L^{-1}$), $Bi(NO_3)_3$ (0.1 mol·$L^{-1}$), $Na_3AsO_4$ (0.1 mol·$L^{-1}$), $Na_2S$ (0.5 mol·$L^{-1}$), $MnSO_4$ (0.02 mol·$L^{-1}$), thioacetamide(5%), iodine water, chlorine water, $Na_3AsO_3$ (0.1 mol·$L^{-1}$), $CCl_4$, hydrogen sulfide gas.

pH test paper, starch-potassium iodide test paper, lead acetate paper.

## Experimental Procedures

**1. Acidity and alkalinity of the oxides and hydroxides of As, Sb and Bi in the quantivalence of $+\text{III}$**

(1) The properties of $As_2O_3$

①Place $As_2O_3$ with a mung bean size (very toxic) solid into 1 mL water (lukewarm) and test the acidity of the solution with a piece of pH test paper.

②Place two grains of $As_2O_3$ with a mung bean size solid in two test tubes and add about 10 drops of 6 mol·$L^{-1}$ HCl solution in a tube, another concentrated HCl, respectively, heat them slightly, observe the dissolving phenomena and write down the reaction equations.

③Place $As_2O_3$ solid with a mung bean size in the test tube and add about 10 drops of 2 mol·$L^{-1}$ NaOH solution, vibrate the tube and observe the dissolving phenomena and write down the equation. Preserve the solution for the next experiment.

(2) The acid property and basic property of hydroxide of Sb and Bi (III)

Add 10 drops of 0.1 mol·$L^{-1}$ $SbCl_3$ and $Bi(NO_3)_3$ solutions in two test tubes and then five drops of 2 mol·$L^{-1}$NaOH solution are dropped respectively, observe the color and state of the products. Divide each precipitate into two portions, drop the first portion with 6 mol·$L^{-1}$ NaOH solution and the second one with 6 mol·$L^{-1}$ HCl solution. Observe the reaction in each portion and write down the reaction equations.

According to above experimental results, compare the acid property of oxides and hydroxides of As, Sb and Bi (Ⅲ) and summarize their transmutation principles.

## 2. The hydrolysis of Sb (Ⅲ) compounds

Place a mung bean size $SbCl_3$ solid in a dry test tube and add water to dissolve it. And drop 6 mol·$L^{-1}$ HCl until the solid is just dissolved then dilute it. Observe and explain the phenomena, write down the reaction equations.

## 3. The reducibility of As (Ⅲ) and Bi (Ⅲ) and the oxidizability of As (V) and Bi (V)

(1) Place five drops of $NaAsO_3$ solution which has been prepared in the procedure(1). ③and then adjust the pH value between 8 and 9 with 2 mol·$L^{-1}$ HCl. Observe the phenomena after dropping one or two drops of iodine water. Acidify the solution with concentrated HCl. The next step is to drop 5 drops of $CCl_4$, and then observe the color of the $CCl_4$ layer. Write down the reaction equations and explain them.

(2) Add 10 drops of $Bi(NO_3)_3$ solution in a test tube before 6 mol·$L^{-1}$ NaOH solution and chlorine water are added. Observe the phenomenon while it is heated. Decant the solution and wash the precipitate. Add concentrated HCl onto the precipitate. Do you know what phenomena will take place? Test what is the gas and then write down the reaction equations.

(3) Add 2 mL 2 mol·$L^{-1}$ $HNO_3$ to one drop of 0.02 mol·$L^{-1}$ $MnSO_4$ and then add a mung bean size $Na_3BiO_3$ solid. Heat it and drop 1~2 mL water to dilute it. Observe the color of the solution and write down the reaction equations.

Summarize the principle of the oxidation-reduction properties of As, Bi and Sb in high quantivalence based on the experiments above.

## 4. Sulfide, and thioarsenate, thiobismuate, thioantimonate

(1) The generation and properties of $As_2O_3$ and sodium thioarsenite

Add 10 drops of $Na_3AsO_3$ solution and 6 mol·$L^{-1}$ HCl into three test tubes respectively and then add 10 drops of thioacetamide (5%), heat it. Observe the color and state of the product. After centrifugalizing, decant the liquid, add concentrated HCl, 2 mol·$L^{-1}$ NaOH and 0.5 mol·$L^{-1}$$Na_2S$ into each test tube respectively. Observe whether the precipitate will be dissolved and write down the reaction equations.

(2) The properties and generation of $Sb_2S_3$ and thioantimonite

Substitute $Na_3AsO_3$ with $SbCl_3$, repeat the procedure 4. (1).

(3) The properties and generation of $Bi_2S_3$

Take place of $Na_3AsO_3$, with $Bi(NO_3)_3$, repeat the procedure 4. (1).

(4) The properties and generation of $As_2S_5$ and thioarsenate

Add 1 mL 0.1 mol·$L^{-1}$ to a centrifugal test tube and add 1 mL concentrated HCl solution into another one. After cooling them down in the ice-water coincidently for 5~10 min, ventilating $H_2S$ gas (in fume hood). Observe the color and state of the product. Centrifugalize it to remove the liquid and divide the precipitate into three portions, repeat the procedure 4. (1). Observe the phenomena and write down the reaction equations.

Summarize all the consequences of the experiments in the table 20-1; compare the properties of As, Sb and Bi's sulfides.

**Table 20-1　Properties of As, Sb, Bi sulfides**

| color and reagents ＼ Sulfides | $As_2S_3$ | $Sb_2S_3$ | $Bi_2S_3$ | $As_2S_5$ |
|---|---|---|---|---|
| color | | | | |
| conc. HCl | | | | |
| $2\ mol \cdot L^{-1}$ NaOH | | | | |
| $0.5\ mol \cdot L^{-1}$ $Na_2S$ | | | | |

## Questions

1. In the experiments, which one will demonstrate the acidity or alkalinity of solution influence the direction of the oxidation-reduction reaction?

2. How to prepare $Bi(NO_3)_3$ solution?

3. Judge whether the reactions below would take place with your knowledge of electrode potential. If it does, point out the product and phenomena.

(1) $Na_3AsO_4 + KI + H^+ \longrightarrow$　　　(2) $Na_3AsO_4 + MnSO_4 + H^+ \longrightarrow$

(3) $NaBiO_3 + MnSO_4 + H^+ \longrightarrow$　　　(4) $NaSb(OH)_6 + KI + H^+ \longrightarrow$

## Notes

1. All of the As, Sb and Bi as well as their compounds are toxic, especially $As_2O_3$ (arsenic) and arsine ($AsH_3$) as well as other soluble arsenide are also destructive. As a result, using them must follow teacher's construction. The quantity must be small and you mustn't contact them with your mouth and some other injured parts of your body. Wash your hands after experiment. If you are hurt by them, go to the infirmary at once, or treat the injured body with $HSCH_2CH_2SH$ and the reaction equation is:

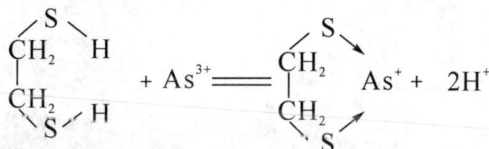

$$\begin{array}{c} \diagup S \diagdown \\ \overset{|}{C}H_2 \quad H \\ | \\ \overset{|}{C}H_2 \\ \diagdown S \diagup H \end{array} + As^{3+} = \begin{array}{c} \diagup S \diagdown \\ \overset{|}{C}H_2 \diagdown \\ | \qquad As^+ \\ \overset{|}{C}H_2 \diagup \\ \diagdown S \diagup \end{array} + 2H^+$$

2. Identification of As invented by Marsh

$AsH_3$ is one of the most toxic inorganic compounds, so that the experiment must be done in the fume hood with special care. The microcrystalline As would be tested by the following way, as the instrument in Fig. 20-1.

1.glass fiber         2.anhydrous calcium chloride

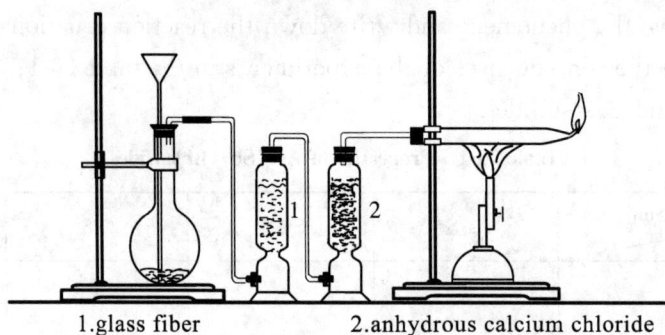

Fig. 20-1    Identification of As invented by Marsh

$20\% H_2SO_4$ would be transferred into a flask with zinc but no As through a long-foot glass funnel and the foot of the funnel is immersed in the acid properly. The produced hydrogen should be burnt after being purified, dried and purity check, and then add 3 mL of any solution containing arsenic compound through the funnel, the $AsH_3$ would come out.

$$2Na_3AsO_3 + 9H_2SO_4 + 6Zn \Longrightarrow 3Na_2SO_4 + 6H_2O + 2AsH_3 + 6ZnSO_4$$

As a result, the color and shape of the flame of the hydrogen would be changed. The shape would be elongated and become light blue with the white smoke of $As_2O_3$. Heat the middle part of the test tube, the $AsH_3$ (arsine) would be decomposed; as a result, in the cool part of the test tube, there will be some black pots(arsenic mirror).

$$2AsH_3 \overset{\triangle}{\Longrightarrow} 2As + 3H_2$$

The instrument mustn't be separated until the whole flask is filled with water in the end of the experiment. Wash the arsenic spots with equal volume of dilute NaOH and $H_2O_2$ solutions.

# 实验二十一 硅、硼、铝

## 实验目的

1. 掌握硅酸盐、硼酸和硼砂的主要性质。
2. 掌握铝及其重要化合物的性质。
3. 练习硼砂珠的操作。

## 药品及材料

$CaCl_2$，$CuSO_4$，$Co(NO_3)_2$，$NiSO_4$，$FeSO_4$，$FeCl_3$，$CrCl_3$，$ZnSO_4$，硼酸，硼砂，$HCl(3.0\ mol\cdot L^{-1}$，$6.0\ mol\cdot L^{-1}$，浓)，$Na_2SiO_3(20\%)$，甘油，$NH_4Cl$(饱和)，乙醇，硼砂(饱和)，$H_2SO_4$(浓)，$NaOH(2.0\ mol\cdot L^{-1}$，$6.0\ mol\cdot L^{-1})$，$Na_2CO_3(0.5\ mol\cdot L^{-1})$，硼酸(饱和)，$HNO_3(0.1\ mol\cdot L^{-1})$，$Al_2(SO_4)_3(0.5\ mol\cdot L^{-1})$，$Na_2S(0.5\ mol\cdot L^{-1})$，$NH_4Ac\ (1.0\ mol\cdot L^{-1})$，$Hg(NO_3)_2(0.1\ mol\cdot L^{-1})$，铝试剂。

镍铬丝，pH 试纸，滤纸，铝片。

## 实验内容

### 1. 硅

(1) 硅酸水凝胶的生成

将 20% $Na_2SiO_3$ 5 mL 加热微沸后滴加 $6.0\ mol\cdot L^{-1}$ HCl，观察产物的颜色和状态。

(2) 硅酸盐的水解

用 pH 试纸测 20%硅酸钠溶液的酸碱性，然后往盛有 5 滴该溶液的试管中滴入 10 滴饱和氯化铵溶液，微热，可用 pH 试纸检验放出的气体为何物，写出反应方程式。

(3) 难溶性硅酸盐的生成——"水中花园"

在一只 50 mL 烧杯中加入约 2/3 体积的 20%(m)水玻璃 (或 20%的硅酸钠)，然后把 $CaCl_2$，$CuSO_4$，$Co(NO_3)_2$，$NiSO_4$，$ZnSO_4$，$FeSO_4$ 和 $FeCl_3$ 固体中任选 3~4 种各黄豆大小的一粒放入烧杯内 (注意：各种固体不能放在一起)，记住它们各自的位置。半小时后观察现象 (实验完毕后，必须立即洗净烧杯，因为 $Na_2SiO_3$ 对玻璃有侵蚀作用)。

### 2. 硼酸的制备、性质和鉴定

(1) 往盛有 5 滴饱和硼砂溶液的小试管中滴加 5 滴浓 $H_2SO_4$，放在冰水中冷却，若无沉淀，可用小玻璃棒摩擦试管壁。观察产物的颜色和状态。

(2) 取 5 滴饱和硼酸溶液，用 pH 试纸测 pH。硼酸溶液中滴入 2 滴甘油，再测溶液的 pH。解释酸度变化原因，硼酸与甘油反应机理应为：

①硼酸溶液

$$B(OH)_3+H_2O \Longrightarrow \left[ HO-B\!\!\overset{OH}{\underset{OH}{\longleftarrow}}OH \right] +H^+$$

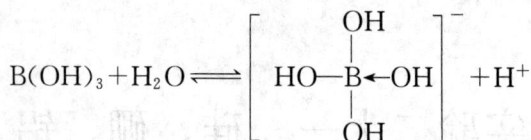

②当甘油与硼酸溶液之比为 1:1 时，反应为：

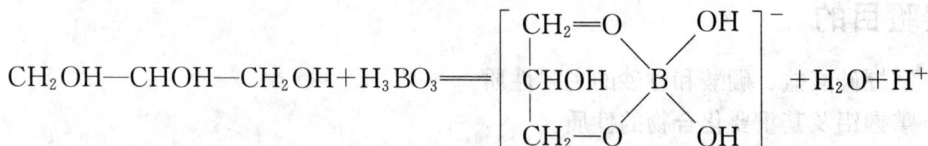

$$CH_2OH-CHOH-CH_2OH+H_3BO_3 \Longrightarrow \left[ \begin{matrix} CH_2=O \\ CHOH \\ CH_2=O \end{matrix} \!\!B\!\!\overset{OH}{\underset{OH}{\longleftarrow}} \right]^- +H_2O+H^+$$

③当甘油与硼酸溶液之比为 2:1 时反应为：

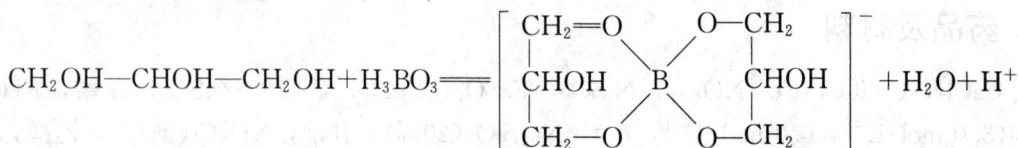

$$CH_2OH-CHOH-CH_2OH+H_3BO_3 \Longrightarrow \left[ \begin{matrix} CH_2=O \\ CHOH \\ CH_2=O \end{matrix} \!\!B\!\! \begin{matrix} O-CH_2 \\ CHOH \\ O-CH_2 \end{matrix} \right]^- +H_2O+H^+$$

（3）在蒸发皿中放入黄豆大小的硼酸晶体，滴入 5 滴酒精和 2 滴浓 $H_2SO_4$，混合后点燃，观察火焰的颜色有何特征。

**3. 硼砂珠的实验**

先将铂丝（或铂镍丝）做如下清洁处理方法：

在一支小试管中注入 1 mL 6.0 mol·L⁻¹ 的盐酸，将铂丝置于氧化焰中灼烧片刻后浸入酸中，取出灼烧后再浸入酸中，如此重复数次即可。

（1）硼砂珠的制备

用按上述处理过的铂丝（或镍铬丝）取一些硼砂固体，在氧化焰中灼烧并熔融成圆球。观察硼砂珠的颜色和状态。

（2）硼砂珠鉴定钴盐和铬盐

用烧热的硼砂珠分别沾上少量硝酸钴，三氯化铬固体，熔融。冷却后观察硼砂珠的颜色。写出反应方程式。

**表 21-1　几种重要金属的硼砂珠颜色**

| 样品元素 | 符号 | 氧化焰 | | 还原焰 | | 容易得到的原料 | 备注 |
|---|---|---|---|---|---|---|---|
| | | 热时 | 冷时 | 热时 | 冷时 | | |
| 钴 | Co | 青色 | 青色 | 青色 | 青色 | $CoCl_2$ | 珠砂的颜色随：①氧化焰和还原焰②热时和冷时③试样的含量不同而不同 |
| 铬 | Cr | 黄色 | 黄绿色 | 绿色 | 绿色 | $CrCl_3$ | |
| 铜 | Cu | 绿色 | 青绿色－淡绿色 | 灰色－绿色 | 红色 | $CuSO_4$ | |
| 铁 | Fe | 黄色－淡褐色 | 黄色－褐色 | 绿色 | 淡绿色 | $FeCl_2$，$FeSO_4$ | |
| 钼 | Mo | 淡黄色 | 无色－白色 | 褐色 | 褐色 | $MoO_3$ | |
| 锰 | Mn | 紫色 | 紫红色 | 无色－灰色 | 无色－灰色 | $MnCl_2$ | |
| 镍 | Ni | 紫色 | 黄褐色 | 无色－灰色 | 无色－灰色 | $NiSO_4$，$NiCl_2$ | |

#### 4. 铝

(1) 铝与水、空气的反应

在试管中放入一铝片，加入 2 mol・L$^{-1}$ HCl 2 mL，加热煮沸 1 min，以清洗表面氧化物。倒出盐酸溶液，用水洗二次，再加入几滴 0.1 mol・L$^{-1}$ Hg(NO$_3$)$_2$，待铝片表面刚变为灰色(约10~20s)，迅速倒出 Hg(NO$_3$)$_2$ 溶液，立即用水洗两次以除去多余的 Hg(NO$_3$)$_2$，然后再加水，观察氢气的放出。最后倒去水，用滤纸吸去表面水分，放置几分钟后可观察到蓬松的 Al$_2$O$_3$ 生成，写出反应式。

(2) 铝盐的水解

取 5 滴 0.1 mol・L$^{-1}$ Al$_2$(SO$_4$)$_3$ 溶液于小试管中，滴入 2~3 滴 0.5 mol・L$^{-1}$ Na$_2$S 溶液，观察现象。设法证明沉淀物是氢氧化铝而不是硫化铝。

(3) Al$^{3+}$ 的鉴定

在 Al$^{3+}$ 溶液中加入 1.0 mol・L$^{-1}$ NH$_4$Ac 至接近中性。然后加入 1~2 滴铝试剂，微热，有红色沉淀生成表示有 Al$^{3+}$。

### 思考题

1. 实验室中为什么不可以用磨砂口玻璃器皿储存碱液？为什么盛过硅酸钠溶液的容器在实验后必须立即洗净？

2. 实验室为何无 Al$_2$(CO$_3$)$_3$ 和 Al$_2$S$_3$ 试剂？

# Exp 21　Silicon, Boron and Aluminum

## Objectives

1. To master the main properties of metasilicate, boric acid (boracid), borax.
2. To master the properties of aluminum and its important compounds.
3. To practice the operation of borax bead.

## Reagents and Materials

$CaCl_2$, $CuSO_4$, $Co(NO_3)_2$, $NiSO_4$, $FeSO_4$, $FeCl_3$, $ZnSO_4$, boric acid, borax, HCl (3.0, 6.0 mol·$L^{-1}$, concentrated), $Na_2SiO_3$(20%), glycerin, $NH_4Cl$(saturated), ethanol, borax (saturated), $H_2SO_4$(concentrated), NaOH(2.0, 6.0 mol·$L^{-1}$), $Na_2CO_3$(0.5 mol·$L^{-1}$), boric acid (saturated), $HNO_3$ (0.1 mol·$L^{-1}$), $Al_2(SO_4)_3$ (0.5 mol·$L^{-1}$), $Na_2S$(0.5 mol·$L^{-1}$), $NH_4Ac$(1.0 mol·$L^{-1}$), $Hg(NO_3)_2$(0.1 mol·$L^{-1}$), aluminum reagent.

nickel-chromium thread, pH test paper, filter paper, aluminum sheet.

## Experimental Procedures

### 1. Silicon

(1) Preparation of silicic acid gel

Heat 20% $Na_2SiO_3$ 5 mL to boil slightly, add 6 mol·$L^{-1}$ HCl , then observe the color and state of the product.

(2) Hydrolysis of silicate

Test the acidity or basicity of 20% $Na_2SiO_3$ by pH test paper, then add 10 drops of saturated $NH_4Cl$ solution to a test tube containing 5 drops of 20% $Na_2SiO_3$, test what gas is given out by pH test paper after tepidity, write down the reaction equation.

(3) Preparation of insoluble silicate— "garden in water"

Add sodium silicate about 2/3 volume that of the beaker of into a beaker, then select 3 ~4 kinds of solid from $CaCl_2$, $CuSO_4$, $Co(NO_3)_2$, $NiSO_4$, $ZnSO_4$, $FeSO_4$ and $FeCl_3$ same as a soybean size, put them into the beaker (note: all of the solid shouldn't be put together), note down each position. Observe the phenomena after half-hour (it is necessary that the beaker should be washed at once after experiment being finished for $Na_2SiO_3$ will erode glass).

### 2. Preparation, properties and identification of boric acid

(1) Add 5 drops of saturated borax solution into a small test tube, add 5 drops of concentrated $H_2SO_4$ into it, cool down in ice water, rub test tube wall by a small glass rod if no

precipitate produced, observe the color and state of product.

(2) Take 5 drops of boric acid solution, then test its pH value by pH test paper. Add 2 drops of glycerol to the boric acid solution, test pH value again. Explain the reason of acidity change, the reaction mechanism of boric acid and glycerol is:

①Boric acid solution

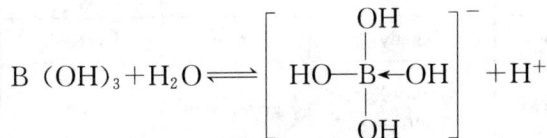

$$B\ (OH)_3+H_2O \Longrightarrow \left[\begin{array}{c} OH \\ | \\ HO-B\leftarrow OH \\ | \\ OH \end{array}\right]^- +H^+$$

②When the ratio of glycerol and boric acid is 1:1, the reaction is:

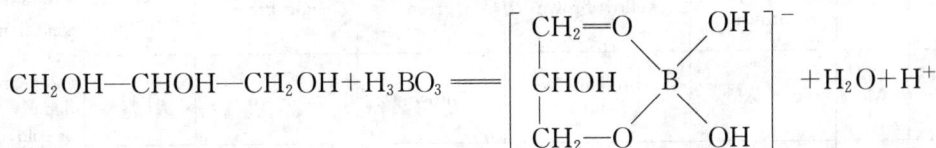

$$CH_2OH-CHOH-CH_2OH+H_3BO_3 == \left[\begin{array}{cc} CH_2=O & OH \\ | & \diagup \\ CHOH & B \\ | & \diagdown \\ CH_2-O & OH \end{array}\right]^- +H_2O+H^+$$

③When the ratio of glycerol and boric acid is 2:1, the reaction is:

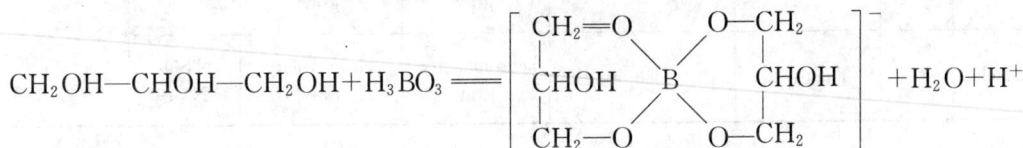

$$CH_2OH-CHOH-CH_2OH+H_3BO_3 == \left[\begin{array}{ccc} CH_2=O & O-CH_2 \\ | & \diagup\diagdown & | \\ CHOH & B & CHOH \\ | & \diagdown\diagup & | \\ CH_2-O & O-CH_2 \end{array}\right]^- +H_2O+H^+$$

(3) Put crystal boric acid as a soybean size into evaporating dish, add 5 drops of alcohol and 2 drops of concentrated $H_2SO_4$, ignite after mixing, what character can be observed on color of flame.

## 3. Experiment of borax bead

Cleaning method of platinum or platinum nickel:

Inject 1 mL 6.0 mol·$L^{-1}$ HCl into a small test tube, burn the platinum on oxidizing flame and then insert it into acid, take out to burn, then insert again. Thus repeat a few times.

(1) Preparation of borax bead

Take some solid borax bead by platinum which has been disposed formerly, burn it in oxidizing flame to fuse pellet. Observe the color and state of borax bead.

(2) Identification of cobalt and chrome salt by borax bead

Stain $Co(NO_3)_2$, $CrCl_3$ solid to melt with hot borax bead. Observe the color of borax bead after cooling down. Write down the reaction equation.

**Table 21-1　Colors of some important metals' borax bead**

| sample element | sym-bol | oxidation flame | | reduction flame | | materials easily got | remark |
|---|---|---|---|---|---|---|---|
| | | hot | cold | hot | cool | | |
| cobalt | Co | cyan | cyan | cyan | cyan | $CoCl_2$ | |
| chromium | Cr | yellow | kelly | green | green | $CrCl_3$ | |
| copper | Cu | green | viridity-light green | gray-green | red | $CuSO_4$ | The colors of pearl bead are different for condition: ① oxidizing flame and reducing flame, ② hot or cold, ③different sample |
| iron | Fe | yellow-lightor brown | yellow-brown | green | light green | $FeCl_2$ $FeSO_4$ | |
| molybde-num | Mo | light yellow | achromaticity-white | brown | brown | $MoO_3$ | |
| manganese | Mn | amethyst | amaranth | achromaticity-gray | achromaticity-gray | $MnCl_2$ | |
| nickel | Ni | amethyst | filemot | achromaticity-gray | achromaticity-gray | $NiSO_4$ $NiCl_2$ | |

## 4. Aluminum

(1) Reaction between aluminum and water, air

Place an aluminum sheet in a test tube in which is added 2 mL 2 mol·L⁻¹ HCl, heat to boil for 1min to remove the oxide on surface. Decant HCl solution, wash it with water twice, then add a few drops of 0.1 mol·L⁻¹ $Hg(NO_3)_2$, decant $Hg(NO_3)_2$ solution rapidly when surface of aluminum become just gray (about 10~20 s), wash with water twice to remove redundant $Hg(NO_3)_2$, then add water, observe hydrogen being given out. Finally pour water out, absorb surface water by filter paper, stand still, observe fluffy $Al_2O_3$ after a few minutes. Write down the reaction equations.

(2) Hydrolysis of aluminum salt

Add 5 drops of 0.5 mol·L⁻¹ $Al_2(SO_4)_3$ to a small test tube, add 2~3 drops of 0.5 mol·L⁻¹ $Na_2S$ solution, observe the phenomena. Try to prove that the precipitate is $Al(OH)_3$ but $Al_2S_3$.

(3) Identification of $Al^{3+}$

Add 1.0 mol·L⁻¹ $NH_4Ac$ to $Al^{3+}$ solution till the solution is near neutral. Then add 1~2 drops of aluminum reagent to it. If red precipitate is formed after tepidity, it indicates the existence of $Al^{3+}$.

## Questions

1. Why the abraded glass can not be used for base in the lab? Why must the container filled sodium with silicate be washed at once after the experiment?

2. Why $Al_2(CO_3)_3$ and $Al_2S_3$ don't exist in the lab?

# 实验二十二　锡和铅

## 实验目的

1. 试验并掌握二价锡和铅的氢氧化物和某些难溶铅盐的重要性质。
2. 试验并掌握锡（Ⅱ）的还原性和铅（Ⅳ）的氧化性。
3. 试验锡和铅的硫化物的性质。

## 预习与思考

1. 复习无机化学教材中有关锡、铅化合物的性质。
2. 思考并回答下列问题。
   (1) 实验室中配制氯化亚锡溶液，往往既加盐酸，又加锡粒，是何原因？
   (2) 在试验氢氧化铅的碱性时，应该用什么酸为宜？
   (3) 比较 $SnS_2$，$SnS$ 和 $PbS$ 在酸、碱以及硫化物水溶液中溶解性的异同点。
   (4) 如何鉴别 $SnCl_4$ 和 $SnCl_2$？如何分离 $PbS$ 和 $SnS$？

## 仪器和药品

试管，离心试管，烧杯，离心机。

$PbO_2(s)$，$Pb_3O_4(s)$，$H_2SO_4(1\ mol\cdot L^{-1}$，$3\ mol\cdot L^{-1})$，$HCl(2\ mol\cdot L^{-1}$，浓），$HNO_3$（$6\ mol\cdot L^{-1}$，浓），醋酸（$6\ mol\cdot L^{-1}$），$NaOH$（$2\ mol\cdot L^{-1}$，$6\ mol\cdot L^{-1}$），$Na_2S$（$1\ mol\cdot L^{-1}$），$(NH_4)_2S_2$（$1\ mol\cdot L^{-1}$），硫化氢水溶液（饱和），$SnCl_2$（$0.1\ mol\cdot L^{-1}$），$SnCl_4$（$0.1\ mol\cdot L^{-1}$），$Pb(NO_3)_2$（$0.1\ mol\cdot L^{-1}$），$FeCl_3$（$0.1\ mol\cdot L^{-1}$），$HgCl_2$（$0.1\ mol\cdot L^{-1}$），$Bi(NO_3)_2$（$0.1\ mol\cdot L^{-1}$），$KI$（$0.1\ mol\cdot L^{-1}$），$MnSO_4$（$0.1\ mol\cdot L^{-1}$），$K_2CrO_4$（$0.1\ mol\cdot L^{-1}$），$NaAc$（饱和）。

## 实验内容

**1. 锡（Ⅱ）和铅（Ⅱ）的氢氧化物的生成和酸碱性**

(1) $Sn(OH)_2$ 的生成和酸碱性

在 $1\ mL\ 0.1\ mol\cdot L^{-1}\ SnCl_2$ 溶液中逐渐滴加入 $2\ mol\cdot L^{-1}\ NaOH$，直至生成的白色沉淀经摇动后不再溶解为止。离心分离，弃去清液，将沉淀分成两份，试验其对稀酸和稀碱的作用。写出反应式。

(2) 试从 $0.1\ mol\cdot L^{-1}Pb(NO_3)_2$ 溶液制得 $Pb(OH)_2$ 沉淀。用实验证明 $Pb(OH)_2$ 是否具有两性（注意：实验其碱性时应该用什么酸?）。写出反应式。

根据上面的实验，对 $Sn(OH)_2$ 和 $Pb(OH)_2$ 的酸碱性作出结论。

**2. 锡（Ⅱ）的还原性和铅（Ⅳ）的氧化性**

(1) 锡（Ⅱ）的还原性

①试验 $0.1\ \mathrm{mol\cdot L^{-1}}$ 的 $SnCl_2$ 溶液与 $0.1\ \mathrm{mol\cdot L^{-1}}$ 的 $FeCl_3$ 溶液的反应。观察现象。写出反应式。

②在 $0.1\ \mathrm{mol\cdot L^{-1}}$ $HgCl_2$ 溶液中，逐滴加入 $0.1\ \mathrm{mol\cdot L^{-1}}$ $SnCl_2$ 溶液，观察有何变化。继续滴加 $SnCl_2$，又有什么变化？反应式为：

$$SnCl_2 + 2HgCl_2 =\!\!= Hg_2Cl_2 \downarrow (白色) + SnCl_4$$

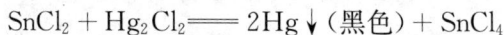
$$SnCl_2 + Hg_2Cl_2 =\!\!= 2Hg \downarrow (黑色) + SnCl_4$$

③在自制的 $[Sn(OH)_4]^{2-}$ 溶液中加入 $0.1\ \mathrm{mol\cdot L^{-1}}$ $Bi(NO_3)_3$ 溶液，立即有黑色沉淀出现。反应式：

$$3[Sn(OH)_4]^{2-} + 2Bi^{3+} + 6OH^- =\!\!= 3[Sn(OH)_6]^{2-} + 2Bi \downarrow (黑色)$$

此反应可用来鉴定 $Sn^{2+}$ 和 $Bi^{3+}$ 离子。

(2) 铅（Ⅳ）的氧化性

①在少量 $PbO_2$ 中加入浓盐酸，观察现象，并检查有无氯气生成。写出反应式。

②在 $2\ \mathrm{mL}$ $3\ \mathrm{mol\cdot L^{-1}}$ $H_2SO_4$ 和 $1$ 滴 $0.1\ \mathrm{mol\cdot L^{-1}}$ $MnSO_4$ 的混合溶液中，加入少量 $PbO_2$，在水浴中微热，观察紫红色的 $MnO_4^-$ 生成。写出反应式。

### 3. 锡和铅的硫化物

(1) $SnS$ 的生成和性质

在 $1\ \mathrm{mL}$ $0.1\ \mathrm{mol\cdot L^{-1}}$ 的 $SnCl_2$ 溶液中，加入几滴硫化氢水溶液（饱和），观察棕色沉淀生成。离心分离，用蒸馏水洗涤沉淀，分别试验沉淀与 $1\ \mathrm{mol\cdot L^{-1}}$ $Na_2S$ 和 $1\ \mathrm{mol\cdot L^{-1}}$ 的多硫化铵（或多硫化钠）溶液的作用。如沉淀溶解，再用稀 $HCl$ 酸化，观察有何变化。反应式：

$$SnS + S_2^{2-} =\!\!= SnS_3^{2-}$$
$$SnS_3^{2-} + 2H^+ =\!\!= SnS_2 \downarrow + H_2S$$

(2) $SnS_2$ 的生成和性质

在 $1\ \mathrm{mL}$ $0.1\ \mathrm{mol\cdot L^{-1}}$ $SnCl_4$ 溶液中加入几滴硫化氢水溶液（饱和），观察黄色 $SnS_2$ 沉淀生成。离心分离，洗涤沉淀，试验沉淀物与 $1\ \mathrm{mol\cdot L^{-1}}$ $Na_2S$ 溶液的作用。如沉淀溶解，再用稀盐酸酸化，观察有何变化。反应式：

$$SnS_2 + S^{2-} =\!\!= SnS_3^{2-}$$
$$SnS_3^{2-} + 2H^+ =\!\!= SnS_2 \downarrow + H_2S$$

(3) $PbS$ 的生成和性质

在 $0.1\ \mathrm{mol\cdot L^{-1}}$ $Pb(NO_3)_2$ 溶液中加入几滴饱和硫化氢水溶液，观察黑色 $PbS$ 生成。分别试验沉淀物与 $1\ \mathrm{mol\cdot L^{-1}}$ $Na_2S$ 和多硫化铵溶液的作用。

根据实验结果，比较 $SnS$ 与 $SnS_2$ 以及 $SnS$ 与 $PbS$ 在性质上的差异。

### 4. 铅（Ⅱ）的难溶盐

(1) 铅（Ⅱ）的卤化物

①$PbCl_2$：在 $1\mathrm{mL}$ 水中加数滴 $0.1\ \mathrm{mol\cdot L^{-1}}$ $Pb(NO_3)_2$ 溶液，再加几滴稀盐酸，即有白色 $PbCl_2$ 沉淀生成。将所得的白色沉淀连同溶液一起加热，沉淀是否溶解？再把溶液冷却，又有什么变化？说明 $PbCl_2$ 的溶解度与温度的关系。

取白色沉淀少许，加入浓盐酸，观察沉淀的溶解。由于在浓盐酸中生成 $[PbCl_4]^{2-}$ 络离子，使溶解度增大：

$$PbCl_2 + 2Cl^- \Longrightarrow [PbCl_4]^{2-}$$

②$PbI_2$：取数滴 $0.1\ mol \cdot L^{-1}\ Pb(NO_3)_2$ 溶液，用水稀释至 1 毫升后，加 1~2 滴 $0.1\ mol \cdot L^{-1}$ KI 溶液，即生成橙黄色 $PbI_2$ 沉淀。试验它在热水和冷水中的溶解度。

（2）铅（Ⅱ）的含氧酸盐

①$PbCrO_4$：观察由 $Pb(NO_3)_2$ 稀释溶液与 $0.1\ mol \cdot L^{-1}\ K_2CrO_4$ 反应而生成 $PbCrO_4$ 的反应。分别试验 $PbCrO_4$ 在 $6\ mol \cdot L^{-1}\ HNO_3$ 和 HAc 的溶解情况。写出反应式。

②$PbSO_4$：在 1 mL 水中加数滴 $0.1\ mol \cdot L^{-1}\ Pb(NO_3)_2$ 溶液，加入几滴稀 $H_2SO_4$，即得白色 $PbSO_4$ 沉淀。离心分离，弃去溶液。分别试验沉淀与 $6\ mol \cdot L^{-1}$ NaOH 和饱和 NaAc 溶液的反应。由于生成可溶性络合离子 $[Pb(OH)_3]^-$ 和弱电解质和 $PbAc_2$ 而使沉淀溶解。

## 5. 铅丹（$Pb_3O_4$）的组成

取少量 $Pb_3O_4$ 固体与浓 $HNO_3$ 反应，不断搅拌，观察固态物的颜色变化，并与 $PbO_2$ 的颜色比较。试检查溶液中有无 $Pb^{2+}$ 离子存在。通过此实验，说明 $Pb_3O_4$ 中铅有哪几种价态。

# Exp 22　　Tin and Lead

## Objectives

1. To test and master the important properties of hydroxide of tin (Ⅱ) and lead (Ⅱ) and some insoluble lead salts.

2. To test and master the reducibility of tin (Ⅱ) and oxidizability of lead (IV).

3. To test the properties of sulfides of tin and lead.

## Previewing and Thinking

1. Review the related properties of tin and lead compounds in inorganic chemistry textbook.

2. Think and answer the following questions:

(1) Add both hydrochloric acid and stannum grain when preparing $SnCl_2$ solution in the lab, what is the reason?

(2) What acid shall be used when testing the alkalinity of $Pb(OH)_2$?

(3) Compare the similarities and differences of dissolution of $SnS_2$, $SnS$, $PbS$ in acid, alkaline, and in aqueous sulfide solution.

(4) How to distinguish $SnCl_4$ from $SnCl_2$? How to separate PbS and SnS ?

## Instruments and Reagents

test tube, centrifugal tube, beaker, centrifugal machine.

$PbO_2$ (solid), $Pb_3O_4$ (solid), $H_2SO_4$ (1, 3 mol·$L^{-1}$), HCl(2 mol·$L^{-1}$, concentrated), $HNO_3$ (6 mol·$L^{-1}$, concentrated), HAc(6 mol·$L^{-1}$), NaOH(2, 6 mol·$L^{-1}$), $Na_2S$ (1 mol·$L^{-1}$), $(NH_4)_2S_2$(1 mol·$L^{-1}$), $H_2S$ (saturated), $SnCl_2$(0. 1 mol·$L^{-1}$), $SnCl_4$(0. 1 mol·$L^{-1}$), $Pb(NO_3)_2$(0. 1 mol·$L^{-1}$), $FeCl_3$(0. 1 mol·$L^{-1}$), $HgCl_2$(0. 1 mol·$L^{-1}$), Bi $(NO_3)_2$(0. 1 mol·$L^{-1}$), KI(0. 1 mol·$L^{-1}$), $MnSO_4$ (0. 1 mol·$L^{-1}$), $K_2CrO_4$ (0. 1 mol· $L^{-1}$), NaAc(saturated).

## Experimental Procedures

### 1. Preparation and the acid or alkaline properties of hydroxide of tin (Ⅱ) and lead (Ⅱ)

(1) Preparation and the acid or alkaline properties of $Sn(OH)_2$

Gradually add 2 mol·$L^{-1}$ NaOH to 1 mL 0. 1 mol·$L^{-1}$ $SnCl_2$ solution, till white precipitate produced is not dissolved anymore. Centrifugalize the solution, discard the clear liquid. The precipitate is divided into two portions, test the reaction between $Sn(OH)_2$ and dilute acid and base. Write down the reaction eqations.

(2) Try to prepare $Pb(OH)_2$ precipitate from $0.1$ $mol \cdot L^{-1}$ $Pb(NO_3)_2$ solution.　Prove whether $Pb(OH)_2$ is amphoteric (note: what acid shall be selected when testing the alkalinity of solution).　Write down reaction equation.

Draw a conclusion on acidity and alkalinity of $Sn(OH)_2$ and $Pb(OH)_2$ according to the above experiments.

## 2.　Reducibility of tin (Ⅱ) and oxidizability lead (Ⅳ)

(1) Reducibility of tin(Ⅱ)

①Test the reaction of $SnCl_2$ with $FeCl_3$ solution.　Observe the phenomena.　Write down the reaction equation.

②Gradually add $0.1$ $mol \cdot L^{-1}$ $SnCl_2$ solution into $0.1$ $mol \cdot L^{-1}$ $HgCl_2$ solution, observe the change.　What change will happen when continuously adding $SnCl_2$ solution, the reaction is as follows.

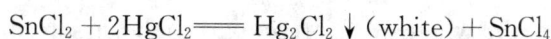

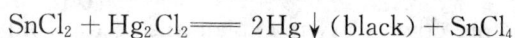

$$SnCl_2 + 2HgCl_2 = Hg_2Cl_2 \downarrow (white) + SnCl_4$$
$$SnCl_2 + Hg_2Cl_2 = 2Hg \downarrow (black) + SnCl_4$$

③Add $0.1$ $mol \cdot L^{-1}$ $Bi(NO_3)_2$ solution to prepared $[Sn(OH)_4]^{2-}$ solution.　black precipitate appears at once.　The reaction equation is:

$$3[Sn(OH)_4]^{2-} + 2Bi^{3+} + 6OH^- = 3[Sn(OH)_6]^{2-} + 2Bi \downarrow (black)$$

The reaction can be used to identify $Sn^{2+}$ and $Bi^{3+}$.

(2) Oxidizability of lead (Ⅳ)

①Add concentrated HCl onto a little $PbO_2(s)$, observe the phenomena, check whether chlorine is produced, write down the reaction equation.

②Add a little $PbO_2(s)$ into the mixture of $2$ mL $3$ $mol \cdot L^{-1}$ $H_2SO_4$ and 1drop of $0.1$ $mol \cdot L^{-1}$ $MnSO_4$, observe the formation of mauve $MnO_4^-$ after heated slightly by water bath, write down the reaction equation.

## 3.　Sulfides of tin and lead

(1) Formation and properties of SnS

Add a few drops of saturated $H_2S$ solution to $1$ mL $0.1$ $mol \cdot L^{-1}$ $SnCl_2$ solution, observe brown precipitate SnS.　Centrifugalize it, distilled water is used to wash the precipitate, test the reaction between the precipitate and $1 mol \cdot L^{-1}$ $Na_2S$, $(NH_4)_2S_X$, $Na_2S_X$ respectively.　If the precipitate can be dissolved, observe the change after being acidified by dilute HCl.　The reaction equations are as follows:

$$SnS + S_2^{2-} = SnS_3^{2-}$$
$$SnS_3^{2-} + 2H^+ = SnS_2 \downarrow + H_2S$$

(2) Formation and the properties of $SnS_2$

Add a few drops of saturated $H_2S$ into $1$ mL $0.1$ $mol \cdot L^{-1}$ $SnCl_4$ solution, observe yellow $SnS_2$, centrifugalize it, distilled water is used to wash the precipitate, test the reaction between the precipitate and $1$ $mol \cdot L^{-1}$ $Na_2S$, if the precipitate can dissolve, observe the change after acidifying by dilute HCl.　The reaction equations are as follows:

$$SnS_2 + S^{2-} = SnS_3{}^{2-}$$
$$SnS_3{}^{2-} + 2H^+ = SnS_2 \downarrow + H_2S$$

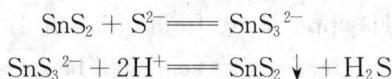

(3) Formation and properties of PbS

Add a few of drops saturated $H_2S(aq)$ into 1 mL 0. 1 $mol \cdot L^{-1} Pb(NO_3)_2$ solution, observe the black precipitate PbS, test the reaction between the precipitate and 1 $mol \cdot L^{-1} Na_2S$ and $(NH_4)_2S_X$ respectively.

Qualitatively compare the differences of SnS and $SnS_2$, SnS and PbS according to the results of above experiments.

### 4. Insoluble salts of Pb( Ⅱ )

(1) Halides of Pb( Ⅱ )

①$PbCl_2$: Add a few drops of 0. 1 $mol \cdot L^{-1} Pb(NO_3)_2$ solution into 1 mL water, add a few drops of dilute HCl, white precipitate $PbCl_2$ will be produced. Heat the white precipitate and its solution, observe whether the precipitate can be dissolved. What change will occur after the solution is cooling down? Try to explain the relationship between the solubility of $PbCl_2$ and temperature.

Take a little white precipitate, adding concentrated HCl, observe the precipitate being dissolved. The solubility of the precipitate is increasing due to the formation of complex $[PbCl_4]^{2-}$ in concentrated HCl.

②$PbI_2$: take a few drops of 0. 1 $mol \cdot L^{-1} Pb(NO_3)_2$ solution, dilute it to 1 mL using distilled water, then add 1~2 drops of 0. 1 $mol \cdot L^{-1}$ KI solution into it, orange-yellow $PbI_2$ precipitate can be observed. Test its solubility in hot water and cold water.

(2) Oxysalt of Pb( Ⅱ )

①$PbCrO_4$: Add $Pb(NO_3)_2$ solution into 0. 1 $mol \cdot L^{-1} K_2CrO_4$, observe the formation of $PbCrO_4$. Test whether $PbCrO_4$ can be dissolved in 6 $mol \cdot L^{-1} HNO_3$ or HAc. Write down the reaction equation.

②$PbSO_4$: Add a few drops of 0. 1 $mol \cdot L^{-1} Pb(NO_3)_2$ solution into 1 mL water, add a few of drops of dilute $H_2SO_4$, white precipitate $PbSO_4$ will be obtained. Centrifugalize it, discard the clear liquid. Test the reaction between the precipitate NaOH and saturated NaAc. The precipitate will dissolve due to the formation of dissolubility coordination compound $[Pb(OH)_3]^-$ and weak electrolyte($PbAc_2$).

### 5. Composition of $Pb_3O_4$

Take a little solid $Pb_3O_4$ to react with concentrated $HNO_3$, with continuous stir, observe the change of solid color and compare it with that of $PbO_2$.

Try to test whether there exists $Pb^{2+}$ in the solution. Explain which kinds of value states of Pb in $Pb_3O_4$ through this experiment.

# 实验二十三　铜、银、醋酸铜制备

## 实验目的

1. 了解铜、银的氧化物，氢氧化物酸碱性。
2. 掌握铜（I）、铜（Ⅱ）重要化合物的性质和相互转化的条件。
3. 了解醋酸铜的制备原理和方法以及铜、银离子的鉴定。
4. 练习抽滤操作。

## 实验仪器、材料和药品

仪器：试管、烧杯、离心机、抽滤装置等

材料：滤纸

液体药品：$NaOH$（2 $mol \cdot L^{-1}$，6 $mol \cdot L^{-1}$），氨水（2 $mol \cdot L^{-1}$，6 $mol \cdot L^{-1}$），$H_2SO_4$（1 $mol \cdot L^{-1}$），$HNO_3$（2 $mol \cdot L^{-1}$），$HCl$（2 $mol \cdot L^{-1}$，浓），$HAc$（2 $mol \cdot L^{-1}$，10％），$CuSO_4$（0.2 $mol \cdot L^{-1}$），$CuCl_2$（0.5 $mol \cdot L^{-1}$），$AgNO_3$（0.2 $mol \cdot L^{-1}$），$KI$（0.1 $mol \cdot L^{-1}$），$Na_2S_2O_3$（0.5 $mol \cdot L^{-1}$），$K_4[Fe(CN)_6]$（0.1 $mol \cdot L^{-1}$），乙醇（95％），乙醚（95％），葡萄糖溶液（10％）

固体药品：铜屑

## 实验内容

### 1. 铜的化合物

（1）氢氧化铜和氧化铜的生成和性质

取三支试管各滴入 2～3 滴 0.2 $mol \cdot L^{-1}$ 硫酸铜溶液，再滴入 1～2 滴 2 $mol \cdot L^{-1}$ 氢氧化钠溶液，观察生成的氢氧化铜的颜色和状态。在第一支试管中加入 1 $mol \cdot L^{-1}$ $H_2SO_4$ 溶液，第二支试管中加入过量的 6 $mol \cdot L^{-1}$ $NaOH$ 溶液，第三支试管加热至固体变黑，再注入 2 $mol \cdot L^{-1}$ $HCl$ 溶液，观察有何现象，写出反应方程式。

（2）氧化亚铜的生成和性质

在两支离心试管中各滴入 5 滴 0.2 $mol \cdot L^{-1}$ $CuSO_4$ 溶液，然后再各注入过量的 6 $mol \cdot L^{-1}$ $NaOH$ 溶液，使生成的沉淀全部溶解，再往此澄清的溶液中各注入 10 滴 10％葡萄糖溶液，观察各有何现象？写出反应方程式。

将上述两支有沉淀的试管分别离心分离，并用蒸馏水洗涤沉淀后，在第一支试管中加入 5 滴 1 $mol \cdot L^{-1}$ $H_2SO_4$ 并加热，观察有何现象？在第二支试管中加入 5～6 滴浓 $NH_3 \cdot H_2O$，振荡后静置 10 min，观察清液颜色。

（3）氯化亚铜的生成和性质

取 2 mL 0.5 $mol \cdot L^{-1}$ $CuCl_2$ 溶液，加 1 mL 浓 $HCl$ 和少量铜屑，加热直到溶液变成深棕色为止。取出 1～2 滴，注入 1 mL 蒸馏水中，如有白色沉淀产生，则迅速把全部溶液倒入 10 mL 蒸馏水中，观察沉淀的生成。等大部分沉淀析出后，静置，倾出上层清液，并用 1

~2 mL 蒸馏水洗涤沉淀，将沉淀分成两份。一份与浓氨水反应，另一份与浓 HCl 反应，观察沉淀是否溶解？写出反应方程式。

（4）碘化亚铜的生成

在一支试管中滴入 5 滴 0.2 mol·L$^{-1}$ 的 CuSO$_4$ 溶液，再滴入 2～3 滴 0.1 mol·L$^{-1}$ KI 溶液，观察有何变化？再滴入 1 滴 0.5 mol·L$^{-1}$ Na$_2$S$_2$O$_3$ 溶液，以除去反应中生成的 I$_2$（加入 Na$_2$S$_2$O$_3$ 不能过量，否则会使 Cu$_2$I$_2$ 溶解，为什么？）。观察生成物 Cu$_2$I$_2$ 的颜色和状态。写出反应方程式。

（5）Cu$^{2+}$ 的鉴定

在试管中滴入 1～2 滴 0.2 mol·L$^{-1}$ CuSO$_4$ 溶液，再滴入 2～3 滴 2 mol·L$^{-1}$ 醋酸和 2～3 滴 0.1 mol·L$^{-1}$ 六氰合亚铁酸钾 K$_4$[Fe(CN)$_6$] 溶液，即生成红棕色的六氰合亚铁酸铜 Cu$_2$[Fe(CN)$_6$] 沉淀。

在沉淀中注入 6mol·L$^{-1}$NH$_3$·H$_2$O，沉淀溶解生成蓝色的溶液，表示有 Cu$^{2+}$ 存在。写出反应方程式。

（6）醋酸铜的制备

慢慢将 4 mL 6 mol·L$^{-1}$NaOH 溶液滴入 50 mL 0.2 mol·L$^{-1}$ 的 CuSO$_4$ 溶液，并不断搅动，静置后抽滤得蓝色沉淀（这是何物？）用少量蒸馏水洗涤沉淀，洗至无可检测的游离 Cu$^{2+}$ 离子（如何检测？）。

将所得沉淀与 12 mL 10％醋酸溶液混合，搅动至完全溶解为止，静置一周，有沉淀生成，抽滤并用少量乙醇洗涤沉淀。沉淀在空气中晾干后称重。

**2. 银的化合物**

（1）氧化银的生成和性质

在两支离心试管中分别滴入 3～5 滴 AgNO$_3$（0.1 mol·L$^{-1}$）溶液，再分别逐滴加入 2 mol·L$^{-1}$ NaOH 溶液，振荡，观察 Ag$_2$O（为什么不是 AgOH？）的颜色和状态。分别将两支试管离心分离，弃去溶液，用蒸馏水洗涤沉淀。在一支试管中加 2 mol·L$^{-1}$ HNO$_3$ 溶液，在另一支试管中加 2 mol·L$^{-1}$ NH$_3$·H$_2$O，观察现象，并写出反应方程式。

（2）银镜反应

在一支洁净的试管中滴入 10 滴 0.1 mol·L$^{-1}$ AgNO$_3$ 溶液，滴入 2 mol·L$^{-1}$ NH$_3$·H$_2$O 至初生成的沉淀恰好溶解为止。再多滴 1 滴。然后滴入 5 滴 10％葡萄糖溶液，摇匀后放在 80～90℃水浴中静置。观察试管内壁上有何变化？写出反应方程式。

## 思考题

1. 土红色的氧化亚铜溶于氨水得到什么配合物？为什么它变成深蓝色呢？

2. 在鉴定 Cu$^{2+}$ 离子的实验中，如果 Cu$^{2+}$ 试液中含有 Fe$^{3+}$ 离子时会干扰鉴定，若有干扰，是为什么？应如何处理？

3. 选用什么试剂来溶解下列沉淀：

氢氧化铜、硫化铜、溴化银、碘化银

4. 在制备氯化亚铜时，能否用氯化铜溶液和铜屑在用盐酸酸化呈微弱的酸性条件下反应？为什么？若不加盐酸，加入浓氯化钠溶液，此反应能否进行？为什么？

# Exp 23   Copper, Silver and Preparation of Copper Acetate

## Objectives

1.  To comprehend acid and basic properties of hydroxides, oxides of copper and silver.

2.  To grasp important properties of Cu (I), Cu (II) and transformation condition.

3.  To know preparative principle and the method of copper acetate, identifying ions of Cu and Ag.

4.  To practice the operation of vaccum filtration.

## Instruments, Materials and Reagents

Instruments: test tube, beaker, centrifugal tube, instrument of filtering under reduced pressure

Materials: filter paper

Liquid reagents: NaOH(2, 6 $mol \cdot L^{-1}$), ammonia water(2, 6 $mol \cdot L^{-1}$), $H_2SO_4$ (1 $mol \cdot L^{-1}$), $HNO_3$(2 $mol \cdot L^{-1}$), HCl(2 $mol \cdot L^{-1}$), HAc(2 $mol \cdot L^{-1}$, 10%), $CuSO_4$(0.2 $mol \cdot L^{-1}$), $CuCl_2$(0.5 $mol \cdot L^{-1}$), $AgNO_3$(0.1 $mol \cdot L^{-1}$), KI(0.1 $mol \cdot L^{-1}$), $Na_2S_2O_3$ (0.5 $mol \cdot L^{-1}$), $K_4[Fe(CN)_6]$(0.1 $mol \cdot L^{-1}$), ethanol(95%), ether(95%), glucose solution(10%)

Solid reagents: copper shavings

## Experimental Procedures

### 1.  Compounds of copper

(1) Formation and properties of $Cu(OH)_2$ and CuO

Add 2~3 drops of 0.2 $mol \cdot L^{-1}$ $CuSO_4$ solution into three test tubes respectively, then add 1~2 drops 2 $mol \cdot L^{-1}$ NaOH solution, observe the color and state of $Cu(OH)_2$. Add 1 $mol \cdot L^{-1}$ $H_2SO_4$ to the first tube, excess 6 $mol \cdot L^{-1}$ NaOH solution to the second one, heat the third one until the solid turns to black, 2 $mol \cdot L^{-1}$ HCl is added, observe the phenomena and write down the reaction equations.

(2) Formation and properties of $Cu_2O$

Add 5 drops of 0.2 $mol \cdot L^{-1}$ $CuSO_4$ solution into two centrifugal test tubes respectively, then excess 6 $mol \cdot L^{-1}$ NaOH solution is added to dissolve all the precipitate, another 10 drops of glucose solution are added to this opaque solution, observe the phenomena, write down the reaction equations.

Centrifugalize the two test tubes, wash the precipitate with distilled water, add 5 drops of 1 $mol \cdot L^{-1}$ $H_2SO_4$ and heat one tube, observe the phenomena; the other one is added 5~6 drops of concentrated ammonia water, shake and stand for 10 min, observe the color of solution.

(3) Formation and properties of CuCl

2 mL of 0.5 mol·L$^{-1}$ CuCl$_2$ solution, 1 mL concentrated HCl and small quantities of copper shaving are mixed, the mixture is heated until it appears dark brown. 1 mL distilled water is added to 1~2 drops of above solution, if white precipitate is produced, pour all the solution to 10 mL distilled water rapidly, observe the phenomena. When bulk of precipitate appears, stand, decant above the layer, wash the precipitate with 1~2 mL distilled water, divided it into two portions. One reacts with concentrated ammonia water, another with concentrated HCl solution, whether does precipitate dissolve? Write down the reaction equations.

(4) Formation and properties of CuI

Add 5 drops of 0.2 mol·L$^{-1}$ CuSO$_4$ solution to a test tube, then 2~3 drops of 0.1 mol·L$^{-1}$ KI solution are added, too. What will you observe? One drop of 0.5 mol·L$^{-1}$ Na$_2$S$_2$O$_3$ is added to eliminate reaction product: I$_2$ (Don't add excess, otherwise Cu$_2$I$_2$ will also be dissolved, why?). Observe the color and state of Cu$_2$I$_2$. Write down reaction equations.

(5) Identification of Cu$^{2+}$

1~2 drops of 0.2 mol·L$^{-1}$ CuSO$_4$ are added to a test tube, 2~3 drops 2 mol·L$^{-1}$ acetic acid and 2~3 drops of 0.1 mol·L$^{-1}$ K$_4$[Fe(CN)$_6$] solution, umber precipitate Cu$_2$[Fe(CN)$_6$] forms.

6 mol·L$^{-1}$ NH$_3$·H$_2$O is added to the precipitate, blue solution appears accompanying precipitate being dissolved, which indicates the existence of Cu$^{2+}$. Write down the reaction equations.

(6) Preparation of copper acetate

4 mL 6 mol·L$^{-1}$ NaOH is added slowly to 50 mL 0.2 mol·L$^{-1}$ CuSO$_4$ solution with stirring, stand, filtering, blue precipitate is obtained (what ?). Wash the precipitate with a small quantity of water until no free Cu$^{2+}$ can be detected (How to do it ?).

Mix the precipitate with 12 mL 10% acetic acid solution, stir until all the precipitate is dissolved, stand for a week, precipitate forms, filter, wash the precipitate with a small quantity of ethanol. Weigh after open-air drying.

## 2. Compound of silver

(1) Formation and properties of Ag$_2$O

3~5 drops of AgNO$_3$ (0.1 mol·L$^{-1}$) solution are added to two test tubes respectively, 2 mol·L$^{-1}$ NaOH solution is then added dropwise, shake, observe the color and state of Ag$_2$O (why is it not AgOH ?). Centrifugalize each tube, remove solution of above layer, wash the precipitate with distilled water. Add 2 mol·L$^{-1}$ HNO$_3$ in another one, 2 mol·L$^{-1}$ NH$_3$·H$_2$O is added, observe the phenomena, write down the reaction equations.

(2) Silver mirror reaction

10 drops of 0.1 mol·L$^{-1}$ AgNO$_3$ is added to one clean test tube, 2 mol·L$^{-1}$ NH$_3$·H$_2$O is added until precipitate is dissolved. One more drop of NH$_3$·H$_2$O is added. Then 5 drops

of 10% glucose solution are added，shake and place it water bath at 80~90℃. What will we see inside of the test tube ? Write down the reaction equations.

## Questions

1. When soil red $Cu_2O$ is dissolved in ammonia water，what coordination compound will we get ? Why does it become a navy blue color ?

2. In identification of $Cu^{2+}$ experiment，if $Fe^{3+}$ exists in $Cu^{2+}$ solution，there will be interference，if it is so，why? And how to deal with it?

3. Which reagents do we choose to dissolve the following precipitates:

$$Cu(OH)_2, CuS, AgBr, AgI$$

4. In preparation of CuCl，whether it can be done in slightly weak acid medium with $CuCl_2$ and Cu，why ? If NaCl is replaced with HCl，whether will this reaction be successful ? Why ?

# 实验二十四　锌、镉、汞

## 实验目的

1. 试验并掌握锌、镉、汞氢氧化物和氧化物的酸碱性，硫化物的溶解性。
2. 试验并熟悉锌、镉、汞的配合能力。
3. 试验并掌握 $Hg_2^{2+}$ 和 $Hg^{2+}$ 离子的转化反应。
4. 学会 $Zn^{2+}$，$Cd^{2+}$，$Hg^{2+}$ 和 $Hg_2^{2+}$ 离子的鉴定反应。

## 实验仪器、材料和药品

仪器：离心试管、小烧杯、离心机

材料：pH 试纸、砂纸、玻棒

固体药品：KI

液体药品：$HCl$（2 mol·$L^{-1}$，浓），$H_2SO_4$（2 mol·$L^{-1}$），$HNO_3$（2 mol·$L^{-1}$），NaOH（2 mol·$L^{-1}$），NaOH（40%），氨水（2 mol·$L^{-1}$），$NH_4Cl$（2 mol·$L^{-1}$），$ZnSO_4$（0.2 mol·$L^{-1}$），$CdSO_4$（0.2 mol·$L^{-1}$），$Hg(NO_3)_2$（0.2 mol·$L^{-1}$），$HgCl_2$（0.2 mol·$L^{-1}$），KI（0.2 mol·$L^{-1}$），$SnCl_2$（0.2 mol·$L^{-1}$），NaCl（0.2 mol·$L^{-1}$），$Na_2S$（1 mol·$L^{-1}$），BaS（0.2 mol·$L^{-1}$），$Zn^{2+}-Cd^{2+}$ 混合液（均含 0.1 mol·$L^{-1}$），金属汞，$AgNO_3$（0.2 mol·$L^{-1}$，不贴标签），$Hg_2(NO_3)_2$（0.2 mol·$L^{-1}$，不贴标签），$Hg(NO_3)_2$（0.2 mol·$L^{-1}$，不贴标签）

## 实验内容

### 1. 锌、镉、汞氢氧化物的生成和性质

（1）锌、镉氢氧化物的生成和性质

在两支试管中分别注入 5 滴 0.2 mol·$L^{-1}$ $ZnSO_4$ 溶液，再分别滴入 2 mol·$L^{-1}$ NaOH 溶液直到沉淀生成为止（不要过量）。在一试管中加 2 mol·$L^{-1}$ $H_2SO_4$，另一试管中加 2 mol·$L^{-1}$ NaOH 溶液，观察现象，写出反应方程式。

用同样的方法试验镉氢氧化物的生成和性质，并与 $Zn(OH)_2$ 比较，写出有关反应方程式。

（2）氧化汞的生成和性质

在两支试管中分别滴入 5 滴 0.2 mol·$L^{-1}$ $Hg(NO_3)_2$ 溶液，再分别滴入 5 滴 2 mol·$L^{-1}$ NaOH 溶液，观察产物的颜色和状态。在一支试管中注入 2 mol·$L^{-1}$ $HNO_3$；另一试管中注入 40% NaOH 溶液，沉淀是否溶解？写出有关反应方程式。

### 2. 锌、镉、汞硫化物的生成和性质

往三支分别盛有 10 滴 0.2 mol·$L^{-1}$ $ZnSO_4$，$CdSO_4$ 和 $Hg(NO_3)_2$ 溶液的试管中，分别滴入 10 滴 1 mol·$L^{-1}$ $Na_2S$ 溶液，观察沉淀的生成和颜色。

将沉淀离心分离，洗涤，往沉淀中分别注入 2 mol·$L^{-1}$ 盐酸，观察沉淀是否溶解。

将不溶的沉淀离心分离，洗涤，往沉淀中注入 6 mol·L$^{-1}$HNO$_3$，观察沉淀是否溶解？

将不溶的沉淀离心分离，洗涤，往沉淀中注入王水（自己配），水浴加热，观察沉淀的溶解情况。

根据实验，对锌、镉、汞硫化物的溶解性作出结论。写出有关的反应方程式。

### 3. 锌、镉、汞的配合物

（1）锌、镉、汞氨合物

在两支试管中分别滴入 5 滴 0.2 mol·L$^{-1}$ZnSO$_4$ 和 0.2 mol·L$^{-1}$CdSO$_4$ 溶液，再分别逐滴滴入 2 mol·L$^{-1}$NH$_3$·H$_2$O，观察沉淀的生成，继续注入过量的 NH$_3$·H$_2$O，又有何现象发生？写出有关反应方程式。用 0.2 mol·L$^{-1}$HgCl$_2$ 溶液做同样的实验，比较 Zn$^{2+}$，Cd$^{2+}$，Hg$^{2+}$ 与 NH$_3$·H$_2$O 反应有什么不同。

（2）汞配合物的生成和应用

往盛有 5 滴 0.2 mol·L$^{-1}$Hg(NO$_3$)$_2$ 溶液中，滴入 0.2 mol·L$^{-1}$KI 溶液，观察沉淀的生成和颜色。再往该沉淀中加入少量固体 KI（直到沉淀刚好溶解为止，不过量）。溶液呈何色？写出有关反应方程。

在所得的溶液中滴 5～6 滴 40％NaOH 溶液并与 2 mol·L$^{-1}$NH$_4$Cl 溶液反应，观察沉淀的颜色。

### 4. 汞（Ⅱ）的氧化性和汞（Ⅰ）与汞（Ⅱ）的转化

（1）Hg$^{2+}$ 的氧化性

在一支试管中滴入 2 滴 0.2 mol·L$^{-1}$Hg(NO$_3$)$_2$ 溶液中，逐滴加入 0.2 mol·L$^{-1}$SnCl$_2$ 溶液，观察沉淀的颜色，再继续加入过量的 0.2 mol·L$^{-1}$ SnCl$_2$ 溶液，观察现象，写出反应方程式。

（2）Hg$^{2+}$ 与 Hg$_2^{2+}$ 的转化

往一支试管中滴入 5 滴 0.2 mol·L$^{-1}$Hg(NO$_3$)$_2$ 溶液，再滴入金属汞 1 滴，振荡。用滴管把清液转入两支小试管中（余下的汞要回收！），在一支试管中滴 1～2 滴 0.2 mol·L$^{-1}$ NaCl 溶液，观察现象。写出反应方程式（另一支试管供下面实验做）。

（3）Hg$_2^{2+}$ 的歧化分解

往上面所得的硝酸亚汞溶液中，滴加 5 滴 2 mol·L$^{-1}$ NH$_3$·H$_2$O，观察现象，写出反应方程式。

### 5. 锌钡白的制备

锌钡白（俗名立德粉）是一种优良的白色颜料，它是 ZnSO$_4$ 同 BaS 共沉淀所形成的混合晶体。制造锌钡白的反应如下：

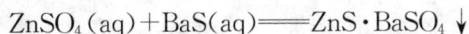

$$ZnSO_4(aq)+BaS(aq)=\!=\!=ZnS·BaSO_4 \downarrow$$

往盛有 1 mL BaS 溶液的试管中，交替加入 ZnSO$_4$ 和 BaS 溶液，而且不断搅拌，制备过程中应维持溶液呈碱性（完全反应时溶液呈微碱性），控制 pH 在 8～9 范围内，静置，用倾析法分离。

### 6. 离子的鉴别

（1）有 Zn$^{2+}$－Cd$^{2+}$ 混合液，试根据其性质进行鉴别，写出实验方法，步骤。

（2）现有三瓶已失去标签的 $Hg(NO_3)_2$，$Hg_2(NO_3)_2$ 和 $AgNO_3$ 溶液，鉴别（至少用两种方法）后，贴上标签。

## 思考题

1. 试从可能含有锌和铝的混合溶液中分离和鉴定这两种离子。

2. 使用汞应注意什么？为什么储存时要用水封？

3. 用平衡原理预测，往 $Hg_2(NO_3)_2$ 溶液中通入 $H_2S$ 气体后，生成的沉淀为何物？并加以解释。再往其中加 KI 溶液有变化吗？为什么？

# Exp 24　Zinc, Cadmium and Mercury

## Objectives

1. To test and master the acid or basic properties of hydroxides and oxides of Zn, Cd and Hg and the soluble property of their sulfides.

2. To test and be familiar with the coordinative ability of Zn, Cd and Hg.

3. To test and master transforming reactions between $Hg^{2+}$ and $Hg_2^{2+}$.

4. To master the identification methods of $Zn^{2+}$, $Cd^{2+}$ and $Hg^{2+}$.

## Instrument, Reagents and Materials

Instrument: centrifugal tube, beaker, centrifugal machine and so on

Materials: pH test paper, sand paper, glass rod

Solid reagents: KI

Liquid reagents: HCl(2 mol·$L^{-1}$, concentrated), $H_2SO_4$(1 mol·$L^{-1}$), $HNO_3$(2 mol·$L^{-1}$), NaOH(2 mol·$L^{-1}$), NaOH(40%), $NH_3·H_2O$(2 mol·$L^{-1}$), $NH_4Cl$(2 mol·$L^{-1}$), $ZnSO_4$(0.2 mol·$L^{-1}$), $CdSO_4$(0.2 mol·$L^{-1}$), $Hg(NO_3)_2$(0.2 mol·$L^{-1}$), $HgCl_2$(0.2 mol·$L^{-1}$), KI(0.2 mol·$L^{-1}$), $SnCl_2$(0.2 mol·$L^{-1}$), NaCl(0.2 mol·$L^{-1}$), $Na_2S$(1 mol·$L^{-1}$), BaS(0.2 mol·$L^{-1}$), $Zn^{2+}-Cd^{2+}$ mixture solution(0.1 mol·$L^{-1}$ each), metal Hg, $AgNO_3$(0.2 mol·$L^{-1}$, without a label), $Hg_2(NO_3)_2$(0.2 mol·$L^{-1}$, without a label), $Hg(NO_3)_2$(0.2 mol·$L^{-1}$, without a label)

## Experimental Procedures

### 1. Formation and properties of hydroxides of Zn, Cd and Hg, ammonia as ligand

(1) Formation and properties of hydroxides of Zn and Cd

Add 5 drops of $ZnSO_4$(0.2 mol·$L^{-1}$) into two test tubes respectively, then add appropriate volume of NaOH(2 mol·$L^{-1}$) into both of them which produce precipitate (the NaOH solution should not be excess). Add $H_2SO_4$(2 mol·$L^{-1}$) into one test tube while NaOH(2 mol·$L^{-1}$)into another one, observe the phenomena and write down the reaction equations. Similarly, the formation and properties of hydroxides of Cd can be tested. Compare it with $Zn(OH)_2$, and write down reaction equations.

(2) Formation and properties of HgO

Add 5 drops of $Hg(NO_3)_2$(2 mol·$L^{-1}$) respectively into two test tubes, followed by another 5 drops of NaOH(2 mol·$L^{-1}$), observe the color change and the state of product. Then add $HNO_3$(2 mol·$L^{-1}$) into one test tube and NaOH(40%) into another, observe whether the precipitate is dissolved?

Write down the reaction equations.

### 2. Formation and properties of Zn, Cd and Hg's sulfides.

Add 10 drops of $Na_2S(1 \ mol \cdot L^{-1})$ respectively into three test tubes containing 10 drops of $ZnSO_4(0.2 \ mol \cdot L^{-1})$, $CdSO_4(0.2 \ mol \cdot L^{-1})$ and $Hg(NO_3)_2(0.2 \ mol \cdot L^{-1})$, observe the color and state of the precipitate.

Add the precipitate to $HNO_3(6 \ mol \cdot L^{-1})$ after it has been separated by centrifugal machine, observe whether the precipitate is dissolved?

Pour the precipitate to aqua regia (prepare by yourself), heat by water bath, observe whether the precipitate is dissolved.

Draw conclusions on soluble abilities of sulfides of Zn, Cd and Hg's according to the above experiments and write down the reaction equations.

### 3. Coordination compounds of Zn, Cd and Hg

(1) Coordination compounds of Zn, Cd and Hg

Add 5 drops of $ZnSO_4(0.2 \ mol \cdot L^{-1})$ and $CdSO_4(0.2 \ mol \cdot L^{-1})$ respectively to two test tubes, then add $NH_3 \cdot H_2O$ (2 $mol \cdot L^{-1}$) in both of them dropwise, observe the generation of precipitate, what will happen if $NH_3 \cdot H_2O$ is excessively added? Write down the reaction equations. Compare the reactions when $HgCl_2(0.2 \ mol \cdot L^{-1})$ is treated in the same way.

(2) Coordination compound of Hg

Add KI (0.2 $mol \cdot L^{-1}$) to a test tube with 5 drops of $Hg(NO_3)_2(0.2 \ mol \cdot L^{-1})$, observe the color of the precipitate formed. Then add small solid KI to it until the precipitate disappears rightly. What color does the solution appear ? Write down the reaction equations.

Add 5~6 drops of NaOH(40%) into the solution, then make it react with $NH_4Cl$ (2 $mol \cdot L^{-1}$), observe the precipitate's color.

### 4. Oxidizability of $Hg^{2+}$ and the transformation of $Hg_2^{2+}$ and $Hg^{2+}$

(1) Oxidizability of $Hg^{2+}$

Add solution of $SnCl_2$ (0.2 $mol \cdot L^{-1}$) dropwise to a test tube with 2 drops of $Hg_2(NO_3)_2$ (0.2 $mol \cdot L^{-1}$), observe the color of the precipitate, and then excess $SnCl_2$ (0.2 $mol \cdot L^{-1}$) is added, observe the phenomena and write down reaction equations.

(2) Transformation of $Hg_2^{2+}$ and $Hg^{2+}$

Add 5 drops of $Hg_2(NO_3)_2(0.2 \ mol \cdot L^{-1})$ and 1 drop of metallic Hg sequentially into one test tube, shake it. Then transfer the upper clean solution to two test tubes (the residual Hg should be recycled), and add 1~2 drops of NaCl (0.2 $mol \cdot L^{-1}$), observe the phenomena and write down the reaction equations (another tube is for the following experiment).

(3) Disproportionate decomposition of $Hg_2^{2+}$

Add 5 drops of $NH_3 \cdot H_2O(2 \ mol \cdot L^{-1})$ into the solution mentioned above. Observe the phenomena and write down the reaction equation.

### 5. Preparation of lithopone

Lithopone is one kind of white pigment，which is the mixed crystal derived from the product of $ZnSO_4$ and BaS.  The preparation equation is listed as below：

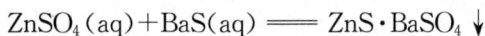

$$ZnSO_4(aq) + BaS(aq) \rightleftharpoons ZnS \cdot BaSO_4 \downarrow$$

Add $ZnSO_4$ and BaS alternatively into a test tube with 1 mL BaS solution，stir the mixture continuously while keep them in alkaline medium （the solution should be kept slightly alkaline）with pH value at 8~9，stand still and separate the precipitate by decantation.

### 6. Identification of ions

(1) Identify the mixed solution containing $Zn^{2+} - Cd^{2+}$ according to their properties. Write down the experimental methods and procedures.

(2) Identify three bottles of reagents without labels：$Hg(NO_3)_2$，$Hg_2(NO_3)_2$ and $AgNO_3$，then label them （At least two methods should be used）.

## Questions

1.  Try to separate $Zn^{2+}$ and $Al^{3+}$ and identify both of them in solution.

2.  What attention should be paid to the use of Hg? Why it should be covered by water for storage?

3.  Predict what precipitate can be obtained when $Hg_2(NO_3)_2$ is ventilated with $H_2S$, and explain it with equilibrium theory.  Would there be something different if KI solution is added continuously? And why?

# 实验二十五　铬、锰、硫酸锰铵的制备

## 实验目的

1. 试验并掌握铬（Ⅲ）和铬（Ⅵ）化合物的性质和它们之间相互转化的条件。
2. 制备硫酸锰铵，练习过滤操作。
3. 试验和了解锰的各种氧化态化合物的重要性质以及它们之间相互转化的条件。

## 实验仪器、药品和材料

仪器：试管、小烧杯、酒精灯、表面皿、天平、吸滤瓶、布氏漏斗等

固体药品：草酸，$MnO_2$，$(NH_4)_2SO_4$，$KClO_3$，$KOH$，$CrO_3$

液体药品：$Cr_2(SO_4)_3(0.5\ mol \cdot L^{-1})$，$H_2SO_4(1.0\ mol \cdot L^{-1}$，浓$)$，乙醇，$HCl$ $(2.0\ mol \cdot L^{-1})$，$NaOH(2.0，6.0\ mol \cdot L^{-1})$，$NH_4Cl($饱和$)$，$MnSO_4(0.1\ mol \cdot L^{-1})$，$NH_3$ $\cdot H_2O(6\ mol \cdot L^{-1})$，$KMnO_4(0.1\ mol \cdot L^{-1})$，$NaClO($浓$)$，$HAc(2\ mol \cdot L^{-1})$，$K_2Cr_2O_7$ $(0.1\ mol \cdot L^{-1})$，$NaNO_2(0.5\ mol \cdot L^{-1})$，$K_2CrO_4(0.5\ mol \cdot L^{-1})$，$AgNO_3(0.1\ mol \cdot L^{-1})$，$BaCl_2(0.2\ mol \cdot L^{-1})$，$Pb(NO_3)_2(0.2\ mol \cdot L^{-1})$，$H_2O_2(3\%)$

材料：冰，滤纸

## 实验内容

### 1. 铬的化合物的性质

（1）氢氧化铬（Ⅲ）的两性

取 5 滴 $0.5\ mol \cdot L^{-1}\ Cr_2(SO_4)_3$ 溶液于试管中，逐滴滴入 $2.0\ mol \cdot L^{-1}\ NaOH$ 溶液（有沉淀生成为止），观察沉淀的颜色。用实验证明沉淀物具有两性性质，并写出有关的反应方程式。

（2）铬（Ⅲ）的还原性

取 $0.5\ mol \cdot L^{-1}\ Cr_2(SO_4)_3$ 溶液 5 滴于一支试管中，注入 $6.0\ mol \cdot L^{-1}\ NaOH$ 溶液，直到沉淀溶解为止。然后往溶液中注入 $4\sim5$ 滴 $3\%H_2O_2$ 溶液，微热，观察溶液颜色的变化。解释现象，写出反应方程式。

（3）铬（Ⅵ）的氧化性

往 5 滴 $0.1\ mol \cdot L^{-1}\ K_2Cr_2O_7$ 溶液中，注入 $1.0\ mol \cdot L^{-1}$ 的 $H_2SO_4$酸化，然后注入 $0.5$ $mol \cdot L^{-1}NaNO_2$ 溶液，将混合溶液加热，观察溶液颜色的变化。写出反应方程式。

（4）铬（Ⅵ）的缩合平衡

取 5 滴 $0.1\ mol \cdot L^{-1}\ K_2Cr_2O_7$ 溶液于试管中，加入 $2.0\ mol \cdot L^{-1}\ NaOH$ 溶液，使溶液呈碱性，观察溶液颜色的变化。再加 $1.0\ mol \cdot L^{-1}\ H_2SO_4$，使溶液呈酸性，观察溶液的颜色又有何变化，写出反应方程式。

（5）重铬酸盐和铬酸盐的溶解性

分别在 $Cr_2O_7^{2-}$ 和 $CrO_4^{2-}$ 溶液中，各加入少量的 $Pb(NO_3)_2$，$BaCl_2$ 和 $AgNO_3$ 溶液，

观察产物的颜色和状态，比较实验结果，写出反应方程式。

（6）三氧化铬的性质

取黄豆大的三氧化铬固体于蒸发皿中，滴入 1～2 滴无水酒精，观察有何现象？写出反应方程式。

## 2. 硫酸锰铵的制备

（1）硫酸锰铵的制备

①硫酸锰的制备

微热盛有 15 mL 1mol·L$^{-1}$硫酸的烧杯，向其中溶解 2.0 g 草酸，再慢慢分次加入 2.0 g MnO$_2$，不断搅拌，使其充分反应。此时发生下列反应：

$$MnO_2 + H_2C_2O_4 + H_2SO_4 \!=\!=\!= MnSO_4 + 2CO_2\uparrow + 2H_2O$$

反应缓慢后，煮沸溶液并趁热过滤，保留滤液。

②硫酸锰铵的制备

往上述溶液中加入 3.5 g 硫酸铵，待硫酸铵全部溶解后，在冰浴（加入适量粗食盐固体）中冷却，即有晶体慢慢析出，30 min 后，抽滤并用 1∶1 的乙醇溶液洗两次，然后用滤纸盛放在表面皿上干燥，称量并计算理论产量和产率。

（2）锰（Ⅱ）化合物的性质

①氢氧化锰的生成和性质

a. 在一支试管中注入 5 滴 0.1 mol·L$^{-1}$ MnSO$_4$ 溶液，滴入 2 滴 2.0 mol·L$^{-1}$NaOH 溶液，观察沉淀的颜色，振荡试管，有何变化？

b. 在一支盛有 5 滴 0.1 mol·L$^{-1}$ MnSO$_4$ 溶液的试管中，滴入 2.0 mol·L$^{-1}$ NaOH 溶液，产生沉淀后，加入过量 NaOH，沉淀是否溶解。

c. 在一支盛有 5 滴 0.1 mol·L$^{-1}$ MnSO$_4$ 溶液的试管中，滴入 2.0 mol·L$^{-1}$ NaOH 溶液后迅速加入 2.0 mol·L$^{-1}$盐酸溶液，有何现象发生。

d. 在一支盛有 5 滴 0.1 mol·L$^{-1}$ MnSO$_4$ 溶液的试管中，滴入 2.0 mol·L$^{-1}$ NaOH 溶液产生沉淀以后，迅速加入 2 mol·L$^{-1}$氯化铵溶液，沉淀是否溶解？

写出上述有关反应方程式。此实验证明氢氧化锰具有哪些性质？

②锰离子的氧化性

试验硫酸锰与次氯酸钠溶液在酸、碱性介质中的反应。比较 Mn$^{2+}$ 在何介质中易氧化？

③硫化锰的生成和性质

往 5 滴 0.1 mol·L$^{-1}$ MnSO$_4$ 溶液中滴入饱和的 H$_2$S 溶液，有无沉淀生成？若用硫化钠代替硫化氢溶液，又有何结果？请用实验说明硫化锰的性质和生成条件。

（3）MnO$_2$ 的生成和性质

①往盛有 1～2 滴 0.1 mol·L$^{-1}$ KMnO$_4$ 溶液中，逐滴滴入 0.5 mol·L$^{-1}$ MnSO$_4$ 溶液，观察沉淀的颜色，往沉淀中加入 1 mol·L$^{-1}$硫酸和 0.1 mol·L$^{-1}$ Na$_2$SO$_3$ 溶液，沉淀是否溶解？写出有关反应方程式。

②在盛有米粒大小的 MnO$_2$ 固体的试管中，加入 2 mL 浓 H$_2$SO$_4$，加热，观察反应前后的颜色。有何气体生成？写出有关反应方程式。

（4）锰酸钾的生成和性质

在干燥的小试管中混合固体 KClO$_3$，MnO$_2$ 和 KOH（其质量分别为 0.1 g，0.2 g 和

0.3 g)，加热熔融，观察产物的颜色。冷却后，加入 2 mL 水，使熔块溶解，取上层清液 10 滴，然后加入 2.0 mol·L$^{-1}$醋酸溶液，观察有何变化？再加入过量的 6.0 mol·L$^{-1}$ NaOH 溶液，又有何变化？写出反应方程式。

通过以上实验说明锰酸钾稳定存在的介质条件和歧化后的产物，并解释之。

（5）高锰酸钾的性质

加热固体高锰酸钾，观察有何现象发生？检验产生的气体，写出反应方程式。

## 思考题

1. 如何实现 $MnO_4^- \rightarrow Mn^{2+}$，$MnO_4^- \rightarrow MnO_2$，$MnO_4^- \rightarrow MnO_4^{2-}$ 的转化？写出反应方程式。

2. 以高锰酸钾为原料制备氯气时，应加浓盐酸，但实验时误加了浓 $H_2SO_4$，加热时引起了爆炸，试解释其原因。

3. 如何实现 Cr（Ⅲ）和 Cr（Ⅵ）之间的相互转化？需要在什么条件下能实现？

# Exp 25   Chromium, Manganese and Preparation of Manganese Ammonium Sulfate

## Objectives

1.  To test and master the properties of Cr(Ⅲ) and Cr(Ⅵ) and the condition of transformation between them.

2.  To prepare $(NH_4)_2Mn(SO_4)_2$ and practice the operation of filtration.

3.  To test and know the important properties of all oxidation states compounds and the condition of transformation between them.

## Instruments, Reagents and Materials

Instruments: test tube, small beaker, alcohol burner, watch glass, balance, suction bottle, büchner funnel and so on

Solid reagents: oxalic acid, $MnO_2$, $(NH_4)_2SO_4$, $KClO_3$, $KOH$, $CrO_3$

Liquid reagents: $Cr_2(SO_4)_3$ (0.5 mol·$L^{-1}$), $H_2SO_4$(1 mol·$L^{-1}$, concentrated), ethanol, $HCl$(2 mol·$L^{-1}$), $NaOH$(2, 6 mol·$L^{-1}$), $NH_4Cl$ (saturated), $MnSO_4$ (0.1 mol·$L^{-1}$), $NH_3·H_2O$(6 mol·$L^{-1}$), $KMnO_4$ (0.1 mol·$L^{-1}$), $NaClO$ (concentrated), $HAc$(2mol·$L^{-1}$), $K_2Cr_2O_7$(2 mol·$L^{-1}$), $NaNO_2$(0.5 mol·$L^{-1}$), $K_2CrO_4$(0.5 mol·$L^{-1}$), $AgNO_3$(0.1 mol·$L^{-1}$), $BaCl_2$(0.2 mol·$L^{-1}$), $Pb(NO_3)_2$(0.2 mol·$L^{-1}$), $H_2O_2$(3%)

Materials: ice, filter paper

## Experimental Procedures

### 1. Properties of compounds of chromium

(1) Amphoteric property of Cr(OH)$_3$(Ⅲ)

Add 5 drops of 0.5 mol·$L^{-1}$ $Cr_2(SO_4)_3$ into a test tube, then add 2 mol·$L^{-1}$NaOH into it (till precipitate is formed), observe the color of precipitate. Try to prove the precipitate is amphoteric by experiments. Write down the related reaction equations.

(2) Reducibility of Cr(Ⅲ)

Add 5 drops of 0.5 mol·$L^{-1}$ $Cr_2(SO_4)_3$ into a test tube, then add 6 mol·$L^{-1}$ NaOH into it (till precipitate is dissolved), add 4~5 drops of 3% $H_2O_2$ solution, observe the change of color in solution after tepidity. Explain the phenomena, write down the related reaction equations.

(3) Oxidizability of Cr(VI)

Add 5 drops of 0.1 mol·$L^{-1}$ $K_2Cr_2O_7$ into the test tube, then add 1 mol·$L^{-1}$ $H_2SO_4$ for acidification, then add 0.5 mol·$L^{-1}$ $NaNO_2$ into it, heat the mixed solution, observe the change of color in solution. Explain the phenomena, write down the related reaction equations.

(4) Condensation balance of Cr(VI)

Add 5 drops of 0. 1 mol·L$^{-1}$ K$_2$Cr$_2$O$_7$ into a test tube, add 2 mol·L$^{-1}$ NaOH to make the solution slightly alkaline, observe the change of color in solution. Then add 1 mol·L$^{-1}$ H$_2$SO$_4$ to make the solution acid, observe the change of color in solution. Explain the phenomena, Write down the related reaction equations.

(5) Soluble properties of dichromate and chromate

Add small quantities of Pb(NO$_3$)$_2$, BaCl$_2$ and AgNO$_3$ solution into Cr$_2$O$_7$$^{2-}$ and CrO$_4$$^{2-}$ solution respectively. Observe the color and state of products. Explain the phenomena, write down the related reaction equations.

(6) The property of CrO$_3$

Take CrO$_3$ solid of a soybean size onto evaporating dish, add 1~2 drops of absolute alcohol to it. Observe the phenomena, write down related reaction equations.

## 2. Preparation of (NH$_4$)$_2$Mn(SO$_4$)$_2$

(1) Preparation of (NH$_4$)$_2$Mn(SO$_4$)$_2$

①Preparation of MnSO$_4$

Add 15 mL 1 mol·L$^{-1}$ H$_2$SO$_4$ into the beaker, add 2. 0 g H$_2$C$_2$O$_4$ after tepidity, and add 2. 0 g MnO$_2$ in portions slowly, mix round continuously to make it react completely. Here the following reaction can be represented as:

$$MnO_2 + H_2C_2O_4 + H_2SO_4 = MnSO_4 + 2CO_2 \uparrow + 2H_2O$$

When the reaction is moderate, heat to boil the solution, filter while it is hot, conserve the filtrate.

②Preparation of (NH$_4$)$_2$Mn(SO$_4$)$_2$

Add 3. 5 g (NH$_4$)$_2$SO$_4$ to the above solution, cool down the solution in ice bath [proper amount of NaCl(s) is added] after (NH$_4$)$_2$SO$_4$ being dissolved completely. Thus crystal will appear slowly, filter and wash twice with 1:1 ethanol after 30 min, dry it with filter paper on watch glass, weigh and calculate theoretical output and yield.

(2) Properties of compounds of Mn(Ⅱ)

①Preparation and properties of Mn(OH)$_2$

a. Add 5 drops of 0. 1 mol·L$^{-1}$ MnSO$_4$ into a test tube, then add 2 drops of 2 mol·L$^{-1}$ NaOH, observe the color of precipitate, what change will happen after shaking the test tube?

b. Add 5 drops of 0. 1 mol·L$^{-1}$ MnSO$_4$ into a test tube, then add 2 mol·L$^{-1}$NaOH, add excess NaOH continuously after precipitate produced, observe whether the precipitate can be dissolved.

c. Add 5 drops of 0. 1 mol·L$^{-1}$ MnSO$_4$ to a test tube, then add 2 mol·L$^{-1}$NaOH into it, add 2 mol·L$^{-1}$ HCl after precipitate produced, observe whether the precipitate can be dissolved.

d. Add 5 drops of 0. 1 mol·L$^{-1}$ MnSO$_4$ into a test tube, then add 2 mol·L$^{-1}$ NaOH into it, add 2 mol·L$^{-1}$ NH$_4$Cl after precipitate produced, observe whether the precipitate can be dissolved.

Write down the related reaction equations. Which properties can be proved through these experiments?

②Oxidizability of Mn(Ⅱ)

Test the reaction between MnSO$_4$ and NaClO in acid and alkalinous medium respectively. Compare them, draw a conclusion in which medium Mn(Ⅱ) is easily oxidized?

③The preparation and properties of MnS

Add 5 drops of 0.1 mol·L$^{-1}$ MnSO$_4$ into a test tube, add saturated H$_2$S solution into it. Observe whether the precipitate can be produced. If H$_2$S is replaced by Na$_2$S, what result will lead? Explain the property and formation condition of MnS.

(3) Preparation and properties of MnO$_2$

①Add 1~2 drops of 0.1 mol·L$^{-1}$ KMnO$_4$ to the test tube, add 0.5 mol·L$^{-1}$ MnSO$_4$ into it dropwise, observe the color of precipitate, then add 1mol·L$^{-1}$ H$_2$SO$_4$ and 0.1 mol·L$^{-1}$ Na$_2$SO$_3$ solution onto precipitate, observe whether the precipitate can be dissolved, write down the related reaction equations.

②Add 2 mL concentrated H$_2$SO$_4$ into the test tube containing MnO$_2$ solid of rice grain size, heat the test tube and observe the change of color, what gas will be produced? Write down the related reaction equations.

(4) Preparation and properties of K$_2$MnO$_4$

Mix round KClO$_3$, MnO$_2$ and KOH (their mass are 0.1 g, 0.2 g and 0.3 g respectively) in a dry small test tube, heat to fuse, observe the color of product. Add 2 mL water to make it dissolve after cooling down. Take 10 drops of above layer clear liquid, add 2 mol·L$^{-1}$ HAc, and observe the change. Then add excess 6 mol·L$^{-1}$ NaOH, what change will happen? Write down the related reaction equations.

Explain medium condition of stabilization for K$_2$MnO$_4$ and the product of disproportionate reaction.

(5) The properties of KMnO$_4$

Heat KMnO$_4$(s), observe the phenomena, test the produced gas, write down the related reaction equation.

## Questions

1. How to realize the mutal transformation of MnO$_4^-$ → Mn$^{2+}$, MnO$_4^-$ → MnO$_2$, MnO$_4^-$ → MnO$_4^{2-}$? Write down the reaction equations.

2. HCl (concentrated) shall be added when chlorine is prepared with KMnO$_4$ as starting material in theory, but if H$_2$SO$_4$ (concentrated) is added by mistake in the lab, thus explosion will occur when heating, explain it.

3. How to realize the mutal transformation between Cr(Ⅲ) and Cr(Ⅵ)? What condition is needed?

# 实验二十六　铁、钴、镍

## 实验目的

1. 试验并掌握铁（Ⅱ）、钴（Ⅱ）、镍（Ⅱ）的还原性和铁（Ⅲ）、钴（Ⅲ）、镍（Ⅲ）的氧化性。

2. 掌握铁、钴、镍的配合物的生成，学会 $Fe^{2+}$，$Fe^{3+}$，$Co^{2+}$ 和 $Ni^{2+}$ 的鉴定方法。

3. 了解金属铁腐蚀的基本原理和防止腐蚀的方法。

## 实验仪器、药品及材料

仪器：试管、离心试管、微型药匙

固体药品：硫酸亚铁铵、硫氰酸钾、氯化铵、锌片、锡片（或锌粒、锡粒）、碘化钾淀粉试纸

液体药品：$H_2SO_4$（1:1，$1.0\ mol\cdot L^{-1}$），HCl（$6.0\ mol\cdot L^{-1}$，5%，浓），NaOH（$6.0\ mol\cdot L^{-1}$，$2.0\ mol\cdot L^{-1}$），$(NH_4)_2Fe(SO_4)_2$（0.2，$0.5\ mol\cdot L^{-1}$），$CoCl_2$（$0.2\ mol\cdot L^{-1}$），$NiSO_4$（$0.2\ mol\cdot L^{-1}$），KI（$0.1\ mol\cdot L^{-1}$），$K_4[Fe(CN)_6]$（$0.5\ mol\cdot L^{-1}$），$NH_3\cdot H_2O$（浓），氯水，碘水，四氯化碳，戊醇，乙醚，$H_2O_2$ 3%（m），丁二酮肟，混合液（在 1 升溶液中含 600 g NaOH 和 60 g $NaNO_2$），KSCN（$0.1\ mol\cdot L^{-1}$），$FeCl_3$（$0.2\ mol\cdot L^{-1}$），$K_3[Fe(CN)_6]$（$0.5\ mol\cdot L^{-1}$）

材料：砂纸、铁钉、回形针、细铁丝

## 实验内容

### 1. 铁（Ⅱ）、钴（Ⅱ）、镍（Ⅱ）的还原性

（1）铁（Ⅱ）的还原性

①在一支试管中滴入 3 滴 1:1 的 $H_2SO_4$ 和 5 滴氯水振荡，再滴入 5 滴 $0.2\ mol\cdot L^{-1}$ $(NH_4)_2Fe(SO_4)_2$ 溶液，观察现象，写出反应方程式。

②在一支试管中滴入 10 滴蒸馏水和 5 滴 $1\ mol\cdot L^{-1}$ $H_2SO_4$ 溶液，煮沸以赶尽溶于其中的空气，然后溶入绿豆大的硫酸亚铁铵晶体。在另一支试管中注入 10 滴 $6\ mol\cdot L^{-1}$ NaOH 溶液，煮沸（为什么？）。冷却后，用一支长滴管吸取 NaOH 溶液，插入前一支试管中的硫酸亚铁铵溶液（直到试管底部）内，慢慢放出 NaOH 溶液（整个操作都要避免将空气带进溶液中），观察产物的颜色和状态。振荡后放置一段时间，观察又有何变化，并留作下面铁（Ⅲ），钴（Ⅲ），镍（Ⅲ）氧化性实验用。写出反应方程式。

（2）钴（Ⅱ）和镍（Ⅱ）的还原性

①往盛有 5 滴 $0.2\ mol\cdot L^{-1}$ $CoCl_2$ 和 5 滴 $0.2\ mol\cdot L^{-1}$ $NiSO_4$ 溶液的试管中注入氯水，观察有何现象。

②在两支盛有 5 滴 $0.2\ mol\cdot L^{-1}$ $CoCl_2$ 溶液的试管中分别注入 5 滴 $2.0\ mol\cdot L^{-1}$ NaOH 溶液，所得沉淀一份置于空气中，一份注入氯水，观察有何变化，第二份留作下面实验用。

③用 $NiSO_4$ 溶液按②实验，观察现象，第二份留作下面实验用。

## 2. 铁（Ⅲ）、钴（Ⅲ）和镍（Ⅲ）的氧化性

（1）在上面保留下来的氢氧化铁（Ⅲ）、氢氧化钴（Ⅲ）和氢氧化镍（Ⅲ）沉淀里各注入浓盐酸，振荡后有何变化？并用 KI-淀粉试纸检验所放出的气体。

$$2CoO(OH) + 6HCl == 2CoCl_2 + Cl_2\uparrow + 4H_2O$$

$$2NiO(OH) + 6HCl == 2NiCl_2 + Cl_2\uparrow + 4H_2O$$

（2）在上述制得的 $FeCl_3$ 溶液中滴入 5 滴 $0.1\ mol\cdot L^{-1}$ KI 溶液，再注入 5 滴四氯化碳，振荡后观察现象。写出反应方程式。

综合上述实验所观察到的现象，总结二价铁（Ⅱ）、钴（Ⅱ）、镍（Ⅱ）化合物的还原性和铁（Ⅲ）、钴（Ⅲ）、镍（Ⅲ）化合物的氧化性的变化规律。

## 3. 配合物的生成以及 $Fe^{2+}$，$Fe^{3+}$，$Co^{2+}$ 和 $Ni^{2+}$ 的鉴定方法

（1）铁的配合物

①往盛有 10 滴 $0.5\ mol\cdot L^{-1}$ $K_4[Fe(CN)_6]$ 溶液的试管中注入 5 滴碘水，振荡试管后滴入 3 滴 $0.5\ mol\cdot L^{-1}$ $(NH_4)_2Fe(SO_4)_2$ 溶液，有何现象发生？此为 $Fe^{2+}$ 的鉴定反应。

$$2[Fe(CN)_6]^{4-} + I_2 == 2[Fe(CN)_6]^{3-} + 2I^-$$

$$2[Fe(CN)_6]^{3-} + 3Fe^{2+} == Fe_3[Fe(CN)_6]_2\downarrow$$

②往盛有 5 滴 $0.5\ mol\cdot L^{-1}(NH_4)_2Fe(SO_4)_2$ 溶液的试管中滴入 3 滴碘水振荡后，将溶液分成两份。并各滴入 3 滴 $0.1\ mol\cdot L^{-1}$ KSCN 溶液，然后往其中一支试管注入 10 滴 3% $H_2O_2$ 溶液，观察现象。此为 $Fe^{3+}$ 的鉴定反应。

$$2Fe^{2+} + 2H^+ + H_2O_2 == 2Fe^{3+} + 2H_2O$$

$$Fe^{3+} + nSCN^- == [Fe(SCN)_n]^{3-n} \quad (n = 1 \sim 6)$$

试从配合物的生成对电极电势的改变来解释为什么$[Fe(CN)_6]^{4-}$能把 $I_2$ 还原成 $I^-$，而 $Fe^{2+}$ 则不能。

③在一支试管中滴入 5 滴 $0.2\ mol\cdot L^{-1}$ $FeCl_3$ 溶液，加入 $1\sim2$ 滴 $0.5\ mol\cdot L^{-1}$ $K_4[Fe(CN)_6]$溶液，观察现象。写出反应方程式。这也是鉴定 $Fe^{3+}$ 的一种常用方法。

④往盛有 5 滴 $0.2\ mol\cdot L^{-1}FeCl_3$ 的试管中，滴入过量浓 $NH_3\cdot H_2O$，观察沉淀是否溶解。

（2）钴的配合物

①往盛有 10 滴 $0.2\ mol\cdot L^{-1}CoCl_2$ 溶液的试管中加入绿豆大的固体 KSCN，观察固体周围的颜色，再注入 5 滴戊醇和 5 滴乙醚，振荡后观察水相和有机相的颜色。这个反应可用来鉴定钴（Ⅱ）离子。

②往 10 滴 $0.2\ mol\cdot L^{-1}CoCl_2$ 溶液中加入绿豆大的固体 $NH_4Cl$，然后滴入浓 $NH_3\cdot H_2O$ 至生成的沉淀刚好溶解为止。静置一段时间后，观察溶液的颜色有何变化。

$$CoCl_2 + NH_3 + H_2O == Co(OH)Cl\downarrow + NH_4Cl$$

$$Co(OH)Cl + H_2O + 7NH_3 == [Co(NH_3)_6](OH)_2 + NH_4Cl$$

$$2[Co(NH_3)_6](OH)_2 + 1/2O_2 + H_2O == 2[Co(NH_3)_6](OH)_3$$

（3）镍的配合物

往盛有 5 滴 $0.2\ mol\cdot L^{-1}NiSO_4$ 溶液的试管中，滴入浓 $NH_3\cdot H_2O$ 至生成的沉淀刚好溶解为止，观察现象。然后滴入 2～3 滴丁二酮肟（二乙酰二肟）试剂，有鲜红色内络盐沉淀生成，这是鉴定 $Ni^{2+}$ 的特征反应。

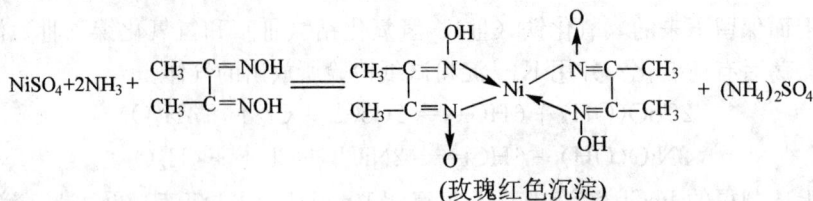

$$NiSO_4 + 2NH_3 + \begin{matrix} CH_3-C=NOH \\ | \\ CH_3-C=NOH \end{matrix} \Longrightarrow \begin{matrix} CH_3-C=N \\ | \\ CH_3-C=N \end{matrix} Ni \begin{matrix} N=C-CH_3 \\ | \\ N=C-CH_3 \end{matrix} + (NH_4)_2SO_4$$

（玫瑰红色沉淀）

### 4. 铁的腐蚀和防腐

（1）铁的腐蚀

在两支试管中各加约 1 mL 水，各滴入 2 滴 HCl 溶液和 2～3 滴 $K_3[Fe(CN)_6]$ 溶液，然后将两根 5 cm 长的细铁丝（铁丝用砂纸擦去氧化膜）分别缠绕在锌粒和锡粒上（均用砂纸擦净），并将其分别投入到这两支试管中。数分钟后，观察试管溶液的颜色，应用 Fe，Zn 和 Sn 的电位顺序分析上述所发生的反应。试说明白口铁（镀锌铁）和马口铁（镀锡铁）的腐蚀机理。

（2）铁的防腐

氧化膜保护区：把擦净了的小铁钉先用 $2\ mol\cdot L^{-1}$ 盐酸浸泡片刻后，取出用水冲洗去铁钉表面的酸。取 1～2 mL 碱性亚硝酸钠混合液（1 升溶液中含有 600 g NaOH 和 60 g $NaNO_2$ 配制成的混合液）于一支试管中加热至沸，将光亮的小铁钉浸在其中，几分钟后取出铁钉，观察现象。其反应方程式为：

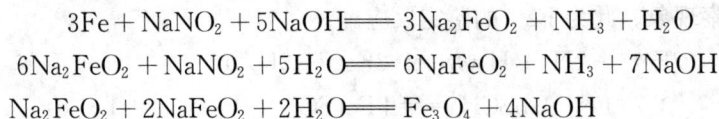

$$3Fe + NaNO_2 + 5NaOH \Longrightarrow 3Na_2FeO_2 + NH_3 + H_2O$$
$$6Na_2FeO_2 + NaNO_2 + 5H_2O \Longrightarrow 6NaFeO_2 + NH_3 + 7NaOH$$
$$Na_2FeO_2 + 2NaFeO_2 + 2H_2O \Longrightarrow Fe_3O_4 + 4NaOH$$

## 思考题

1. 今有一瓶含有 $Fe^{3+}$，$Cr^{3+}$ 和 $Ni^{2+}$ 离子的混合液，设计出分离示意图，将它们分离并加以鉴定。

2. 有一浅绿色的晶体 A，可溶于水制得 B，将 B 注入饱和 $NaHCO_3$ 溶液，有白色沉淀 C 和气体 D 生成。C 在空气中逐渐变为棕色，将气体 D 通入澄清的石灰水会变浑浊。

若将溶液 B 加以酸化，再滴加一滴紫红色溶液 E，则得到浅黄色溶液 F，于 F 中注入黄血盐溶液，立即产生深蓝色的沉淀 G。

若溶液 B 中注入 $BaCl_2$ 溶液，有白色沉淀 H 析出，此沉淀不溶于强酸。

问 A、B、C、D、E、F、G、H 分别是什么物质？写出分子式，并写出有关的反应方程式。

# Exp 26　Iron, Cobalt and Nickel

## Objectives

1. To test and master the reducibility of iron (+2), cobalt (+2) and nickel (+2) and the oxidizability of iron (+3), cobalt (+3) and nickel (+3).

2. To master the preparative methods of coordination compounds of iron, cobalt and nickel and learn the identification methods of $Fe^{2+}$, $Fe^{3+}$, $Co^{2+}$ and $Ni^{2+}$.

3. To know the principle of iron eroding and the method of protecting it from eroding.

## Instruments, Reagents and Materials

Instruments: test tube, centrifuge tube, mini reagent spoon

Solid reagents: ammonium ferrous sulfate, potassium thiocyanide, ammonium chloride, zinc sheet, tin sheet (zinc grain or tin grain), potassium iodide-starch test paper.

Liquid reagents: $H_2SO_4$ (1:1, 1 mol·$L^{-1}$), HCl(6 mol·$L^{-1}$, 5%, concentrated), NaOH (6 mol·$L^{-1}$, 2 mol·$L^{-1}$), $(NH_4)_2Fe(SO_4)_2$ (0.2 mol·$L^{-1}$, 0.5 mol·$L^{-1}$), $CoCl_2$ (0.2 mol·$L^{-1}$), $NiSO_4$(0.2 mol·$L^{-1}$), KI(0.1 mol·$L^{-1}$), $K_4[Fe(CN)_6]$(0.5 mol·$L^{-1}$), $NH_3$·$H_2O$(concentrated), chlorine water, iodine water, carbon tetrachloride, pentanol, ether, $H_2O_2$(3%), diacetyldioxime, mixed liquid(containing 600 g NaOH and 60 g $NaNO_2$ in one liter of solution), KSCN(0.1 mol·$L^{-1}$), $FeCl_3$(0.2 mol·$L^{-1}$), $K_3[Fe(CN)_6]$(0.5 mol·$L^{-1}$)

Materials: sand paper, iron hail, clip, thin iron wire

## Experimental Procedures

**1. The reducibility of iron ( Ⅱ ), cobalt ( Ⅱ ) and nickel ( Ⅱ )**

(1) Reducibility of iron ( Ⅱ )

①Add 3 drops of 1:1 $H_2SO_4$ and 5 drops of chlorine water to a test tube, shake it, then add 5 drops of 0.2 mol·$L^{-1}$ $(NH_4)_2Fe(SO_4)_2$ solution, observe the phenomena, write down the related reaction equation.

②Add 10 drops of distilled water and 5 drops of 1 mol·$L^{-1}$ $H_2SO_4$ into a test tube, heat to boil in order to drive off the air, then add $(NH_4)_2Fe(SO_4)_2$ crystal of a mung bean size to it. Add 10 drops of 6mol·$L^{-1}$ NaOH solution in another test tube, heat to boil (why?). Extract NaOH solution with a long burette after cooling down, then insert it into $(NH_4)_2Fe(SO_4)_2$ solution of the former test tube (till to bottom of test tube), slowly drop NaOH (preventing air from entering the solution in the whole procedure). Observe the color and state of product. Stand for a while after shaking the tube, observe the change. Write down the related reaction equation. Conserve the product for the following experiment.

(2) Reducibility of cobalt（Ⅱ）, nickel（Ⅱ）

①Add 5 drops of 0.2 $mol \cdot L^{-1}$ $CoCl_2$ and 5 drops of 0.2 $mol \cdot L^{-1}$ $NiSO_4$ into a test tube, then add chlorine water into it, observe the phenomenon.

②Add 5 drops of 2 $mol \cdot L^{-1}$ NaOH to two test tubes containing 5 drops of 0.2 $mol \cdot L^{-1}$ $CoCl_2$ respectively, one portion of precipitate obtained is laid in air, the other is added to chlorine water, observe the phenomena. Conserve the latter product for the following experiment.

③Repeat the experiment, replacing $CoCl_2$ with $NiSO_4$ solution according to experiment procedure ② observe the phenomena. Conserve the latter product for the following experiment.

## 2. Oxidizability of iron（Ⅲ）, cobalt（Ⅲ）, nickel（Ⅲ）

(1) Add concentrated HCl to the above conserved precipitate of $Fe(OH)_3$, $Co(OH)_3$ and $Ni(OH)_3$, what change will take place after shaking? Test the gas by potassium iodide-starch test paper.

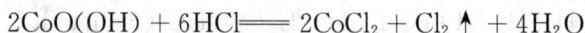

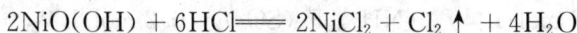

$$2CoO(OH) + 6HCl = 2CoCl_2 + Cl_2 \uparrow + 4H_2O$$
$$2NiO(OH) + 6HCl = 2NiCl_2 + Cl_2 \uparrow + 4H_2O$$

(2) Add 5 drops of 0.1 $mol \cdot L^{-1}$ KI solution to above $FeCl_3$ solution prepared, add 5 drops of $CCl_4$, observe the phenomena after shaking.

Sum up the change rule of the reducibility of $Fe^{2+}$, $Co^{2+}$ and $Ni^{2+}$ and the oxidizability of $Fe^{3+}$, $Co^{3+}$ and $Ni^{3+}$ on the basis of above experimental phenomena.

## 3. The preparation of coordination compounds and the identification methods of $Fe^{2+}$, $Fe^{3+}$, $Co^{2+}$ and $Ni^{2+}$

(1) Coordination compounds of iron

①Add 10 drops of 0.5 $mol \cdot L^{-1}$ $K_4[Fe(CN)_6]$ to a test tube, inject 5 drops of iodine water into it, then add 3 drops of 0.5 $mol \cdot L^{-1}$ $(NH_4)_2Fe(SO_4)_2$ solution after shaking, observe the phenomenon. This is the identification reaction of $Fe^{2+}$.

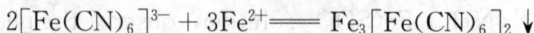

$$2[Fe(CN)_6]^{4-} + I_2 = 2[Fe(CN)_6]^{3-} + 2I^-$$
$$2[Fe(CN)_6]^{3-} + 3Fe^{2+} = Fe_3[Fe(CN)_6]_2 \downarrow$$

②Add 5 drops of 0.5 $mol \cdot L^{-1}$ $(NH_4)_2Fe(SO_4)_2$ solution into a test tube, inject 3 drops of iodine water into it, shake it, then divide the solution into two portions, add 3 drops of 0.1 $mol \cdot L^{-1}$ KSCN into the two test tubes, inject 10 drops of 3% $H_2O_2$ solution in a test tube, observe the phenomena.

This is the identification reaction of $Fe^{3+}$.

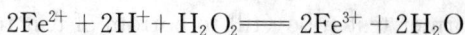

$$2Fe^{2+} + 2H^+ + H_2O_2 = 2Fe^{3+} + 2H_2O$$
$$Fe^{3+} + nSCN^- = [Fe(SCN)_n]^{3-n} \quad (n = 1 \sim 6)$$

Try to explain why can $[Fe(CN)_6]^{4-}$ reduce $I_2$ to $I^-$, but $Fe^{2+}$ can't reduce $I_2$ with the knowledge of electrode potential changed for formation of coordination compounds.

③Add 5 drops of 0. 2 mol·L$^{-1}$ FeCl$_3$ into a test tube, then add 1~2 drops of 0. 5 mol·L$^{-1}$ K$_4$[Fe(CN)$_6$] to it, observe the phenomena. This is also the identification method of Fe$^{3+}$.

④Add 5 drops of 0. 1 mol·L$^{-1}$ FeCl$_3$ into a test tube, then add excess concentrated NH$_3$·H$_2$O into it, observe whether precipitate can be dissolved.

(2) Coordination compounds of cobalt

①Add solid KSCN as a mung bean to the test tube containing 10 drops of 0. 2 mol·L$^{-1}$ CoCl$_2$. Observe the color of solid's surrounding, inject 5 drops of pentanol and ether. Observe the color of water and organic phase after shaking. This reaction can be used to identify Co (Ⅱ).

②Add 10 drops of 0. 2 mol·L$^{-1}$ CoCl$_2$ into a test tube, add solid NH$_4$Cl as a mung bean into it. Then add concentrated NH$_3$·H$_2$O till precipitate just being dissolved. Observe the color of the solution after laying for a period of time.

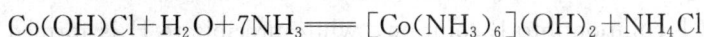

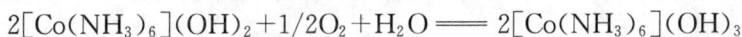

$$CoCl_2 + NH_3 + H_2O \rightleftharpoons Co(OH)Cl\downarrow + NH_4Cl$$
$$Co(OH)Cl + H_2O + 7NH_3 \rightleftharpoons [Co(NH_3)_6](OH)_2 + NH_4Cl$$
$$2[Co(NH_3)_6](OH)_2 + 1/2O_2 + H_2O \rightleftharpoons 2[Co(NH_3)_6](OH)_3$$

(3) Coordination compounds of nickel

Add 5 drops of 0. 2 mol·L$^{-1}$ NiSO$_4$ to a test tube, add concentrated NH$_3$·H$_2$O till precipitate just being dissolved, observe the phenomena. Then add 2~3 drops of diacetyldioxime reagent, observe the forming of turkey red coordination compound precipitate. This reaction can be used to identify Ni$^{2+}$.

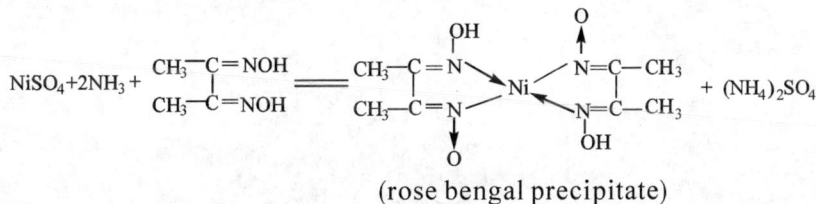

(rose bengal precipitate)

## 4. Iron eroding and protecting from eroding

(1) Iron eroding

Add 1 mL water, then 2 drops of 2 mol·L$^{-1}$ HCl solution and 2~3 drops of K$_3$[Fe(CN)$_6$] solution to the two test tubes. Then plunge two stips of thin iron with 5 cm length (oxidation film is scraped by sand paper) wrapped zinc grain and lead grain (oxidation film is scraped by sand paper) into the two test tubes respectively. Observe the color of solution in the test tube, analyze above reaction through electric potential order of Fe, Zn and Sn after a few minutes. Try to explain the iron eroding mechanism of white iron (galvanization iron) and tin plate (tinning iron).

(2) Corrosion protection of iron

Protection section of oxidation film: dip a small scraped iron nail in 2 mol·L$^{-1}$ HCl for a minute, take out and clear up the acid on its surface. Heat 1~2 mL alkaline NaNO$_2$ mixture (the mixed solution prepared with 600 g NaOH and 60 g NaNO$_2$ in one liter solution) to boil. Dip the bright iron nail into it, take out after a few minutes, observe the phenomena.

Its reaction equations are as follows:

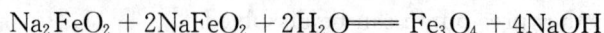

$$3Fe + NaNO_2 + 5NaOH = 3Na_2FeO_2 + NH_3 + H_2O$$
$$6Na_2FeO_2 + NaNO_2 + 5H_2O = 6NaFeO_2 + NH_3 + 7NaOH$$
$$Na_2FeO_2 + 2NaFeO_2 + 2H_2O = Fe_3O_4 + 4NaOH$$

## Questions

1. Design the separating sketch map for a mixed solution containing $Fe^{3+}$, $Cr^{3+}$ and $Ni^{2+}$. Separate and identify them.

2. Light green crystal A can dissolve in water to form B, white precipitate C and gas D is formed when adding saturated $NaHCO_3$ solution to B. C gradually becomes brown in air, limewater will become muddy when gas D is ventilated.

Add a drop of mauve solution E to acidified solution B, light yellow solution F is obtained, inject potassium ferrocyanide solution into F, bring dark blue precipitation G.

Inject $BaCl_2$ solution into B solution, white precipitate H is formed which is indissolvable in strong acid.

What are A, B, C, D, E, F, G, H respectively? Write down molecular formulas and related reaction equations.

# 实验二十七　配位化合物

## 实验目的

1. 试验并了解配位化合物的形成和组成。
2. 试验并了解中心离子、配合剂对配位化合物稳定性的影响。
3. 了解配合平衡与沉淀反应、氧化还原反应以及溶液的酸碱度的关系。
4. 练习生成配合物及检验其性质的操作。

## 实验原理

配位化合物是由形成体（中心离子或原子）和一定数目的配位体（阴离子或分子）以配位键相结合形成的具有一定的组成和空间构型的复杂化合物。在配离子溶液中存在配合—离解平衡：

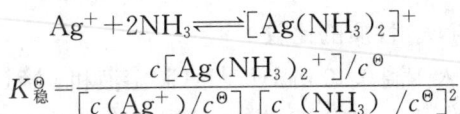

$$Ag^+ + 2NH_3 \rightleftharpoons [Ag(NH_3)_2]^+$$

$$K_{稳}^\ominus = \frac{c[Ag(NH_3)_2^+]/c^\ominus}{[c(Ag^+)/c^\ominus][c(NH_3)/c^\ominus]^2}$$

式中 $K_{稳}^\ominus$ 称为稳定常数，不同的配合物具有不同的稳定常数，对于同种类型的配合物，$K_{稳}^\ominus$ 越大，表示配离子越稳定。

根据平衡移动原理，条件改变后，配合平衡将发生移动，如加入沉淀剂、改变溶液的浓度以及改变溶液的酸碱性，配合平衡都将发生移动。

由简单物质形成配合物时，物质的性质将发生改变，如稳定性、颜色、溶解性、酸碱性以及氧化还原性等。

## 实验仪器、药品

仪器：离心机，离心试管，酒精灯，小试管（规格 mm：10×100），小玻棒等

酸：$H_2SO_4$（1:1），HCl（2.0 mol·L$^{-1}$，浓），$H_3BO_3$（0.10 mol·L$^{-1}$）

碱：NaOH（2.0 mol·L$^{-1}$），$NH_3 \cdot H_2O$（1.0，2.0，6.0 mol·L$^{-1}$）

盐：$CuSO_4$（0.10 mol·L$^{-1}$），$BaCl_2$（0.10 mol·L$^{-1}$），$Hg(NO_3)_2$（0.10 mol·L$^{-1}$），KI（0.10 mol·L$^{-1}$），$FeCl_3$（0.10 mol·L$^{-1}$），KSCN（0.10，1.0 mol·L$^{-1}$），NaF（0.10，1.0 mol·L$^{-1}$），$NH_4Cl$（1.0 mol·L$^{-1}$），$CrCl_3$（0.10 mol·L$^{-1}$），NaCl（0.10，1.0 mol·L$^{-1}$），$AgNO_3$（0.10 mol·L$^{-1}$），KBr（0.10 mol·L$^{-1}$），$Na_2S_2O_3$（0.10 mol·L$^{-1}$），$K_3[Fe(CN)_6]$（0.10 mol·L$^{-1}$），$ZnSO_4$（0.10 mol·L$^{-1}$），$CdSO_4$（0.10 mol·L$^{-1}$），$FeSO_4$（0.10 mol·L$^{-1}$），$K_4[Fe(CN)_6]$（0.10 mol·L$^{-1}$），$Na_2S$（0.10 mol·L$^{-1}$），明矾（s），$Na_3[Co(NO_2)_6]$（饱和溶液），$CoCl_2$（0.10，0.50 mol·L$^{-1}$），EDTA 二钠盐（0.10 mol·L$^{-1}$），$NiCl_2$（或 $NiSO_4$）（0.10 mol·L$^{-1}$）

其他药品：$CCl_4$，无水乙醇，戊醇，丁二酮肟溶液，邻菲罗啉（25%），甲基橙溶液（0.1%），甘油，硬水

## 实验内容

### 1. 配离子的生成和组成

(1) 在三支小试管中分别加入 3 滴 $0.10\ mol \cdot L^{-1} CuSO_4$ 溶液，分别小心逐滴滴加 $1.0\ mol \cdot L^{-1} NH_3 \cdot H_2O$，观察浅蓝色 $Cu_2(OH)_2SO_4$ 沉淀的生成。继续滴加氨水，直至沉淀完全溶解，再加 1 滴氨水，观察溶液的颜色。

在其中一份溶液中加入 1 滴 $0.10\ mol \cdot L^{-1} BaCl_2$ 溶液，另一份加入 1 滴 $0.1\ mol \cdot L^{-1}$ NaOH 溶液，观察现象。在第三支试管中加入 2 滴无水乙醇，摇动试管，观察现象。离心分离，观察晶体的性状。

(2) 在小试管中加入 1 滴 $0.10\ mol \cdot L^{-1} Hg(NO_3)_2$ 溶液，再加入 $1\sim2$ 滴 $0.10\ mol \cdot L^{-1}$ KI 溶液，观察红色 $HgI_2$ 沉淀的生成，继续滴加 KI，直至沉淀完全溶解，观察现象。

根据以上实验，分析说明两种配合物溶液的内界和外界的组成。

### 2. 配离子和简单离子性质的比较

(1) $FeCl_3$ 与 $K_3[Fe(CN)_6]$ 性质的比较

往两支小试管中分别加入 1 滴 $0.10\ mol \cdot L^{-1} FeCl_3$ 溶液和 1 滴 $0.10\ mol \cdot L^{-1}\ K_3[Fe(CN)_6]$ 溶液，然后各加入 1 滴 $0.10 mol \cdot L^{-1} KSCN$ 溶液，观察有何变化？两种化合物中都有 $Fe^{3+}$，为什么实验结果不同？

(2) $FeSO_4$ 与 $K_4[Fe(CN)_6]$ 性质的比较

在两支小试管中分别加入 1 滴 $0.10\ mol \cdot L^{-1} FeSO_4$ 溶液和 1 滴 $0.10\ mol \cdot L^{-1}\ K_4[Fe(CN)_6]$ 溶液，然后各加入 1 滴 $0.10\ mol \cdot L^{-1} Na_2S$ 溶液，是否都有 FeS 沉淀生成？为什么？

(3) 取少量明矾 $[K_2SO_4 \cdot Al_2(SO_4)_3 \cdot 24H_2O]$ 晶体，放入试管中用蒸馏水溶解，分别用 $Na_3[Co(NO_2)_6]$ 饱和溶液、$2.0\ mol \cdot L^{-1} NaOH$ 溶液、$0.10\ mol \cdot L^{-1} BaCl_2$ 溶液检出其中的 $K^+$，$Al^{3+}$ 和 $SO_4^{2-}$ 离子。

比较以上实验结果，配离子和简单离子之间，配合物和复盐之间有何区别？

### 3. 配合物的稳定性

(1) 中心离子对配合物稳定性的影响

在两支小试管中分别加入 2 滴 $0.10\ mol \cdot L^{-1} ZnSO_4$ 和 $0.10\ mol \cdot L^{-1} CdSO_4$ 溶液，再滴加 $2.0\ mol \cdot L^{-1} NH_3 \cdot H_2O$，直到最初生成的沉淀溶解为止，比较所需氨水量的多少。每支试管中再加入过量氨水 1 滴，然后各加入 1 滴 $2.0\ mol \cdot L^{-1} NaOH$ 溶液，是否都有沉淀生成？

(2) 配位剂对配合物稳定性的影响

在三支小试管中各加入 1 滴 $0.10\ mol \cdot L^{-1} FeCl_3$ 溶液，在第一支试管中加入 1 滴 $0.10 mol \cdot L^{-1}$ 的 NaF 溶液，在第二支试管中加入 1 滴 $0.10\ mol \cdot L^{-1} KSCN$ 溶液，在第三支试管中加入 1 滴 $0.10\ mol \cdot L^{-1} KSCN$ 溶液后再滴加 $0.10\ mol \cdot L^{-1} NaF$ 溶液，直至红色褪去，解释观察到的现象。

**4. 配合物的水合异构现象**

（1）在小试管中加入 $2\sim3$ 滴 $0.10\ mol\cdot L^{-1}CrCl_3$ 溶液，加热试管，观察溶液颜色的变化，然后将溶液冷却，溶液的颜色又有何变化？

（2）在另一支小试管中加入 2 滴 $0.50\ mol\cdot L^{-1}CoCl_2$ 溶液，重复以上实验，观察溶液颜色的变化。

$[Cr(H_2O)_6]^{3+}$ 和 $[Co(H_2O)_6]^{2+}$ 离子的水合异构反应为：

$$[Cr(H_2O)_6]^{3+}+2Cl^-\rightleftharpoons[Cr(H_2O)_4Cl_2]^++2H_2O$$

$$[Co(H_2O)_6]^{2+}+4Cl^-\rightleftharpoons[Co(H_2O)_2Cl_4]^{2-}+4H_2O$$

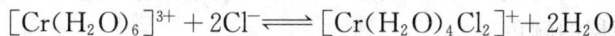

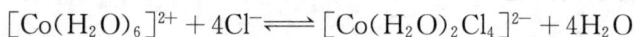

**5. 配位平衡的移动**

（1）配位平衡与沉淀反应

在小试管中加入 1 滴 $0.10\ mol\cdot L^{-1}NaCl$ 溶液和 1 滴 $0.10\ mol\cdot L^{-1}AgNO_3$ 溶液，振荡试管，观察沉淀，再滴加 $2.0\ mol\cdot L^{-1}\ NH_3\cdot H_2O$ 至沉淀刚好溶解，然后加 1 滴 $0.10\ mol\cdot L^{-1}NaCl$ 溶液，看有无沉淀生成？再加 1 滴 $0.1\ mol\cdot L^{-1}KBr$，观察有无 $AgBr$ 沉淀生成？加入几滴 $0.1\ mol\cdot L^{-1}Na_2S_2O_3$ 溶液，沉淀是否溶解？然后加入 1 滴 $0.10\ mol\cdot L^{-1}KBr$，观察有无 $AgBr$ 生成。再加 1 滴 $0.10\ mol\cdot L^{-1}KI$ 溶液，有无沉淀生成？解释以上沉淀生成及溶解的原因。

（2）配位平衡与氧化还原反应

在两支小试管中分别加入 2 滴 $0.10\ mol\cdot L^{-1}FeCl_3$ 溶液和几滴 $CCl_4$，振荡，观察 $CCl_4$ 层颜色。在第一支试管中滴加 $0.10\ mol\cdot L^{-1}KI$ 溶液，振荡后观察 $CCl_4$ 层颜色，解释现象。在第二支试管中滴加 $0.10\ mol\cdot L^{-1}NaF$ 溶液至无色，然后加入与第一支试管加的滴数相同的 $0.10\ mol\cdot L^{-1}KI$ 溶液，振荡后观察 $CCl_4$ 层颜色，比较两支试管有何不同？

（3）配位平衡与介质的酸碱性

在两支试管中各加入 1 滴 $0.10\ mol\cdot L^{-1}FeCl_3$ 溶液，再加入 $5\sim6$ 滴 $0.10\ mol\cdot L^{-1}NaF$ 至溶液变为无色，然后在一支试管中加 1 滴 $0.1\ mol\cdot L^{-1}NaOH$，另一支试管中加 1 滴 $0.10\ mol\cdot L^{-1}KSCN$ 后加 1 滴 $1:1H_2SO_4$，观察现象。

（4）配位平衡与配合剂的浓度

在试管中加入 1 滴 $0.10\ mol\cdot L^{-1}CoCl_2$ 溶液，再加入几滴浓 $HCl$，观察溶液颜色的变化，然后加水稀释，至溶液刚变为粉红色时，停止加水，再滴加浓 $HCl$，观察溶液颜色变化。

**6. 螯合物的形成和应用**

（1）在试管中加入硬水约 $1\ mL$，在酒精灯上加热煮沸约 $1\ min$，观察水中有无沉淀或混浊。取另一支试管，加入硬水 $1\ mL$ 后，再加入几滴 $0.10\ mol\cdot L^{-1}EDTA$ 二钠盐溶液，加热煮沸约 $1\ min$，是否有浑浊现象？为什么？

（2）在试管中加入 $0.10\ mol\cdot L^{-1}H_3BO_3$ 溶液 2 滴，加入甲基橙指示剂 1 滴，观察颜色变化，再加入 2 滴甘油，摇匀后，颜色又有何变化？

**7. 配合物在分析化学中的应用**

（1）利用形成带色配合物来鉴定某些离子

①$Fe^{2+}$的鉴定：在试管中加入 5 滴 0.10 mol·$L^{-1}Fe^{2+}$离子溶液，再加入 2 滴 25％邻菲罗啉溶液，观察现象。

②$Ni^{2+}$离子的鉴定：在试管中加入 1 滴 0.10 mol·$L^{-1}Ni^{2+}$溶液及 5 滴水，再加入 2.0 mol·$L^{-1}$氨水，使溶液呈碱性，然后加入 5 滴丁二酮肟溶液，观察现象。

（2）利用配合物掩蔽干扰离子

在试管中加入 0.10 mol·$L^{-1}FeCl_3$溶液 1 滴和 0.50 mol·$L^{-1}CoCl_2$溶液 1 滴，1.0 mol·$L^{-1}KSCN$溶液 10 滴，再逐滴加入 0.10 mol·$L^{-1}NaF$溶液至无色，再加戊醇 5 滴，振荡试管，静置，观察戊醇层的颜色。

## 思考题

1. 怎样根据实验来推测配合物的结构？结合本实验举例说明。

2. Cu 能从 $Hg^{2+}$ 盐中置换出 Hg，能否从 $[Hg(CN)_4]^{2-}$ 中置换出 Hg？为什么？

3. 影响配合平衡的因素有哪些？

# Exp 27　Coordination Compounds

## Objective

1. To test and learn about the formation and the composition of the coordination compounds.

2. To learn the influence of the central ion and compounding agent on coordination compounds' stability.

3. To learn the relationship between the coordination balance and precipitate reaction, oxidation-reduction reaction, and solution's pH value.

4. To practice the operation of the formation of the coordination compound and test its property.

## Principles

Coordination compounds is complex compound with certain formation and space configuration, which is formed by coordination bond from organisator (central ion or atom) and certain number of ligands (anion or molecular). There exists a balance of coordination-dissociation in the solution of coordination ion.

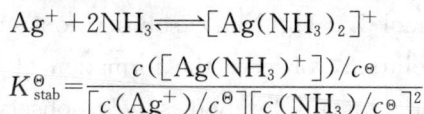

$$Ag^+ + 2NH_3 \rightleftharpoons [Ag(NH_3)_2]^+$$

$$K_{stab}^{\ominus} = \frac{c([Ag(NH_3)^+])/c^{\ominus}}{[c(Ag^+)/c^{\ominus}][c(NH_3)/c^{\ominus}]^2}$$

$K_{stab}^{\ominus}$ is the stability constant, different coordination compounds have different stability constants, for the same type of coordination compounds, the higher the $K_{stab}^{\ominus}$ is, the more stable the coordination ion is.

According to the principle of equilibrium shifting, coordination balance is broken when the condition changes, for example , add precipitant to it, change the concentration or pH value of the solution will make the coordination balance shift.

The property of the compound will change a lot when the coordination compounds are formed by simple materials, such as the stability, color, solubility, acidity or alkalinity, oxidation-reduction quality, *etc*.

## Instruments and Reagents

Instruments: centrifugal machine, centrifuge test tube , alcohol burner, small test tube (Spec: $10 \times 100$ mm) , small glass rod and so on.

Acid: $H_2SO_4$(1:1), HCl(2. 0 mol·$L^{-1}$, conc. ), $H_3BO_3$(0. 1 mol·$L^{-1}$)

Base: NaOH(2. 0 mol·$L^{-1}$), $NH_3 \cdot H_2O$(1. 0, 2. 0, 6. 0 mol·$L^{-1}$)

Salt : $CuSO_4$(0. 10 mol·$L^{-1}$), $BaCl_2$(0. 10 mol·$L^{-1}$), $Hg(NO_3)_2$(0. 10 mol·$L^{-1}$), KI(0. 10 mol·$L^{-1}$), $FeCl_3$(0. 10 mol·$L^{-1}$), KSCN(0. 10, 1. 0 mol·$L^{-1}$), NaF(0. 10, 1. 0

mol·L$^{-1}$), NH$_4$Cl(1.0 mol·L$^{-1}$), CrCl$_3$ (0.10 mol·L$^{-1}$), NaCl(0.10, 1.0 mol·L$^{-1}$), AgNO$_3$(0.10 mol·L$^{-1}$), KBr(0.10 mol·L$^{-1}$), Na$_2$S$_2$O$_3$(0.10 mol·L$^{-1}$), K$_3$[Fe(CN)$_6$] (0.10 mol·L$^{-1}$), ZnSO$_4$(0.10 mol·L$^{-1}$), CdSO$_4$(0.10 mol·L$^{-1}$), FeSO$_4$(0.10 mol·L$^{-1}$), K$_4$[Fe(CN)$_6$](0.10 mol·L$^{-1}$), Na$_2$S(0.10 mol·L$^{-1}$), alum crystal(s), Na$_3$[Co(NO$_2$)$_6$] (saturated), CoCl$_2$(0.10, 0.50 mol·L$^{-1}$), alum crystal(s)(saturated solution), EDTA disodium salt

Others: carbon tetrachloride, absolute alcohol, pentanol, diacetyldioxime solution, o-phenanthroline (25%), Lung's indicator solution (0.1%), glycerol, hard water

## Experimental Procedures

### 1. Formation and composition of coordination compounds

(1) Add 3 drops of 0.10 mol·L$^{-1}$ CuSO$_4$ solution to three small test tubes respectively. Then drop 1.0 mol·L$^{-1}$ NH$_3$·H$_2$O carefully, observe the formation of light blue Cu$_2$ (OH)$_2$SO$_4$ precipitate. Go on dropping ammonia water until the precipitate is dissolved completely, add one more drop of ammonia water in it, observe the color of the solution.

Add 1 drop of 0.1 mol·L$^{-1}$ BaCl$_2$ solution to the first test tube, add 1 drop of 0.1 mol· L$^{-1}$ NaOH solution to the second test tube and observe the phenomenon.

Add 2 drops of absolute ethanol to the third test tube, shake the test tube, observe the phenomenon. Then filter, observe the appearance of the crystal.

(2) Add 1 drop of 0.1 mol·L$^{-1}$ Hg(NO$_3$)$_2$ solution to a small test tube, then add 1~2 drops of 0.1 mol·L$^{-1}$ KI solution, observe the formation of red HgI$_2$ precipitate, go on dropping KI until the precipitate is absolutely dissolved, observe the phenomenon.

According to above experiments, please analyze and illustrate the composition of metal state and outside of the two kinds of coordination compound.

### 2. Comparison the property of complex ion with simple ion

(1) Comparison between the property of FeCl$_3$ with K$_3$[Fe(CN)$_6$]

Add 1 drop of 0.1 mol·L$^{-1}$ FeCl$_3$ solution and 1 drop of 0.1 mol·L$^{-1}$ K$_3$[Fe(CN)$_6$]solution to two small test tubes respectively, then add 1 drop of 0.1 mol·L$^{-1}$KSCN solution respectively, observe what changes does it have.

In these two compounds both have Fe (Ⅲ) ion, but why the experiment results are different?

(2) Comparison between the property of FeSO$_4$ with K$_4$[Fe(CN)$_6$]

Add 1 drop of 0.1 mol·L$^{-1}$ FeSO$_4$ solution and 1 drop of 0.1 mol·L$^{-1}$ K$_4$[Fe(CN)$_6$] solution to two small test tubes respectively, then add 1 drop of 0.1 mol·L$^{-1}$ Na$_2$S solution respectively, whether there is FeS precipitate from both solutions or not. Why?

(3) Dissolve a little alum crystal [K$_2$SO$_4$·Al$_2$(SO$_4$)$_3$·24H$_2$O] with distilled water in a test tube, identify the K$^+$, Al$^{3+}$, SO$_4{}^{2-}$ by saturated Na$_3$[Co(NO$_2$)$_6$] solution, 2.0 mol· L$^{-1}$ NaOH solution, 0.10 mol·L$^{-1}$BaCl$_2$ respectively.

　　Compare the above experimental results, what are the differences between the complex ion and simple ion, coordination compound and the complex salt ?

### 3. The stability of coordination compounds

　　(1) The influence of central ion on the stability of coordination compounds

　　Add 2 drops of 0. 10 mol·L$^{-1}$ ZnSO$_4$ solution and 0. 10 mol·L$^{-1}$ CdSO$_4$ solution into two test tubes respectively, then add 2. 0 mol·L$^{-1}$ NH$_3$·H$_2$O until the newly formed precipitate is absolutely dissolved which was formed at first. Compare the amount of NH$_3$·H$_2$O needed. Add one excess drop of NH$_3$·H$_2$O respectively, then add 1 drop of 2. 0 mol·L$^{-1}$ NaOH solution respectively, whether there are precipitate in the solutions?

　　(2) The influence of chelant on the stability of coordination compounds

　　Add 1 drop of 0. 10 mol·L$^{-1}$ FeCl$_3$ to three small test tubes respectively, add 1 drop of 0. 10 mol·L$^{-1}$ NaF solution to the first test tube, add 1 drop of 0. 10 mol·L$^{-1}$ KSCN, to the second test tube, add 1 drop of 0. 10 mol·L$^{-1}$ KSCN to the third test tube, and then add 0. 10 mol·L$^{-1}$ NaF until the red color fades, observe the phenomenon.

### 4. The hydrate isomerism of coordination compounds

　　(1) Add 2~3 drops of 0. 10 mol·L$^{-1}$ CrCl$_3$ solution to the small test tubes, heat the test tubes, observe the color of the solutions , then make the solution cool down, observe the color change.

　　(2) Add 2 drops of 0. 50 mol·L$^{-1}$ CoCl$_2$ solution to another small test tube, repeat the above experiment, observe the color change. The hydrate isomerism reactions between [Cr(H$_2$O)$_6$]$^{3+}$ and [Co(H$_2$O)$_6$]$^{2+}$ are like following:

$$[Cr(H_2O)_6]^{3+} + 2Cl^- \rightleftharpoons [Cr(H_2O)_4Cl_2]^+ + 2H_2O$$
　　（blue purple）　　　　　　　　　　（green）
$$[Co(H_2O)_6]^{2+} + 4Cl^- \rightleftharpoons [Co(H_2O)_4Cl_2]^{2-} + 4H_2O$$
　　（violet red）　　　　　　　　（blue purple）

### 5. The movement of complexation equilibrium

　　(1) Complexation equilibrium and precipitate reaction

　　Add 1 drop of 0. 10 mol·L$^{-1}$ NaCl solution and 1 drop of 0. 10 mol·L$^{-1}$ AgNO$_3$ to a small test tube, shake the test tube, observe the precipitate, then add 2. 0 mol·L$^{-1}$ NH$_3$·H$_2$O until the precipitate is exactly dissolved completely, then add 1 drop of 0. 10 mol·L$^{-1}$ NaCl, whether is there precipitate formed? Add 1 drop of 0. 1 mol·L$^{-1}$ KBr solution to the above solution, whether the precipitate is dissolved, then add 1 drop of 0. 10 mol·L$^{-1}$ KBr, and observe whether there is precipitate AgBr in the solution. Add 1 drop of 0. 10 mol·L$^{-1}$ KI solution, whether there is precipitate? Explain the reason that the precipitate formed and dissolved.

　　(2) Complexation equilibrium and oxidation-reduction reaction

　　Add 2 drop of 0. 10 mol·L$^{-1}$ FeCl$_3$ solution and a few drops of CCl$_4$ to two small test tubes respectively, shake and observe the color of the $CCl_4$ layer. Add $0.10$ mol·$L^{-1}$ KI solution to the first test tube until the solution become umber, shake and observe the color of $CCl_4$ layer, explain the phenomenon. Add $0.10$ mol·$L^{-1}$ NaF solution until the solution become colorless, then add $0.10$ mol·$L^{-1}$ KI solution (the amount of KI is the same as the first test tube), shake and observe the color of $CCl_4$ layer, compare the differences between the two above test tubes.

(3) Complexation equilibrium and pH value of solution

Add 1 drop of $0.10$ mol·$L^{-1}$ $FeCl_3$ solution to two small test tubes respectively, then add $5\sim6$ drops of $0.10$ mol·$L^{-1}$ NaF until the solution become colorless, after adding 1 drops of $0.1$ mol·$L^{-1}$ NaOH solution to one test tube, add 1 drop of $0.10$ mol·$L^{-1}$ KSCN to another one, add $1:1$ $H_2SO_4$, observe the phenomenon.

(4) Relationship between complexation equilibrium and the concentration of ligand

Add 1 drop of $0.10$ mol·$L^{-1}$ $CoCl_2$ solution to a test tube, then add a few drops of conc. HCl, observe the color change, then dilute with water until the solution just turning pink, stop adding water into it, add more conc. HCl into it, observe the color change.

### 6. The formation and application of chelate

(1) Add about 1 mL hard water to a test tube, heat it until it boils by alcohol burner for about 1 min, observe whether there is precipitate from the water and whether it becomes muddy. Take another test tube, after adding 1 mL hard water, then add some drops of $0.10$ mol·$L^{-1}$ EDTA disodium salt solution, heat until it boils for about 1 min, whether the muddy phenomenon appears again? Why?

(2) Add 2 drops of $0.10$ mol·$L^{-1}$ $H_3BO_3$ to the test tube, add 1 drop of Lung's indicator, observe the color change, then add 2 drops of glycerol, shake and observe the color change.

### 7. The application of coordination compounds in analytical chemistry

(1) Identifying some ions by the formation of colorful coordination compounds

①Identification of $Fe^{2+}$

Add 5 drops of $0.10$ mol·$L^{-1}$ $Fe^{2+}$ solution to a test tube, then add 2 drops of $25\%$ $o$-phenanthroline solution, observe the phenomenon.

②Identification of $Ni^{2+}$

Add 1 drop of $0.10$ mol·$L^{-1}$ $Ni^{2+}$ solution and 5 drops of water to the test tube, then add $2.0$ mol·$L^{-1}$ ammonia water, make the solution alkalinous, then add 5 drops of diacetyldioxime solution, observe the phenomenon.

(2) Making use of coordination compounds to mask interference ion

Add 1 drop of $0.10$ mol·$L^{-1}$ $FeCl_3$ solution, 1 drop of $CoCl_2$ solution and 10 drops of $1.0$ mol·$L^{-1}$ KSCN to the test tube, then add $0.10$ mol·$L^{-1}$ NaF to the above solution dropwise until it becomes colorless, next add 5 drops of pentanol, shake the test tube, stand, observe the color of pentanol layer.

## Questions

1.  How to deduce the coordination compound's structure according to experiment? Give an example according to this experiment.

2.  Cu can displace Hg from salt of $Hg^{2+}$, whether or not it can displace Hg from $[Hg(CN)_4]^{2-}$? And why?

3.  What factors will influence the complexation equilibrium?

# 实验二十八　三草酸合铁（Ⅲ）酸钾的制备

## 实验目的

1. 了解三草酸合铁（Ⅲ）酸钾的制备方法。
2. 理解制备过程中化学平衡原理的应用。
3. 熟悉合成 $K_3[Fe(C_2O_4)_3]\cdot 3H_2O$ 的操作方法。

## 实验原理

三草酸合铁（Ⅲ）酸钾 $K_3[Fe(C_2O_4)_3]\cdot 3H_2O$ 是一种亮绿色的单斜晶体，溶于水而不溶于乙醇，光照易分解。它是制备负载型活性铁催化剂的主要原料，也是一些有机反应的良好催化剂。

本实验首先用硫酸亚铁铵与草酸反应制备出难溶于水的 $FeC_2O_4\cdot 2H_2O$，反应式为：

$$(NH_4)_2Fe(SO_4)_2\cdot 6H_2O + H_2C_2O_4 =\!=\!= FeC_2O_4\cdot 2H_2O\downarrow + (NH_4)_2SO_4 + H_2SO_4 + 4H_2O$$

草酸亚铁在草酸钾和草酸的存在下，被过氧化氢氧化得到三草酸合铁（Ⅲ）酸钾配合物。加入乙醇后便有三草酸合铁（Ⅲ）酸钾晶体析出。

$$2FeC_2O_4\cdot 2H_2O + H_2O_2 + 3K_2C_2O_4 + H_2C_2O_4 =\!=\!= 2K_3[Fe(C_2O_4)_3]\cdot 3H_2O + H_2O$$

## 实验仪器和药品

电子天平，温度计，烧杯，量筒，漏斗，抽滤瓶，布氏漏斗。

$(NH_4)_2Fe(SO_4)_2\cdot 6H_2O$，$H_2SO_4(3.0\ mol\cdot L^{-1})$，$H_2C_2O_4$（饱和溶液），$K_2C_2O_4$（饱和溶液），$H_2O_2(3\%)$，乙醇$(95\%, m)$。

## 实验步骤

### 1. 草酸亚铁的制备

在 100 mL 烧杯中加入 5.0 g $(NH_4)_2Fe(SO_4)_2\cdot 6H_2O$ 固体，15 mL 蒸馏水和几滴 3.0 mol·L$^{-1}$ H$_2$SO$_4$，加热溶解后再加入 25 mL 饱和 $H_2C_2O_4$ 溶液，加热至沸，搅拌片刻，停止加热，静置。待黄色晶体 $FeC_2O_4\cdot 2H_2O$ 沉降后倾析弃去上层清液，加入 10~20 mL 蒸馏水，搅拌并温热，静置冷却，倾析弃去上层清液。

### 2. 三草酸合铁（Ⅲ）酸钾的制备

在上述沉淀中加入 10 mL 饱和 $K_2C_2O_4$ 溶液。水浴加热至 313K，并恒温在 313K 左右，用滴管慢慢加入 20 mL 3% $H_2O_2$，边加边搅拌（此时有什么现象?），然后将溶液加热至沸，并分两次加入 8 mL 饱和 $H_2C_2O_4$ 溶液，第一次加 5 mL，第二次慢慢加入 3 mL，趁热过滤。滤液中加入 10 mL 95%（m）乙醇，温热溶液使析出的晶体再溶解后用表面皿盖好烧杯，自然冷却，避光静置，待晶体完全析出后抽滤，称重，计算产率，产品保留作测定用。

## 思考题

1. 影响三草酸合铁（Ⅲ）酸钾产量的主要因素有哪些？

2. 三草酸合铁（Ⅲ）酸钾见光易分解，应如何保存？

3. 用 $FeSO_4$ 为原料合成 $K_3[Fe(C_2O_4)_3]\cdot 3H_2O$，用 $HNO_3$ 代替 $H_2O_2$ 作氧化剂有什么缺点？

# Exp 28　Preparation of Potassium Trioxalatoferrate（Ⅲ）

## Objectives

1.　To learn how to prepare potassium trioxalatoferrate（Ⅲ）.
2.　To comprehend the usage of chemical equilibrium in preparative process.
3.　To be familiar with the operation of preparative procedure of potassium trioxalatoferrate（Ⅲ）.

## Principles

Potassium trioxalatoferrate（Ⅲ）is a kind of green monoclinic crystal, which is soluble in water but insoluble in ethanol, readily decomposes when exposed to light. It is the main starting material for preparation of active loading iron catalyst, it also is a kind of good catalyst for organic reaction.

In this experiment $FeC_2O_4 \cdot 2H_2O$ firstly is prepared between ferrous ammonium sulfate and oxalic acid:

$$(NH_4)_2Fe(SO_4)_2 \cdot 6H_2O + H_2C_2O_4 \longrightarrow FeC_2O_4 \cdot 2H_2O \downarrow + (NH_4)_2SO_4 + H_2SO_4 + 4H_2O$$

In the presence of ferrous oxalate and oxalic acid, ferrous oxalate is oxidized by hydrogen peroxide produces potassium trioxalatoferrate（Ⅲ）coordination compound, the crystal crystallizes after the addition of ethanol.

$$2FeC_2O_4 \cdot 2H_2O + H_2O_2 + 3K_2C_2O_4 + H_2C_2O_4 \longrightarrow 2K_3[Fe(C_2O_4)_3] \cdot 3H_2O + H_2O$$

## Instrument and Reagents

electronic balance, thermometer, beaker, measuring cylinder, funnel, suction bottle, büchner funnel.

$(NH_4)_2Fe(SO_4)_2 \cdot 6H_2O$, $H_2SO_4(3.0 \ mol \cdot L^{-1})$, $H_2C_2O_4$ (saturated), $K_2C_2O_4$ (saturated), $H_2O_2(3\%)$, ethanol 95%(m).

## Experimental Procedures

### 1.　Preparation of ferrous oxalate

5.0 g $(NH_4)_2Fe(SO_4)_2 \cdot 6H_2O$ are added to a 100 mL beaker, then 15 mL distilled water and several drops of 3.0 $mol \cdot L^{-1}$ $H_2SO_4$. Warm the solution to dissolve the solid, then 25 mL saturated oxalic acid are added, heat to boil, stir 1~2 minutes, stop heating, stand still. The above layer is discarded by decantation while all yellow solid precipitate remains, 10~20 mL $H_2O$ are added, stir, warm and stand still, the above layer is discarded.

### 2.　Preparation of potassium trioxalatoferrate

10 mL saturated potassium oxalate is added to the solution mentioned above. The mix-

ture is heated to 313K in water bath, retain the mixture at 313K. 20 mL 3% (m) $H_2O_2$ is added dropwise (observe the phenomena) with stirring. Heat the solution to boil, 8 mL saturated oxalic acid are added in two portions, first 5 mL, then 3 mL, filter while it is hot. 10 mL ethanol are added to the filtrate, if crystal appears, warm the solution till the crystal is dissolved. The beaker is covered with watch-glass, stand still, cool down naturally (keep it away from light overnight), when the crystal precipitates completely, filter under reduced pressure, weigh and calculate the yield, the product is kept for analysis.

## Questions

1. What main factors will influence the yield of potassium trioxalatoferrate (Ⅲ)?

2. Potassium trioxalatoferrate (Ⅲ) readily decomposes, how should we keep store it?

3. If $FeSO_4$ is chosen as the starting material for preparatsion of $K_3[Fe(C_2O_4)_3] \cdot 3H_2O$, or $HNO_3$ is replaced by $H_2O_2$, what are shortcomings?

# 实验二十九　三草酸合铁（Ⅲ）酸钾的组成测定

## 实验目的

1. 学习确定化合物化学式的基本原理的方法。
2. 学习热重分析法，由热重曲线了解 $K_3[Fe(C_2O_4)_3] \cdot 3H_2O$ 的热分解过程。

## 基本原理

配离子的组成可通过化学分析确定，其中 $C_2O_4{}^{2-}$ 含量直接由 $KMnO_4$ 标准溶液在酸性介质中滴定测得。$Fe^{3+}$ 离子含量可先用过量锌粉将其还原为 $Fe^{2+}$ 离子，然后再用 $KMnO_4$ 标准溶液滴定而测得。反应式：

$$5C_2O_4{}^{2-} + 2MnO_4{}^- + 16H^+ =\!=\!= 10CO_2 + 2Mn^{2+} + 8H_2O$$

$$5Fe^{2+} + MnO_4{}^- + 8H^+ =\!=\!= 5Fe^{3+} + Mn^{2+} + 4H_2O$$

$K_3[Fe(C_2O_4)_3] \cdot 3H_2O$ 加热到 100℃ 脱去结晶水，加热到 230℃ 时分解，其质量也随之变化。在程序控制温度下测定物质的质量与温度的函数关系，即进行热重分析。由热重曲线可确定 $K_3[Fe(C_2O_4)_3] \cdot 3H_2O$ 的结晶水和热分解过程。

## 实验仪器与药品

热分析仪。

$K_3[Fe(C_2O_4)_3] \cdot 3H_2O$，$H_2SO_4(1\ mol \cdot L^{-1})$，$KMnO_4(0.1000\ mol \cdot L^{-1})$。

## 实验步骤

### 1. 测定结晶水和热分解过程（在热重天平上进行测试操作）

操作条件：

　　样品重量：~10 mg

　　升温速度：10℃ /min

　　热重量程：25 mg

　　走纸速度：60 cm/h

由得到的热重（TG）曲线（图 29-1）确定化合物的结晶水，并写出它的热分解反应式。

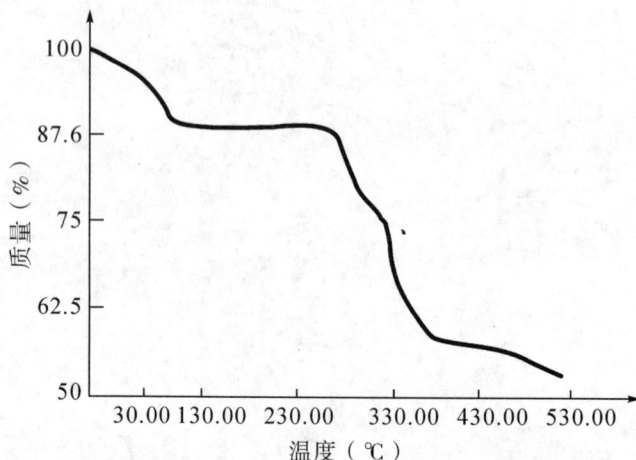

图 29-1　三草酸合铁（Ⅲ）酸钾的 TG 曲线

### 2. 草酸根含量的测定

把制得的 $K_3[Fe(C_2O_4)_3] \cdot 3H_2O$ 在 110℃ 烘箱中干燥 1h，在干燥器中冷却至室温。精确称取 0.10~0.20 g 样品 3 份，分别放入三个锥形瓶中，加 30 mL 1 mol·

$L^{-1}$ $H_2SO_4$ 溶解，加热溶液到 70～80℃（不高于 85℃）。用 0.1000 mol·$L^{-1}$ $KMnO_4$ 标准溶液滴定热溶液（$KMnO_4$ 溶液用 $Na_2C_2O_4$ 进行标定），滴定至微红色在 30 s 内不消失为终点。记下消耗的 $KMnO_4$ 的体积，计算 $K_3[Fe(C_2O_4)_3]\cdot3H_2O$ 中草酸根的浓度，并换算成物质的量。滴定后的溶液保留待用。

在上述用 $KMnO_4$ 滴定过草酸根的保留液中加锌粉还原，至黄色消失。加热 2～3 min，使 $Fe^{3+}$ 转变为 $Fe^{2+}$，抽滤除去多余锌粉，并用温水洗涤沉淀。将二次滤液合并放入锥形瓶，再用 $KMnO_4$ 标准溶液滴定至微红色，计算 $K_3[Fe(C_2O_4)_3]\cdot3H_2O$ 中含铁量，并换算成物质的量。

将测定结果记录于表 29-1 和 29-2 中。

**表 29-1 $C_2O_4^{2-}$ 含量测定数据**

| 实验序号 | 1 | 2 | 3 |
|---|---|---|---|
| 样品质量（g） | | | |
| $KMnO_4$ 浓度（mol·$L^{-1}$） | | | |
| $KMnO_4$ 溶液滴数 | | | |
| $KMnO_4$ 溶液体积（mL） | | | |
| $KMnO_4$ 物质的量（mol） | | | |
| $C_2O_4^{2-}$ 物质的量（mol） | | | |
| 1mol 样品中 $C_2O_4^{2-}$ 的物质的量（mol） | | | |

**表 29-2 $Fe^{3+}$ 含量测定数据**

| 实验序号 | 1 | 2 | 3 |
|---|---|---|---|
| 样品质量（g） | | | |
| $KMnO_4$ 浓度（mol·$L^{-1}$） | | | |
| $KMnO_4$ 溶液滴数 | | | |
| $KMnO_4$ 溶液体积（mL） | | | |
| $KMnO_4$ 物质的量（mol） | | | |
| $Fe^{3+}$ 物质的量（mol） | | | |
| 1mol 样品中 $Fe^{3+}$ 的物质的量（mol） | | | |

结论：在 1mol 产品中含结晶水_____mol，$C_2O_4^{2-}$ 离子_____mol，$Fe^{3+}$ 离子_____mol。该物质的化学式为_____。

## 思考题

1. 滴定 $C_2O_4^{2-}$ 时要加热，但又不能太高，为什么？

2. 根据三草酸合铁（Ⅲ）酸钾的合成过程及它的 TG 曲线，你认为该化合物应如何保存。

# Exp 29　Analysis of Potassium Trioxalatoferrate（Ⅲ）

## Objectives

1.  To study principles of determination the formula of compound.

2.  To study analytical method of thermogravimetry and know the decomposing procedures of potassium trioxalatoferrate（Ⅲ）through TG curve.

## Principles

Composition of coordination ions can be confirmed by chemical analysis, the content of $C_2O_4^{2-}$ can be measured in acid medium by standard potassium permanganate. $Fe^{3+}$ can be reduced by zinc powder to $Fe^{2+}$ and then the content of $Fe^{3+}$ can be measured by standard potassium permanganate. The reaction equations are as following:

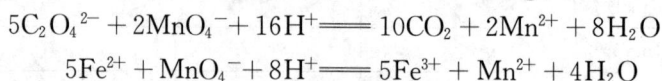

$$5C_2O_4^{2-} + 2MnO_4^- + 16H^+ = 10CO_2 + 2Mn^{2+} + 8H_2O$$

$$5Fe^{2+} + MnO_4^- + 8H^+ = 5Fe^{3+} + Mn^{2+} + 4H_2O$$

Potassium trioxalatoferrate will lose its crystal water at 100℃, decomposes at 230℃, its mass will change at the same time. Measure the function relation between quality and temperature under procedures controls, that is TG analysis. The crystal water and decomposing procedures of $K_3[Fe(C_2O_4)_3] \cdot 3H_2O$ can be dertermined.

## Instruments and Reagents

thermal analyzer.

$K_3[Fe(C_2O_4)_3] \cdot 3H_2O$, $H_2SO_4(1 \ mol \cdot L^{-1})$, $KMnO_4(0.1000mol \cdot L^{-1})$.

## Procedures

**1.  Analysis of crystal water and thermolysis procedures（operating on TGA balance）**

Operation conditions:

Sample weight: ~10 mg

Velocity of heating: 10℃/min

Thermogravimetry rang: 25 mg

Paper speed: 60 cm/h

Identifying its crystal water with TG curve (Fig. 29-1) and write down its decomposition reaction equation.

**2.  Determination of content of $C_2O_4^{2-}$**

Prepared $K_3[Fe(C_2O_4)_3] \cdot 3H_2O$ is laid in oven at 110℃ for 1h, cool down in desiccator

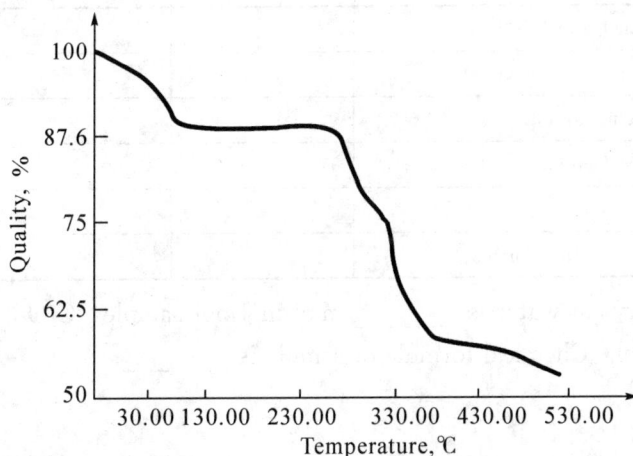

Fig. 29-1 TG curve of potassium trioxalatoferrate（Ⅲ）

to room temperature. Weigh 3 portions of precisely 0. 10~0. 20 g samples to three 25 mL conical flasks respectively, 30 mL 0. 1 mol·L$^{-1}$ H$_2$SO$_4$ are added to dissolve the solid, then the mixture is heated to 70~85℃, titrate them with 0. 1000 mol·L$^{-1}$ KMnO$_4$ standard solution （concentration of KMnO$_4$ solution must be demarcated by Na$_2$C$_2$O$_4$ following instruction of analytical handbook）. Titrate hot solution until red lilac doesn't disappear within 30s. Record the volume of KMnO$_4$ solution, calculate the concentration of C$_2$O$_4^{2-}$ in the K$_3$[Fe(C$_2$O$_4$)$_3$]·3H$_2$O, then convert it into gram-molecule. The titrated solution is kept for later use.

Add zinc powder to mixture mentioned above until the color of yellow disappears, heat for 2~3 min, convert Fe$^{3+}$ to Fe$^{2+}$, filter so as to remove redundant zinc powder, wash the precipitate with warm water. Transfer filtrate into 5 mL conical flask, titrate with standard solution to red lilac, calculate content of Fe$^{3+}$, convert it to gram-molecule.

Data in the experiment are recorded in table 29-1 and table 29-2.

**Table 29-1    Data assaying of C$_2$O$_4^{2-}$ content**

| Serial number | 1 | 2 | 3 |
| --- | --- | --- | --- |
| Sample quality, g | | | |
| Concentration of KMnO$_4$, mol·L$^{-1}$ | | | |
| Drop numbers of KMnO$_4$ | | | |
| Volume of KMnO$_4$ solution, mL | | | |
| Gram-molecule of KMnO$_4$, mol | | | |
| Gram-molecule of C$_2$O$_4^{2-}$, mol | | | |
| Gram-molecule of C$_2$O$_4^{2-}$ in 1mol sample | | | |

**Table 29-2    Data Assaying of Fe$^{3+}$ content**

| Serial number | 1 | 2 | 3 |
| --- | --- | --- | --- |
| Sample quality, g | | | |

| | | | |
|---|---|---|---|
| Concentration of $KMnO_4$, $mol \cdot L^{-1}$ | | | |
| Drop numbers of $KMnO_4$ | | | |
| Volume of $KMnO_4$ solution, mL | | | |
| Gram-molecule of $KMnO_4$, mol | | | |
| Gram-molecule of $Fe^{3+}$, mol | | | |
| Gram-molecule of $Fe^{3+}$ in 1mol sample | | | |

Conclusion: crystal water is _____ mol in 1mol sample, $C_2O_4{}^{2-}$ is _____ mol, $Fe^{3+}$ is _____ mol. Chemical formula of sample is _____.

## Questions

1. Heating is needed when $C_2O_4{}^{2-}$ is titrated, but the temperature should not be too high, why?

2. On the basis of preparative procedures of potassium trioxalatoferrate (Ⅲ) and its TG curve, how do you think it should be kept properly?

# 实验三十 硫代硫酸钠的制备

## 实验目的

1. 学习用溶剂法提纯工业硫化钠和用提纯的硫化钠制备硫代硫酸钠的方法。
2. 练习冷凝管的安装和回流操作。
3. 练习抽滤，气体发生和器皿连接操作。

## 实验原理

### 1. 非水溶剂重结晶法提纯硫化钠

纯硫化钠为含有不同数目结晶水的无色晶体（如 $Na_2S \cdot 5H_2O$，$Na_2S \cdot 9H_2O$）。工业硫化钠由于含有大量的杂质，如重金属硫化物、煤粉等而呈红褐色或棕黑色。本实验是利用硫化钠能溶解于热的酒精中，其他杂质或在趁热过滤时除去，或在冷却后硫化钠结晶析出时留在母液中除去达到使硫化钠纯化的目的。

### 2. 硫代硫酸钠的制备

用硫化钠制备硫代硫酸钠的反应大致可分为三步进行[1]。

（1）碳酸钠与二氧化硫中和而生成亚硫酸钠

$$Na_2CO_3 + SO_2 = Na_2SO_3 + CO_2$$

（2）硫化钠与二氧化硫反应而生成亚硫酸钠和硫

$$2Na_2S + SO_2 = 2Na_2SO_3 + 3S$$

（3）亚硫酸钠与硫反应而生成硫代硫酸钠

$$Na_2SO_3 + S = Na_2S_2O_3$$

总反应如下：$2Na_2S + Na_2CO_3 + 4SO_2 = 3Na_2S_2O_3 + CO_2$

含有硫化钠和碳酸钠的溶液，用二氧化硫气体饱和。反应中碳酸钠的用量不宜过少。如用量过少，则中间产物亚硫酸钠量少，使析出的硫不能全部生成硫代硫酸钠。硫化钠和碳酸钠以 2:1 的摩尔比取量较为合适。

反应完毕，过滤得到硫代硫酸钠的溶液，然后浓缩蒸发，冷却，析出晶体为 $Na_2S_2O_3 \cdot 5H_2O$，干燥后即得产品。

## 实验仪器和药品

仪器：圆底烧瓶(250 mL)，水浴锅，300 mm 球型冷凝管，抽滤瓶(250 mL)，布氏漏斗，烧杯(250 mL)，锥形瓶(250 mL)，分液漏斗，橡皮塞，蒸馏烧瓶(250 mL)，洗气瓶，磁力搅拌器

固体药品：硫化钠(工业级)，亚硫酸钠(无水)，碳酸钠

液体药品：乙醇(95%)，硫酸(浓)，$NaOH(6\ mol \cdot L^{-1})$

## 实验步骤

### 1. 硫化钠的提纯

取粉碎的工业级硫化钠 18 g，装入 250 mL 的烧瓶中，再加入 75 mL 95％酒精。将烧瓶放在水浴锅上，烧瓶上装一支长 300 mm 的球形冷凝管，并向冷凝管中通入冷却水装置如图 30-1 所示。水浴锅的水保持沸腾，回流约 40 min。

停止加热并使烧瓶在水浴锅上静置 5 min，然后取下烧瓶，用两层滤纸趁热抽滤，以除去不溶性杂质。将滤液转入一只 250 mL 的烧杯中，不断搅拌以促使硫化钠晶体大量析出。再放置一段时间，冷却至室温。冷却后倾析出上层母液。硫化钠晶体用少量 95％酒精在烧杯中用倾析法洗涤 1～2 次，然后抽滤。抽干后，再用滤纸吸干。母液装入指定的回收瓶中。按本方法制得的产品组成相当于 $Na_2S \cdot 5H_2O$。

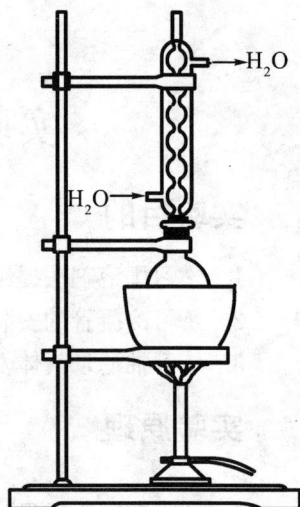

图 30-1　硫化钠提纯装置

### 2. 硫代硫酸钠的制备[2]

按图 30-2 安装好制备装置。将 2.7 g $Na_2CO_3$ 放入 150 mL 的锥形瓶中，加入 75 mL 蒸馏水使其溶解，再加入 12.0 g $Na_2S \cdot 5H_2O$ 溶解得到反应液。吸收瓶中放入 6 mol·$L^{-1}$ NaOH 用于吸收多余的 $SO_2$。在蒸馏烧瓶中加入 16 g 无水亚硫酸钠，并加入 4～5 mL 蒸馏水润湿。在滴液漏斗中加入比理论量稍多些的浓硫酸，并慢慢打开滴液漏斗的活塞，以 13 秒 1 滴的速度将浓硫酸缓慢滴入蒸馏烧瓶中。反应生成的二氧化硫通入锥形瓶与其中的硫化钠和碳酸钠溶液反应，在不断搅拌下反应约 30 min 后，至锥形瓶中溶液呈无色，当其中有淡黄色不溶的硫磺颗粒出现，且瓶壁上也附有黄色的硫时，再反应 3～5 min 待溶液 pH 约为 7 时可停止反应。把锥形瓶中的液体过滤，得到的滤液经水浴浓缩为原反应液体积的四分之一时（约 20 mL 左右），冷却，在乙醇中结晶，抽滤，用滤纸吸干后称重，计算产率。

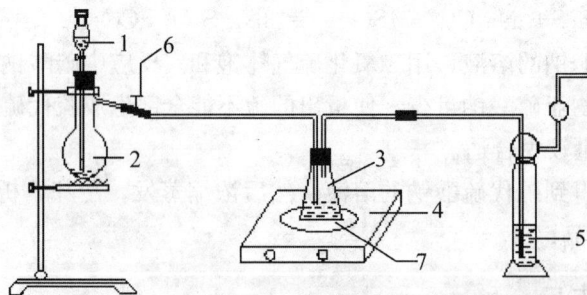

1.滴液漏斗（内装硫酸）2.蒸馏烧瓶（内装亚硫酸钠）

3.锥形瓶 4.电磁搅拌器 5.碱吸收瓶 6.螺旋夹

7. 磁力搅拌子

图 30-2　制备硫代硫酸钠的装置

## 思考题

1. 在 $Na_2S-Na_2CO_3$ 溶液中通 $SO_2$ 的反应是放热反应，还是吸热反应？为什么？
2. 停止通入 $SO_2$ 时，为什么必须控制溶液的 pH 约为 7 而不能使其小于 7？

## 参考文献

[1] 北京师范大学无机化学教研室. 无机化学实验（第三版），北京：高等教育出版社，2005.
[2] 李芳，郑典慧，陈红. 硫代硫的室内制备方法的优化. 实验技术与原理. 2005，22（10）：57-59.

# Exp 30　Preparation of Sodium Hyposulfite

## Objectives

1. To learn how to purify technically pure sodium sulfide and prepare sodium hyposulfite with purified sodium sulfide.

2. To practice operation of installation and reflux.

3. To practice operation of filter under reduced pressure, gas producing and linkage of apparatus.

## Principles

### 1. Purification of $Na_2S$ by recrystallization in non-aqueous solvent

Pure $Na_2S$ is a kind of colorless crystal with different crystal water (e. g $Na_2S \cdot 5H_2O$, $Na_2S \cdot 9H_2O$). Industrial $Na_2S$ appears red-brown or brown-black for it is mixed with too many impurities, such as heavy metal sulfide, coal. In this experiment, we draw on that crude $Na_2S$ containing a lot of impurities which are easily dissolved in hot ethanol, other impurities can be filtered while the solution is hot, so we can purify it by recrystallization.

### 2. Preparing $Na_2S_2O_3$ abides by following method

First, sodium carbonate reacts with sulfur dioxide to prepare sodium sulfite.

$$Na_2CO_3 + SO_2 =\!=\!= Na_2SO_3 + CO_2$$

Second, sodium sulfide reacts with sulfur dioxide to prepare sodium sulfite and sulfur.

$$2Na_2S + SO_2 =\!=\!= 2Na_2SO_3 + 3S$$

The solution being mixed with $Na_2S$ and $Na_2CO_3$ is saturated by gaseous $SO_2$. The ammount of $Na_2CO_3$ should not be too small. If does, then the intermediate $Na_2SO_3$ is little. The preupitated sulfur can not be converted into sodium hyposulfite thoroughly. The best mole ratio of $Na_2S$ to $Na_2CO_3$ is 2:1.

Third, sodium sulfite reacts with sulfur to prepare sodium hyposulfite.

$$Na_2SO_3 + S =\!=\!= Na_2S_2O_3$$

overall reaction equation: $2Na_2S + Na_2CO_3 + 4SO_2 =\!=\!= 3Na_2S_2O_3 + CO_2$

When the reaction is over, fitter, concentrate the filtrate, cool. The crystal which appears is $Na_2S_2O_3 \cdot 5H_2O$, the product is obtained after drying.

## Instruments and Reagents

Instruments: round bottom flask (250 mL), water bath, straight (or ball) condenser, suction bottle (250 mL), büchner funnel, beaker (250 mL), conical flask, rubber stopple, distilled flask (250 mL), gas wash bottle, magnetic stirrer

Solid reagents: sodium sulfide (industrial grade), sodium sulfite, sodium carbonate

Liquid reagents: ethanol (95%), sulfuric acid (concentrated), sodium hydroxide (6 mol·L$^{-1}$)

## Experimental Procedures

### 1. Purification of Na$_2$S

Charge 18 g industrial pure Na$_2$S in 250 mL beaker, add 75 mL 95% ethanol. Then put the flask equipped a 300 mm ball condenser in the water bath (Fig. 30-1), keep the water of the water bath boiling, reflux about 40 minutes.

Stop heating and keep the flask in water bath for 5 minutes, filter under reduced pressure with double layer filter paper while it is hot in order to remove insoluble impurities. Transfer the filtrate to a 250 mL beaker, stir continuously to hasten crystal appear. Keep it longer and cool down to room temperature. Decant the mother liquid of above layer. Na$_2$S was washed 1~2 times with 95% ethanol by decantation in the beaker. Filter, the water on the Na$_2$S was absorbed with filter paper. Mother liquid is transferred to a appointed flask.

The product prepared by this method is equivalently Na$_2$S pentahydrate.

### 2. Preparation of Na$_2$S$_2$O$_3$

Install preparative instrument according Fig. 30-2. Charge 2.7 g Na$_2$CO$_3$ in 150 mL conical flask, add 75 mL H$_2$O to make it

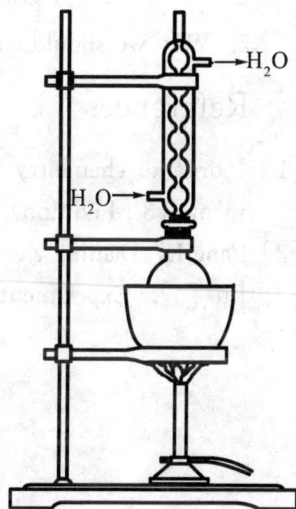

Fig. 30-1 Instrument of Na$_2$S purification

dissolve, after that, 12.0 g Na$_2$S·5H$_2$O are added at the same time. 6 mol·L$^{-1}$ NaOH is added to absorb excess SO$_2$. Add 16 g anhydrous Na$_2$S$_2$O$_3$ and 4~5 mL distilled water are used to moisten Na$_2$S$_2$O$_3$. Add conc. H$_2$SO$_4$ slightly more than theoretical amount to dropping funnel and turn the pistons of funnel into add conc. H$_2$SO$_4$ to the flask at the rate of 1 drop per 13 seconds, generated SO$_2$ is ventilated to the solution. Continuous stirring is kept for about 30 min until the solution in conical flask appears colorless and indissolvable sulfur appears on wall of conical flask and solution, SO$_2$ is ventilated for more than 3~5 min to pH value reaching to 7. Filter, the filtrate is evaporated to 1/4 volume of the original volume (about 20 mL) in water-bath, cool down and recrystallize in cold ethanol, filter, the redundant solvent is absorbed by filter paper, weigh and calculate the yield.

Equip the apparatus according to following figure for preparation of sodium hyposulfite.

## Questions

1. Is the reaction exothermic one or endothermic one which SO$_2$ is ventilated in Na$_2$S-Na$_2$CO$_3$ solution? And why?

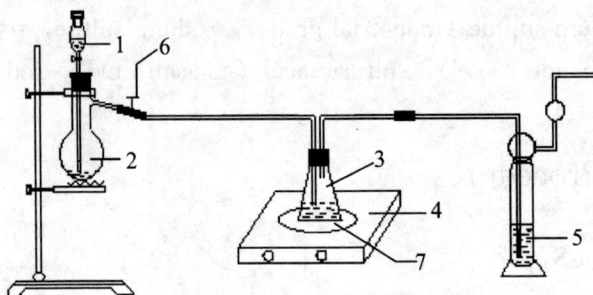

1.dropping funnel($H_2SO_4$ in it)    2.distilled flask($Na_2SO_3$ in it )    3.conical flask

4.magnetic stirrer    5.alkali solution absorption bottle    6.spiral clip    7.stirrer

Fig. 30-2    Instrument for preparation of sodium hyposulfite

2. Why we should control the pH of the solution at about 7 and not less than 7?

# References

[1] Inorganic chemistry college, Beijing normal university. Imorganic chemistry experiments (3 rd edition). Bei Jing: Higher Education Press, 2005.

[2] Fang L, Dianhui Z, hong C, et al. Aoptimized preparation method of sodium hyposulfite [J]. Experimental technology and management. 2005, 22 (10): 57~59.

# 实验三十一　三氯化六氨合钴（Ⅲ）的合成和组成的测定

## 实验目的

1. 了解三氯化六氨合钴的制备原理及其组成的测定方法。
2. 加深理解配合物的形成对三价钴稳定性的影响。

## 实验原理

根据标准电极电势，在酸性介质中钴（Ⅱ）的化合物比钴（Ⅲ）的化合物稳定，而在它们的配合物中，大多数的钴（Ⅲ）配合物比钴（Ⅱ）配合物稳定，所以常采用空气或过氧化氢氧化钴（Ⅱ）配合物来制备钴（Ⅲ）配合物。

氯化钴（Ⅲ）的氨合物有许多种。主要有三氯化六氨合钴（Ⅲ）$[Co(NH_3)_6]Cl_3$（橙黄色晶体），三氯化一水·五氨合钴（Ⅲ）$[Co(NH_3)_5H_2O]Cl_3$（砖红色晶体），二氯化一氯·五氨合钴（Ⅲ）$[Co(NH_3)_5Cl]Cl_2$（紫红色晶体）等。它们的制备条件各不相同。例如，在没有活性炭存在时，由氯化亚钴在过量氨、氯化铵中被氧化的主要产物是二氯化一氯·五氨合钴（Ⅲ），有活性炭存在时被氧化的主要产物是三氯化六氨合钴（Ⅲ）。

本实验用活性炭作催化剂，用过氧化氢作氧化剂，氯化亚钴溶液与过量氨和氯化铵作用制备三氯化六氨合钴（Ⅲ）。其总反应式如下：

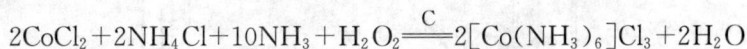

$$2CoCl_2+2NH_4Cl+10NH_3+H_2O_2 \xrightarrow{C} 2[Co(NH_3)_6]Cl_3+2H_2O$$

三氯化六氨合钴（Ⅲ）溶解于酸性溶液中，通过过滤可以将混在产品中的大量活性炭除去，然后在高浓度盐酸中使三氯化六氨合钴（Ⅲ）结晶。

三氯化六氨合钴（Ⅲ）为橙黄色单斜晶体。固态的$[Co(NH_3)_6]Cl_3$在 488 K 转变为$[Co(NH_3)_5Cl]Cl_2$，高于 523 K 则被还原为$CoCl_2$。

$[Co(NH_3)_6]Cl_3$可溶于水不溶于乙醇，在 293K 水中的溶解度 0.26 $mol \cdot L^{-1}$。在强碱作用下（冷时）或强酸的作用下基本不被分解，只有在沸热条件下才被强碱分解：

$$2[Co(NH_3)_6]Cl_3+6NaOH \xrightarrow{沸热} 2Co(OH)_3+12NH_3+6NaCl$$

用过量的标准盐酸溶液吸收分解逸出的氨。剩余的盐酸用标准氢氧化钠溶液回滴，便可分析出氨的含量，从而确定出配合物组成中氨的百分比。

然后用碘量法测定蒸除氨后的样品溶液中的 Co（Ⅲ），确定出配合物组成中中心离子的百分比。

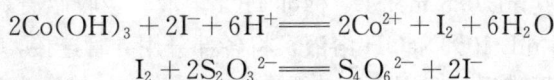

$$2Co(OH)_3 + 2I^- + 6H^+ = 2Co^{2+} + I_2 + 6H_2O$$
$$I_2 + 2S_2O_3{}^{2-} = S_4O_6{}^{2-} + 2I^-$$

用沉淀滴定法（莫尔法）测定样品中氯离子的含量，确定出配合物组成中 $Cl^-$ 的百分比。通过组分分析，可确定出配合物的实验式。

## 实验仪器和药品

电子天平，分析天平，酸式滴定管(50 mL)，碱式滴定管(50 mL)，玻璃管，碱封管。

CoCl$_2$·6H$_2$O(s)，NH$_4$Cl(s)，KI(s)，活性炭，HCl(6 mol·L$^{-1}$、浓)，标准 HCl 溶液 (0.5 mol·L$^{-1}$)，标准 NaOH 溶液 (0.5 mol·L$^{-1}$)，H$_2$O$_2$ 6％ (m)，氨水（浓），NaOH 溶液 10％(m)，标准 Na$_2$S$_2$O$_3$ 溶液(0.1 mol·L$^{-1}$)，标准 AgNO$_3$ 溶液（0.1 mol·L$^{-1}$)，K$_2$CrO$_4$ 溶液 5％(m)，冰。

## 实验步骤

### 1. 制备三氯化六氨合钴（Ⅲ）

在 100 mL 锥形瓶中加入 6 g 研细的氯化亚钴 CoCl$_2$·6H$_2$O，4 g 氯化铵和 7 mL 蒸馏水。加热溶解后加入 0.3 g 活性炭。冷却，加 14 mL 氨水，冷却至 283 K 以下，缓慢加入 14 mL 6％(m)的过氧化氢，水浴加热至 333 K 左右并恒温 20 min（适当摇动锥形瓶）。取出，先用自来水冷却，后用冰水冷却，抽滤。将沉淀溶解于含有 2 mL 浓盐酸的 30 mL 酸溶液中，抽滤。在滤液中慢慢地加入 30 mL 浓盐酸，三氯化六氨合钴（Ⅲ）结晶析出。抽滤。沉淀用冷却过的 6 mol·L$^{-1}$ 盐酸洗涤后取出，产品置于烘箱中，在 100℃下干燥 1~2 h。称重，计算产率。

### 2. 三氯化六氨合钴（Ⅲ）组成的测定

（1）氨的测定（图 31-1）

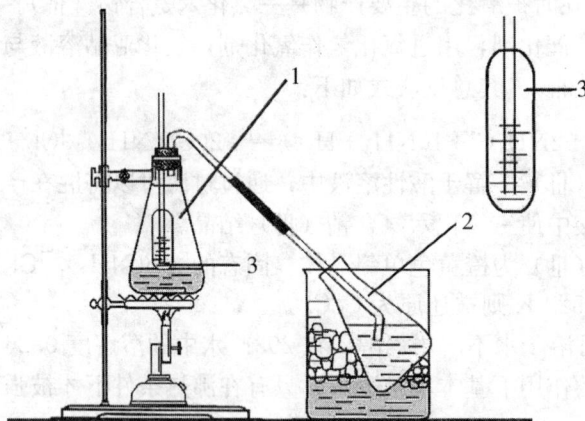

1. 反应瓶　2. 接收瓶　3. 碱封瓶
图 31-1 测定氨的装置

在 250 mL 锥形瓶 1 中加入 0.2 g（准确至 0.1 mg）待测的三氯化六氨合钴（Ⅲ）晶体，加入 80 mL 蒸馏水，摇动溶解，然后加入 10 mL 10％NaOH 溶液。在接收瓶（锥形瓶 2）中加入 30.00~35.00 mL 0.5 mol·L$^{-1}$ 标准 HCl 溶液，接收瓶浸入冰水槽中。在锥形瓶中的碱封管内注入 3~5 mL 10％NaOH 溶液，将各部分用导管连接，按图 31-1 安装好。

大火加热样品溶液至沸后，改用小火，微沸 50~60 min，使氨全部蒸出，并被导入标准 HCl 溶液中吸收，停止加热，取出接收瓶，用少量蒸馏水将导管内外可能沾附的盐酸溶液冲洗入接收瓶内，用 0.5 mol·L$^{-1}$ 标准 NaOH 液滴定剩余的盐酸，记录数据。

（2）钴的测定

取下装样品溶液的锥形瓶 1，用少量蒸馏水将塞子、碱封管上沾附的溶液冲洗回锥形瓶内。待样品溶液冷却后加入 1 g 固体 KI，振荡溶解，再加入 12 mL 左右的 6 mol·L$^{-1}$盐酸酸化后放在暗处静置 10 min，然后，加入 60～70 mL 蒸馏水，用 0.1 mol·L$^{-1}$标准 Na$_2$S$_2$O$_3$溶液滴定，开始滴定可以快些，滴定至溶液淡黄色时加入几滴淀粉溶液，继续慢慢滴加 Na$_2$S$_2$O$_3$ 溶液，滴定至终点（终点溶液颜色？），记录数据。

（3）氯的测定

在锥形瓶中加入样品 0.2 g（准确至 0.1 mg），加适量蒸馏水溶解，用沉淀滴定法（莫尔法）测定氯的含量。

（4）记录与结果

①以表格形式记录有关数据。

②计算出样品中氨、钴和氯的百分含量。

③确定出产品的实验式。

## 思考题

1. 制备过程中水浴加热 333 K 并恒温 20 min 的目的是什么？能否加热至沸？
2. 制备三氯化六氨合钴过程中加 H$_2$O$_2$ 和浓盐酸各起什么作用？要注意什么问题？
3. 碘量法测定钴（Ⅲ）离子时要注意什么问题？

# Exp 31  Preparation and Component Analysis of $[Co(NH_3)_6]Cl_3$

## Objectives

1. To know the preparation principle and structure determination method.
2. To learn well the influence of the coordination compound formation on the stability of Co (Ⅲ) further.

## Principles

According to standard electrode potential, cobaltous salt is more stable than trivalent cobalt salt in acid medium, but for their coordinates, most of cobalt (Ⅲ) complex are more stable than cobalt (Ⅱ) complex, so the preparation method of Co (Ⅲ) coordination compound, air or hydrogen peroxide is often used to oxide Co (Ⅱ) complex to prepare Co (Ⅲ) coordination compound from Co (Ⅱ) complex.

There are many kinds of ammoniate of cobalt (Ⅲ) chloride. The main kinds are $[Co(NH_3)_6]Cl_3$ (Ⅲ)  (orange crystal), $[Co(NH_3)_5H_2O]Cl_3$ (brick red crystal), $[Co(NH_3)_5Cl]Cl_2$ (Ⅲ) (purplish red crystal) $etc$. Their preparation conditions are different. For example, in the case of no active carbon existence, the main products are $[Co(NH_3)_5Cl]Cl_2$ (Ⅲ) reacting between cobalt dichloride and extra ammonia, ammonium chloride, in the case of active carbon existence, the main products are components of $[Co(NH_3)_6]Cl_3$ (Ⅲ).

In this experiment, active carbon is used as catalyst, hydrogen peroxide is used as oxidizer, cobalt dichloride reacts with excessive ammonia and ammonium chloride to prepare aqueous $[Co(NH_3)_6]Cl_3$ (Ⅲ). The main reaction is shown as follow:

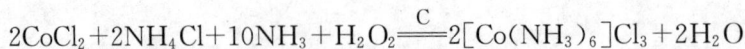

$$2CoCl_2+2NH_4Cl+10NH_3+H_2O_2 \xrightarrow{C} 2[Co(NH_3)_6]Cl_3+2H_2O$$

Components of $[Co(NH_3)_6]Cl_3$ (Ⅲ) can dissolve in acid solution, a great deal of active carbon mixed in products can be removed by filtrating. Then crystal components of $[Co(NH_3)_6]Cl_3$ is crystallized in concentrated hydrochloric acid.

$[Co(NH_3)_6]Cl_3$ (Ⅲ) is dark orange monoclinic crystal. Solid $[Co(NH_3)_6]Cl_3$ change into $[Co(NH_3)_5Cl]Cl_2$ at 488K, it be reduced to $CoCl_2$ above 523K.

$[Co(NH_3)_6]Cl_3$ can be dissolved in water but not dissolved in alcohol, solubility is 0.26mol·L$^{-1}$ in water at 293K. It generally not is decompose in existence of alkali (cool) or strong acid, but it can be decomposed in boiling strong alkali:

$$2[Co(NH_3)_6]Cl_3+6NaOH \xrightarrow{\triangle} 2Co(OH)_3+12NH_3+6NaCl$$

The decomposed ammonia can be absorbed with extra standard hydrochloric acid solution. The left hydrochloric acid can be titrated back with standard sodium hydroxide solution, then the content of ammonia can be determined, then the ammonia percentage in the coordination compound can be calculated.

The Co (Ⅲ) in the sample solution ammonia vaporized can be determined by iodometry method:

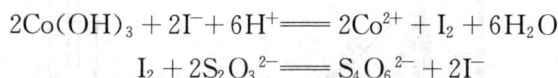

$$2Co(OH)_3 + 2I^- + 6H^+ = 2Co^{2+} + I_2 + 6H_2O$$

$$I_2 + 2S_2O_3^{2-} = S_4O_6^{2-} + 2I^-$$

Determine the chlorine ion content of sample by precipitate titration method (Mohr's method). The percentage of the Cl$^-$ in the coordination compound once being determined, the conventional formula of the coordination compound can be determined through composition analysis.

## Instruments and Reagents

electronic balance, analytical balance, acid burette (50 mL), base burette (50 mL), glass pipe, alkali sealed tube.

$CoCl_2 \cdot 6H_2O(s)$, $NH_4Cl(s)$, $KI(s)$, active carbon, HCl(6 mol·L$^{-1}$, concentrated), standard HCl solution (0.5 mol·L$^{-1}$), standard NaOH solution (0.5 mol·L$^{-1}$), $H_2O_2$ 6%(m), ammonia (dense), NaOH solution 10%(m), standard $Na_2S_2O_3$ solution (0.1 mol·L$^{-1}$), standard AgNO$_3$ solution (0.1 mol·L$^{-1}$), $K_2CrO_4$ solution 5%(m), ice.

## Experimental Procedures

### 1. Preparation of components of $[Co(NH_3)_6]Cl_3$ (Ⅲ)

Add 6 g grinded $CoCl_2 \cdot 6H_2O$, 4 g ammonium chloride and 7 mL distilled water in 100 mL Erlenmeyer flask. Add 0.3 g active carbon after heating and dissolving. Cool, add 14 mL concentrated ammonia, cool to below 283K, add 14 mL 6% (m) hydrogen peroxide slowly, heat by water bath to about 333K and keep constant temperature for 20 min (shaking Erlenmeyer flask properly). Take out, cool by tap water first, and then cool by ice-water. Filter under reduced pressure, dissolve the precipitate in 50 mL boiling water which containing 2 mL concentrated hydrochloric acid, and filter while it is hot. Add 7 mL concentrated hydrochloric acid slowly into filtrate, cool by using ice water, filter, wash (what reagent can be used?), take out, dry it in vacuum desiccator or bake it at below 378K, weigh, calculate the yield.

### 2. Determination of components of $[Co(NH_3)_6]Cl_3$ (Ⅲ)

(1) Determination of ammonia

Add 0.2 g(accurate to 0.1 mg) sample of $[Co(NH_3)_6]Cl_3$ (Ⅲ) crystal in 250 mL Erlenmeyer flask 1, add 80 mL distilled water, shake, dissolve, then add 10 mL 10%(m) NaOH solution. Add 30.00~35.00 mL 0.5 mol·L$^{-1}$ standard HCl solution to receivers (Erlenmeyer flask 2), immerge receiver in ice-water gutter. Inject 3~5 mL10%(m) NaOH solution into alkali sealed tube in Erlenmeyer flask, connect all parts by vessel, fix them according to Fig. 31-1.

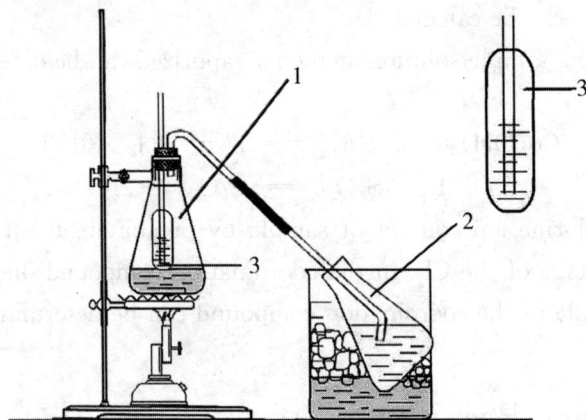

1. reaction bottle　2. accepter　3. alkali sealed tube

Fig. 31-1　Instrument for determining ammonia

Heat up sample solution to boiling, then reduce heating, simmering for $50 \sim 60$ min, vapor ammonia out completely, and ammonia be absorbed by standard HCl solution through pipe, stop heating, take out receiver, wash the hydrochloric acid solution which perhaps adhered to inside and outside pipe into accepter by a little distilled water, titrate the left hydrochloric acid by $0.5 \ \mathrm{mol \cdot L^{-1}}$ standard NaOH solution, record data.

(2) Determination of cobalt

Put down the Erlenmeyer flask that contains sample solution, wash the solution which is adhered to alkali sealed tube and plug back into Erlenmeyer flask with a little distilled water. Add 1 g solid KI after cooling the sample, surge to make it dissolve, then add about 12 mL $6 \ \mathrm{mol \cdot L^{-1}}$ hydrochloric acid, lay aside for 10 min in dark, then add $60 \sim 70$ mL distilled water, titrate it by $0.1 \ \mathrm{mol \cdot L^{-1}}$ standard $Na_2S_2O_3$ solution, it can be titrated quickly at the beginning, titrate until solution primrose yellow, add drops of phenolphthalein solution, add slowly $Na_2S_2O_3$ solution continuously, until to end point (what is the color of solution at end point?), record data.

(3) Determination of chlorine

Add 0.2 g sample (accurate to 0.1 mg) into Erlenmeyer flask, add proper distilled water to dissolve it, determine content of chlorine by precipitate titration method (Mohr's method).

(4) Data record and result

①Record the data in form.

②To calculate the percentage of nitrogen, cobalt, chlorine of sample.

③Determine the empirical formula of products.

## Thinking the Following Questions

1. What is the purpose of heating at 333K in water bath and keeping constant temperature for 20 min? Whether can it be heated to boil?

2.　What are the role of $H_2O_2$ and concentrated hydrochloric acid in the course of preparation of $[Co(NH_3)_6]Cl_3$? What should be noticed?

3.　What should be paid attention to in determining cobalt (Ⅲ) ion by iodometry method?

# 实验三十二　配合物的离子交换树脂分离

## 实验目的

1. 学习离子交换树脂分离的一般原理。
2. 掌握用离子交换树脂分离配合物离子的基本操作方法。

## 实验原理

离子交换树脂分离是常用的分离方法之一。特别是对于那些性质很相似或者含量很低的元素进行分离，离子交换树脂分离法更显示出它的重要性。

离子交换树脂是一种高分子聚合物，它是由苯乙烯和二乙烯苯等单体聚合而成的高分子聚合物母体，然后引入可交换的活性基团。根据活性基团性质的不同可以分为阳离子交换树脂和阴离子交换树脂两大类。根据交换基团酸碱性的强弱，也可以分为强酸型、弱酸型、强碱型和弱碱型等类型。

本实验所用的阳离子交换树脂，其结构为：

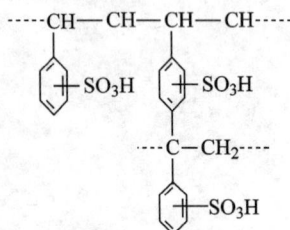

$$----CH—CH—CH—CH----$$

整个树脂可用简式 $RSO_3^- H^+$ 来表示。

离子交换树脂的性质与它的交换性能有着密切的关系。树脂颗粒的大小对树脂的交换能力、水通过树脂层的压力降，以及交换和淋洗时的流失都有很大的影响。树脂颗粒小、总面积大，有利于交换，但颗粒过细，树脂层的压力降大，淋洗时的流失也大。所以树脂颗粒大小的选择需视分离程度的要求而定。在能达到所要求分离程度的前提下，颗粒尽可能选择大些。这样有利于操作并能提高效率。树脂的交联度对交换性能也有影响。交联度是表示树脂结构中交联程度的大小，是指树脂中的乙烯苯的百分重量。交联度大，树脂网眼就小，对交换反应选择性好，但达到平衡的时间增加。目前生产上采用的聚苯乙烯型树脂的交联度一般是 8%～10%。树脂的交换性能和分离效果还与具体的操作条件有关。淋洗速度直接影响树脂的交换性能和分离效果，淋洗速度慢，交换反应进行得完全，分离效果好，但速度太慢，离子向其他方向扩散的机会增加，反而降低分离效果。适当的淋洗速度主要是通过实践来确定。离子交换柱的柱长和直径之比对分离效果也有影响。一般说，柱长和直径比越大，分离效果越好。但柱长太长，直径太小，则会增加吸附层的厚度，使阻力增大，淋洗速度变慢，并增加淋洗液的消耗。实践表明，分离柱的柱长与直径之比一般为 20:1 左右。另外，淋洗液的浓度，操作温度等对树脂的交换性能和分离效果也有一定的影响。

含有阳离子 $M^+$ 的溶液通过上述树脂 $RSO_3^- H^+$ 时，$M^+$ 对树脂 $RSO_3^-$ 有一定的亲和力，并将置换 $H^+$，置换的程度取决于 $M^+$ 的性质和浓度，可用以下的平衡式表示：

$$RSO_3^- \cdot H^+ + M^+ \rightleftharpoons RSO_3^- \cdot M^+ + H^+$$

对于给定的 $M^+$，都有一个特定的平衡常数，而平衡位置将由溶液中 $M^+$ 和 $H^+$ 的相对浓度来决定。如果溶液中 $H^+$ 浓度低，$M^+$ 就将在最大程度上与 $SO_3^-$ 基团结合，增加 $H^+$ 浓度就可以将树脂中结合的 $M^+$ 置换出来，对于不同的阳离子 $M_1^+$ 和 $M_2^+$ 来说，与树脂亲和力较弱的阳离子 $M^+$ 可在 $H^+$ 浓度相对较低时被置换，而为了置换与树脂亲和力较强的 $M^+$ 则要求较高的 $H^+$ 浓度。

本实验中要分离的配合物离子是 $[Cr(H_2O)_4Cl_2]^+$，$[Cr(H_2O)_5Cl]^{2+}$ 和 $[Cr(H_2O)_6]^{3+}$，在 $CrCl_3 \cdot 6H_2O$ 的弱酸性的溶液中由于始终存在着 $Cr^{3+}$ 的水合作用，因此在溶液中会存在 $[Cr(H_2O)_4Cl_2]^+$，$[Cr(H_2O)_5Cl]^{2+}$ 和 $[Cr(H_2O)_6]^{3+}$ 配离子，其相对数量决定溶液的放置时间和温度。当含有着三种配离子的弱酸性溶液（$2 \times 10^{-3}$ mol·L$^{-1}$ HClO$_4$）通过 $RSO_3^-$·H$^+$ 树脂的交换柱时，则与树脂结合最弱的 $[Cr(H_2O)_4Cl_2]^+$ 被淋洗下来，如将 $H^+$ 浓度增加至 $1.0$ mol·L$^{-1}$HClO$_4$ 的酸溶液通过交换柱时，$[Cr(H_2O)_5Cl]^{2+}$ 被淋洗下来，最后用 $3.0$ mol·L$^{-1}$ HClO$_4$ 可把与树脂结合得最牢固的 $[Cr(H_2O)_6]^{3+}$ 淋洗下来，这样就可分离得到三种配离子，分别测定这三种配离子溶液的紫外—可见光谱，对其进行鉴定，确定出各配离子的含量。

## 实验仪器和试剂

仪器：紫外-可见分光光度计，烧杯（100 mL），玻璃交换柱（0.5 cm×15 cm），容量瓶（50 mL），刻度移液管（10 mL），容量瓶（100 mL）

试剂：三氯化铬（CrCl$_3$·6H$_2$O），高氯酸，732 树脂

## 实验步骤

### 1. 树脂的预处理和装柱

（1）树脂的预处理

将市售的树脂用水洗涤多次，除去可溶性杂质，然后用蒸馏水浸泡几小时，使其充分溶涨，再用蒸馏水洗两次，随后用 5 倍树脂体积的 $2\sim3$ mol·L$^{-1}$HCl 浸泡半天，并不断搅拌。使树脂转为 $H^+$ 型。最后用蒸馏水洗去余下的酸，一直洗到使洗涤水的 pH 约等于 3 为止。

（2）装柱

将处理好的树脂和蒸馏水一起慢慢地装入交换柱中，在树脂间不要有空隙和气泡，也不能让树脂干涸，以免影响交换效率，一共装有 $20\sim25$ mL 树脂。

### 2. 溶液的配制

（1）淋洗液的配制

量取一定量高氯酸（70%）分别配制 $0.1$ mol·L$^{-1}$，$1.0$ mol·L$^{-1}$ 和 $3.0$ mol·L$^{-1}$ 的高氯酸溶液各 100 mL。

（2）三氯化铬溶液的配制

称取一定量 CrCl$_3$·6H$_2$O，加入一定量的高氯酸，配制成为 100 mL 含铬为 $0.35$ mol·L$^{-1}$，含 HClO$_4$ 为 $0.002$ mol·L$^{-1}$ 的溶液。本溶液即为 $0.35$ mol·L$^{-1}$ $[Cr(H_2O)_4Cl_2]^+$ 溶液。

### 3. 不同电荷铬配离子溶液的制备及其紫外—可见光谱测定

(1) $[Cr(H_2O)_4Cl_2]^+$ 溶液

将 5 mL 0.35 mol·L$^{-1}$ $[Cr(H_2O)_4Cl_2]^+$ 溶液加入到离子交换柱中，然后排除多余的溶液直至和树脂高度相同。向柱内加入 0.1 mol·L$^{-1}$ HClO$_4$ 溶液淋洗 $[Cr(H_2O)_4Cl_2]^+$ 配离子，淋洗速度约为每秒 2 滴，当流出液出现绿色时开始收集在 50 mL 容量瓶中，至流出液绿色消失为止。用 0.1 mol·L$^{-1}$ HClO$_4$ 溶液稀释到刻度，立即测定其紫外—可见光谱，用 1 cm 比色皿在 350~700 nm 波长进行测定。

(2) $[Cr(H_2O)_5Cl]^{2+}$ 溶液

$[Cr(H_2O)_4Cl_2]^+$ 溶液在加热时会大量转化为 $[Cr(H_2O)_5Cl]^{2+}$，将 5 mL 0.35 mol·L$^{-1}$ 的 $[Cr(H_2O)_4Cl_2]^+$ 溶液在 50℃~60℃ 的水浴中放置 2~3 min，立即把此溶液加入到交换柱中，排除溶液直至其高度与树脂相同，用 0.1 mol·L$^{-1}$ HClO$_4$ 淋洗除去可能还未转化的 $[Cr(H_2O)_4Cl_2]^+$，然后用 1.0 mol·L$^{-1}$ HClO$_4$ 淋洗所需的 $[Cr(H_2O)_5Cl]^{2+}$。用同样的方法收集淋洗液并测定 $[Cr(H_2O)_5Cl]^{2+}$ 的紫外—可见光谱。

(3) $[Cr(H_2O)_6]^{3+}$ 溶液

将 5 mL $[Cr(H_2O)_4Cl_2]^+$ 溶液使其沸腾 5 min，冷却到室温后加入到交换柱中，排除溶液高度与树脂高度相同，先用 1.0 mol·L$^{-1}$ HClO$_4$ 淋洗除去可能未转化的 $[Cr(H_2O)_4Cl_2]^+$ 或 $[Cr(H_2O)_5Cl]^{2+}$，然后用 3.0 mol·L$^{-1}$ HClO$_4$ 淋洗 $[Cr(H_2O)_6]^{3+}$，用同样方法收集蓝色的淋洗液，并测定 $[Cr(H_2O)_6]^{3+}$ 的紫外—可见光谱。

### 4. 三氯化铬溶液中不同配合物离子的分离和鉴定

将 10 mL 放置若干小时的 CrCl$_3$·6H$_2$O 溶液加入到交换柱中，当排除液高度与树脂高度相同时，先用 0.1 mol·L$^{-1}$ HClO$_4$ 溶液淋洗交换柱，接收其蓝色溶液，测定其紫外—可见光谱，接着用 1.0 mol·L$^{-1}$ HClO$_4$ 溶液淋洗交换柱，同样方法接收其蓝色溶液，测定其紫外—可见光谱，最后用 3.0 mol·L$^{-1}$ HClO$_4$ 溶液淋洗交换柱，同样接收其蓝色溶液，测定其紫外—可见光谱。

## 实验结果和处理

1. 由各配合物离子的紫外—可见光谱，确定其特征吸收峰波长 $\lambda$ 和摩尔消光系数 $\varepsilon$。

$[Cr(H_2O)_4Cl_2]^+$: $\lambda$ _____，$\varepsilon$ _____；

$[Cr(H_2O)_5Cl]^{2+}$: $\lambda$ _____，$\varepsilon$ _____；

$[Cr(H_2O)_6]^{3+}$: $\lambda$ _____，$\varepsilon$ _____；

2. 由三氯化铬溶液的离子交换淋洗液的紫外—可见光谱确定其配离子种类及其相对含量。

## 思考题

1. 为什么用高氯酸而不是用盐酸来淋洗交换柱中的 Cr（Ⅲ）配合物离子？

2. 试从配合物离子可见光谱中吸收峰位置的变化说明 Cl$^-$ 和 H$_2$O 的相对配体场强度。

# Exp 32   Separation of Coordination Compound by Ion Exchange Resin

## Objectives

1. To know the principle of ion exchange separation.
2. To grasp the operation methods to separate ion complex by ion exchange resin.

## Principles

Ion exchange separation is one of the most common separation methods. It is most important for those who have similar properties or element with less content, especially for the application of ion exchange resin.

Ion exchange resin is a kind of polymer mixture, it is a high molecular weight matrix polymer composed of monomer polymerizes i. e. styrene and divinylbenzene, then introduce exchangeable active group. It can be divided into cation exchange resin and anion exchange resin according to the properties of active group. According to the acid-base intensity of ion exchange group, it can be divided into strong base, weak acid, strong-base, and weak base and so on.

The structure of cation exchange resin used in this experiments show as follow:

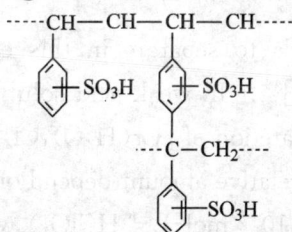

$$\cdots\text{CH}\!-\!\text{CH}\!-\!\text{CH}\!-\!\text{CH}\cdots$$

$$\text{SO}_3\text{H} \qquad \text{SO}_3\text{H}$$

$$\cdots\text{C}\!-\!\text{CH}_2\cdots$$

$$\text{SO}_3\text{H}$$

The structural formulas of resin can be expressed as $RSO_3^- \ H^+$.

The properties of ion exchange resin have close relationship with its ion exchange capacity. The particle size of resin has big effects on exchange capacity of resin, the pressure drop for water pass resin and loss in exchanging and washing. Small resin particles, big total area is beneficial to exchange, but if particle is too small and the pressure drop of resin layer is high, then the loss in eluting is big too. The choice for resin particle size depends on the demand for separation degree. On the premise of degree of separation, particle should be as big as possible. Thus it is convenient to operate and improve the efficiency. The crosslinking degree of resin has effects on exchange capacity. Crosslinking degree is the degree of crosslinking in resin structure; it is mass percentage of styrene in resin. Big crosslinking degree, small mesh of resin, more good selectivity for exchange reaction, but the time to reach balance increases. The general degree of crosslinking of resin of polystyrene style in produce is $8\%\sim10\%$. The exchange capacity and separation effect of resin are also related to the operation condition. The elution rate influences directly on the exchange properties and separation effect of resin; slow elution speed, perfectly the reaction process and separate effect, but as the speed is slow, the chance for ion diffusion to other direction increases, on the contrary the separation effect decreases. The proper elution speed is mainly decided by practice. The length/diameter ratio of column of ion exchange column has effect the separation effect too. Generally speaking, bigger the length/diameter ratio of column is, better the separation effect. But too long column and small diameter would increase the thickness of adsorption layers, and the resistance increase, the elution rate decrease, and increase the dosage of eluting agent. Practice has shown the general column length is in proportion to diameter 20 to 1. In addition, the concentration of elution, operative temperature also influence the exchange capacity and separate effect, too.

If the solution which contains cation $M^+$ passes through the above resin $RSO_3^- H^+$, $M^+$ will attract resin $RSO_3^-$, and the degree of substitution depends on the properties and concentration of $M^+$ the process can be expressed as follow:

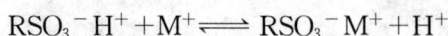

$$RSO_3^- H^+ + M^+ \rightleftharpoons RSO_3^- M^+ + H^+$$

For a certain $M^+$, it has a special equilibrium constant, but the equilibrium point will depend on the relative concentration of $M^+$ and $H^+$ in solution. If the concentration of $H^+$ is small, $M^+$ will combine with $SO_3^-$ group to the greatest degree. Increasing the concentration of $H^+$ will lead to the $M^+$ combined in resin be replaced. For different cation $M_1^+$ and $M_2^+$, the cation $M^+$ that has weak appetency to resin can be replaced in low relative concentration, if want to exchange the $M^+$ that has strong affinity to resin, then the higher concentration of $H^+$ is needed.

The coordinate ion that needs to separate in this experiment is $[Cr(H_2O)_4Cl_2]^+$, $[Cr(H_2O)_5Cl]^{2+}$ and $[Cr(H_2O)_6]^{3+}$. In weak acid solution of $CrCl_3 \cdot 6H_2O$, hydration of $Cr^{3+}$ always exist, so the coordinate ion of $[Cr(H_2O)_4Cl_2]^+$, $[Cr(H_2O)_5Cl]^{2+}$, $[Cr(H_2O)_6]^{3+}$ would exist, the relative amount depend on deposited time and temperature. When the weak acid solution ($2 \times 10^{-3} mol \cdot L^{-1} HClO_4$) which has the three coordinate ions pass through exchange column of resin containing $RSO_3^- H^+$, the weakly combined $[Cr(H_2O)_4Cl_2]^+$ will be eluted. If pass $1.0 mol \cdot mol \cdot L^{-1} HClO_3$ solution through exchange column, $[Cr(H_2O)_5Cl]^{2+}$ is eluted, then by using $3.0 mol \cdot L^{-1} HClO_3$ solution elute down the $[Cr(H_2O)_6]^{3+}$ that combine well with resin, thus the three coordinate ions can be separated, identify the ultraviolet-visible spectrum of three coordinate ion respectively, determine the content of coordinate ions.

## Instruments and Reagents

Instruments: UV-visible spectrophotometer, beaker(100 mL), glass exchange column (0.5 cm×15 cm), volumetric flask(50 mL), graduated pipette(10 mL), volumetric flask (100 mL)

Reagents: chromium chloride ($CrCl_3 \cdot 6H_2O$) (C.P), perchloric acid (C.P), 732-type resin

## Experimental Procedures

### 1. Pretreatment and column packing of resin

(1) Pretreatment of resin

Wash commercial resin several times by water, remove soluble impurity, then dip it in distilled water for a few hours to make it swell well, wash it two times by distilled water, then dip it in 5 times the volume of resin $2\sim3$ $mol \cdot L^{-1}$ HCl for half of the day, and keep stirring to change resin to $H^+$ style. At last wash out the acid left with water, until the pH value of washing water is 3, then resin can be used.

(2) Loading column

To load the well managed resin and distilled water into exchange column, neither the space and air bubble in resin nor drying up resin are permitted, in case it effect exchange efficiency, the total amount of resin is $20\sim25$ mL.

### 2. Preparation of solution

(1) Preparation of elution solution

Fetch an amount of perchloric acid (70%) to prepare 100 mL of 0.1 $mol \cdot L^{-1}$, 1.0 mol $\cdot L^{-1}$ and 3.0 $mol \cdot L^{-1}$ respectively.

(2) Preparation of chromium chloride

Weigh some amount of $CrCl_3 \cdot 6H_2O$, add certain amount of perchloric acid to prepare 100 mL solution which containing 0.35 $mol \cdot L^{-1}$ chromium and 0.002 $mol \cdot L^{-1}$ $HClO_4$. This solution is the solution of 0.35 $mol \cdot L^{-1}[Cr(H_2O)_4Cl_2]^+$.

### 3. Preparation of different charged chromium complex ion and determination of its ultraviolet-visible spectrum

(1) $[Cr(H_2O)_4Cl_2]^+$ solution

Add 5 mL 0.35 $mol \cdot L^{-1}[Cr(H_2O)_4Cl_2]^+$ into ion exchange column, then remove excess solution until its height same as resin. Pour 0.1 $mol \cdot L^{-1}HClO_4$ into column to elute$[Cr(H_2O)_4Cl_2]^+$ coordinate ion, eluting speed is 2 drops per second, begin to collect solution in 50 mL volumetric flask from the color of solution flown out appearing itself green to green disappearing. Dilute the solution to 50 mL by using 0.1 $mol \cdot L^{-1}$ $HClO_4$, determine immediately its ultraviolet-visible spectrum, and determine the wavelength in $350\sim700$ nm by using 1cm sampling cells.

(2) $[Cr(H_2O)_5Cl]^{2+}$ solution

$[Cr(H_2O)_4Cl_2]^+$ solution will transfer into $[Cr(H_2O)_5Cl]^{2+}$ while heating, lay 5 mL 0.35 $mol \cdot L^{-1}[Cr(H_2O)_4Cl_2]^+$ in $50\sim60℃$ water bath for $2\sim3$ min, add this solution immediately into exchange column, remove the solution until its height is same with resin, elute the under transferred $[Cr(H_2O)_4Cl_2]^+$ by using 0.1 $mol \cdot L^{-1}HClO_4$, then elute the wanted $[Cr(H_2O)_5Cl]^{2+}$ with 1.0 $mol \cdot L^{-1}$ $HClO_4$. Collect the elute by the same method and determine ultraviolet-visible spectrum of $[Cr(H_2O)_5Cl]^{2+}$.

(3) $[Cr(H_2O)_6]^{3+}$ solution

Boil the solution of 5 mL $[Cr(H_2O)_4Cl_2]^+$ for 5 min, cool to room temperature and add it into exchange column, the height of eluting is same as that of resin, elute the under transferred $[Cr(H_2O)_4Cl_2]^+$ or $[Cr(H_2O)_5Cl]^{2+}$ with 1.0 mol·$L^{-1}$ $HClO_4$, then elute $[Cr(H_2O)_6]^{3+}$ with 3.0 mol·$L^{-1}$ $HClO_4$, collect the blue elution by using the same method, determine ultraviolet-visible spectrum of $[Cr(H_2O)_6]^{3+}$.

### 4. Separation and identification of chromium complex ion in chromium chloride

Add 10 mL $CrCl_3 \cdot 6H_2O$ solution that lay aside several hours into exchange column. When the height of elute is same as that of resin, elute exchange column firstly with 0.1 mol·$L^{-1}$ $HClO_4$ solution, collect green solution with the method with step 3, determine its ultraviolet-visible spectrum immediately, then elute exchange column with 1.0 mol·$L^{-1}$ $HClO_4$, collect blue solution by same method, determine the ultraviolet-visible spectrum, at last elute exchange column by using 3.0 mol·$L^{-1}$ $HClO_4$, also collect its blue solution, determine the ultraviolet-visible spectrum.

## Data Process

1. By ultraviolet-visible spectrum of coordinate ion, determine the wavelength $\lambda$ of characteristic absorption peak and its molar extinction coefficient $\varepsilon$.

$[Cr(H_2O)_4Cl_2]^+$: $\lambda$ _____, $\varepsilon$ _____;

$[Cr(H_2O)_5Cl]^{2+}$: $\lambda$ _____, $\varepsilon$ _____;

$[Cr(H_2O)_6]^{3+}$: $\lambda$ _____, $\varepsilon$ _____;

2. Determine the kind of coordinate ion and it relative content in ion exchange eluting solution of chromium chloride solution by ultraviolet-visible spectrum.

## Questions

1. Why do we use not hydrochloric acid but perchloric acid to elute Cr (Ⅲ) coordinate ion in exchange column?

2. Try to explain the ligand field intensity of $Cl^-$ and $H_2O$ from the change of the location of absorption peak in visible spectrum of coordinate ion.

# 实验三十三　生物中几种元素的定性鉴定

## 实验目的

1. 通过实验初步了解动植物体内某些重要元素的简单检验方法。
2. 进一步练习溶液配制操作。

## 实验仪器、药品及材料

仪器：小试管、漏斗、石棉网、坩埚、泥三角、燃烧勺

固体药品：红磷、石灰石

液体药品：$HNO_3$（$0.1\ mol\cdot L^{-1}$，$6\ mol\cdot L^{-1}$，浓）、$(NH_4)_2MoO_4$ 溶液、$K_4[Fe(CN)_6]$ 溶液、KSCN 溶液、$(NH_4)_2C_2O_4$ 溶液、（浓）$NH_3\cdot H_2O$

材料：树叶、棉花、骨头（小块）、鸡蛋黄

## 实验步骤

1. 按照附注 2 的方法配制钼酸铵、六氰合亚铁酸钾、硫氰化钾和草酸铵溶液。

2. 原料的灰化

准备几枚树叶（枯叶、青叶皆可），春天青叶取 6 g，枯叶 2.5 g。用镊子夹取树叶，直接在酒精灯上（或煤气灯上）加热燃烧，待碳化后，将已碳化的叶子放在石棉网上或坩埚中，继续加热至灰化完全。

3. 硝化和分解

将灰分移入试管中，加入浓 $HNO_3$ 3 mL。灰分中的磷转化成磷酸，铁转化成铁（Ⅲ）离子，钙转化成钙（Ⅱ）离子。再加入 2～3 mL 水，过滤，用 1 mL 水洗涤滤纸。

4. 测定

将滤液分成 4 份，分别加入 $0.1\ mol\cdot L^{-1}$ $(NH_4)_2MoO_4$（A 管），$0.1\ mol\cdot L^{-1}$ $K_4[Fe(CN)_6]$（B 管），$0.1\ mol\cdot L^{-1}$ KSCN（C 管）和饱和 $(NH_4)_2C_2O_4$（D 管）试剂。观察现象，判断四支试管中各检出何物，写出反应方程式。

5. 对照实验

取绿豆大的红磷于燃烧勺中，加热使其燃烧，单质磷变成 $P_2O_5$。加 1 mL 水加热沸腾，再加入 2～3 滴 $6\ mol\cdot L^{-1}$ $HNO_3$ 和 $0.1\ mol\cdot L^{-1}$ $(NH_4)_2MoO_4$ 试剂，观察颜色。并与 A 管颜色比较。

取 1mL $6\ mol\cdot L^{-1}$ $HNO_3$，加入小指头大小的一块棉花，加热后再加入 2～3 mL 水，过滤，滤液分成两份。一份中加入 $0.1\ mol\cdot L^{-1}$ $K_4[Fe(CN)_6]$，另一份加入 $0.1\ mol\cdot L^{-1}$ KSCN，与 B，C 管颜色比较。

取一小块石灰石，加入 $0.1\ mol\cdot L^{-1}$ $HNO_3$ 溶解，加入 2 mL 水，再加 $NH_3\cdot H_2O$ 呈碱性后，加入 $(NH_4)_2C_2O_4$ 与 D 管比较。

6. 取小指头大小的一块动植物骨头，1/3 个鸡蛋黄（放在坩埚中）灰化，用硝酸处理，然后按上述方法分别进行钙、铁、磷元素的鉴定。

## 思考题

原料在灰化时燃烧不完全，对实验结果有何影响？

## 附注

1. 该实验鉴定方法还可用来检测以下原材料

肉类(磷、钙)，筋头(磷、铁)，血液(磷、铁)，贝壳(磷、钙)，鸟肉(磷、钙、铁)，DNA(脱氧核糖核酸)，RNA(核糖核酸，含 P)，植物种子(铁、钙、磷)，红萝卜(钙、磷、铁)，牛奶(钙、磷)，藻类(磷、钙、铁) 等。

2. 试剂配制

$(NH_4)_2MoO_4$ 溶液：1 g $(NH_4)_2MoO_4$ 固体加 10 mL 水。

$K_4[Fe(CN)_6]$ 溶液：1 g $K_4[Fe(CN)_6]$ 固体加 10 mL 水。

KSCN 溶液：1g KSCN 固体加 10 mL 水。

$(NH_4)_2C_2O_4$ 溶液：1 g $(NH_4)_2C_2O_4$ 固体加 10 mL 水。

# Exp 33   Qualitative Analysis of a Few Elements in Organism

## Objectives

1. To know the simple identification method of a few elements in animal and plant body through experiment

2. To further practice the operation of formulating method of solution

## Instruments, Reagents and Materials

Instruments: small tube, funnel, asbestos net, crucible, mud triangle, burning spoon

Solid reagents: red phosphor, limestone

Liquid reagents: $HNO_3$ (0.1 mol·$L^{-1}$, 6 mol·$L^{-1}$ and concentrated), $(NH_4)_2MoO_4$, $K_4[Fe(CN)_6]$, KSCN, $(NH_4)_2C_2O_4$, (concentrated) $NH_3 \cdot H_2O$

Materials: leaves, cotton, bone (nubble), yolk

## Experimental Procedures

1. Prepare $(NH_4)_2MoO_4$, $K_4[Fe(CN)_6]$, KSCN, $(NH_4)_2C_2O_4$ solutions following additional information 2.

2. Ashing of material

Prepare several leaves (dried leaves or leafiness), for leafiness in spring, 6 g. dried leaves 2.5 g. Take leaves with forceps, burn with alcohol burner directly, and then lay carbonized leaves on a asbestos net or a crucible, continuing to carbonize them completely.

3. Nitration and decomposing

Transfer ashes to a test tube, 3 mL $HNO_3$ are added. Phosphor in ashes is transformed to phosphoric acid, iron to $Fe^{3+}$, and calcium to $Ca^{2+}$. Add 2~3 mL $H_2O$, filter, wash filter paper with 1 mL $H_2O$.

4. Analysis

Divide the filtrate into 4 portions, 0.1 mol·$L^{-1}$ $(NH_4)_2MoO_4$ (A), 0.1 mol·$L^{-1}$ $K_4[Fe(CN)_6]$ (B), 0.1 mol·$L^{-1}$ KSCN (C) and saturated $(NH_4)_2C_2O_4$ (D) are added respectively. Observe the phenomena, judge what elements have been identified in 4 test tubes, write out reaction equations.

5. Control experiment

Take red phosphor (a mung bean size) in a burning spoon, heat it to burn, phosphor is transformed to $P_2O_5$. Add 1 mL $H_2O$ to boil, another 2~3 drops of 6 mol·$L^{-1}$ $HNO_3$ and 0.1 mol·$L^{-1}$ $(NH_4)_2MoO_4$ are added, observe the color of solution, compare the color with tube A.

Take 1 mL 6 mol·$L^{-1}$ $HNO_3$, a piece of cotton (a little finger size) is added, heat and

2~3 mL $H_2O$ is added, filter, divide the filtrate into 2 portions. 0.1 mol·$L^{-1}$ $K_4[Fe(CN)_6]$ is added to one, another 0.1 mol·$L^{-1}$ KSCN, compare the color with tubes B and C.

Take a bit of limestone, 0.1 mol·$L^{-1}$ $HNO_3$, 2 mL $H_2O$ and $NH_3 \cdot H_2O$ are added to alkaline, $(NH_4)_2C_2O_4$ is added last, compare the color with tube D.

6. Take an animal bone (a little finger size), 1/3 of egg yolk in a crucible, transform them to ashes, dispose with $HNO_3$, following above methods to identify element Ca, Fe and P.

## Questions

How does incomplete combustion of material influence result of experiment?

## Additional information

1. Those following starting material can also be identified with just this method mentioned in above experiment.

Flesh(Ca, P), tendon(P, Fe), blood(P, Fe), shells(P, Ca), bird flesh(P, Ca, Fe), DNA(deoxyribose nucleic acid), RNA(ribonucleic acid, containing P), plant seeds (Fe, Ca, P), garden radish(Ca, P, Fe), milk(Ca, P), alga(P, Ca, Fe) *etc*.

2. Preparation of solutions:

$(NH_4)_2MoO_4$ solution: 10 mL $H_2O$ per 1 g $(NH_4)_2MoO_4$

$K_4[Fe(CN)_6]$ solution: 10 mL $H_2O$ per 1 g $K_4[Fe(CN)_6]$

KSCN solution: 10 mL $H_2O$ per 1 g KSCN

$(NH_4)_2C_2O_4$ solution: 10 mL $H_2O$ per 1 g $(NH_4)_2C_2O_4$

# 实验三十四　水热法制备纳米 $SnO_2$ 微粉

## 实验目的

1. 了解水热法制备纳米粒子的反应原理。
2. 学习如何控制对产物微晶性状产生影响的水热反应条件。

## 实验原理

纳米粒子（nanosized particles）通常是指粒径大约 $1{\sim}100$ nm 的超微颗粒。物质处于纳米尺度状态时，其许多性质既不同于原子、分子，又不同于大块体相物质，而是构成物质的一种"新状态"——介观态（mesoscopic state）。处于介观态的纳米粒子，其中电子的运动受到颗粒边界的束缚而被限制在纳米尺度内，当粒子的尺寸可以与其中电子（或空穴）的 De Broglie 波长相比时，电子运动呈现显著的波粒二象性，此时材料的光、电、磁性质出现许多新的特征和效应。例如量子尺寸效应使半导体的带隙能增大，光吸收带边蓝移。磁性材料中出现由多畴到单畴，铁磁性到超顺磁性的转变等。从化学角度来看，在纳米材料中，位于表、界面上的原子数足以与粒子内部的原子数相抗衡，因而总表面积能大大增加，粒子的表、界面化学性质异常活泼，此特征通常称为表、界面效应。另外还将会产生宏观量子隧道效应，介电限域效应等。纳米粒子的这些新的特性为物理学、电子学、化学和材料科学等开辟了全新的研究领域[1-2]。

$SnO_2$ 是一种半导体氧化物，它在传感器、催化剂和透明导电薄膜等方面具有广泛用途。纳米 $SnO_2$ 具有很大的比表面积，是一种很好的气敏与湿敏材料[3]。制备超细 $SnO_2$ 微粉的方法很多，如 Sol−Gel 法、激光分解法、水热法等。水热法制备纳米氧化物微粉具有产物直接为晶态，无需经过焙烧晶化过程，采用适当措施能尽量减少颗粒团聚，粒度比较均匀，形态比较规则的优点。因此，水热法是制备纳米氧化物微粉的好方法之一[4]。

水热法是指在温度超过 $100{}^\circ\!\mathrm{C}$ 和高于常压条件下利用水溶液（溶剂介质也可不是水）中物质间的化学反应合成化合物的方法[5]。在水热条件下（相对高的温度和压力）下，水的反应活性提高，其蒸气压上升、离子积增大，而密度、表面张力及粘度下降。体系的氧化−还原电势发生改变。物质在水热条件下的热力学性质均不同于常态。因此，水热条件下，由于反应物和溶剂活性的提高，有利于某些特殊中间态及特殊物相的形成，可能合成具有某些特殊结构的新化合物。在水热条件下有利于某些晶体的生长，可以获得纯度高、取向规则、非平衡态缺陷尽可能少的晶体材料；通过改变水热反应条件，可能形成具有不同晶体结构和结晶形态的产物，也有利于低价、中间价态与特殊价态化合物的生成。水热合成化学日益受到化学与材料科学界的重视。

本实验以水热反应制备纳米微晶 $SnO_2$，适当的条件下，$SnCl_4$ 在逐渐碱化时发生水解，形成无定形的 $Sn(OH)_4$。反应方程式：

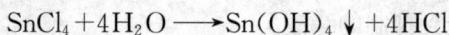

$$SnCl_4 + 4H_2O \longrightarrow Sn(OH)_4 \downarrow + 4HCl$$

生成的 $Sn(OH)_4$ 紧接着产生脱水缩合和晶化作用，形成 $SnO_2$ 纳米微晶。

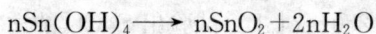

$$nSn(OH)_4 \longrightarrow nSnO_2 + 2nH_2O$$

## 仪器和试剂

仪器：100 mL 不锈钢压力釜（具有聚四氟乙烯衬里），管式电炉套及温控装置，电动搅拌器，抽滤水泵，pH 计

试剂：$SnCl_4 \cdot 5H_2O$（分析纯），KOH（分析纯），乙酸（分析纯），乙酸铵（分析纯），$\omega = 95\%$ 乙醇（分析纯）

## 实验步骤

### 1. 原料液的配制

用蒸馏水配制 $1.0 \ mol \cdot L^{-1}$ 的 $SnCl_4$ 溶液，$10 \ mol \cdot L^{-1}$ 的 KOH 溶液。

每次取 50 mL 的 $1.0 \ mol \cdot L^{-1}$ 的 $SnCl_4$ 溶液于 100 mL 烧杯中，在电磁搅拌下逐滴加入 $10 \ mol \cdot L^{-1}$ 的 KOH 溶液，调节反应液的 pH 至所要求值（例如 1.45），制得的原料液待用。观察记录反应液状态随 pH 的变化。

### 2. 水热反应

将配制好的原料液倾入具有聚四氟乙烯衬里的不锈钢压力釜内，用管式电炉套加热压力釜。用控温装置控制压力釜的温度，使水热反应在所要求的温度下进行一定时间（~2h）。为保证反应的均匀性，水热反应应在搅拌下进行。反应结束，停止加热，待压力釜冷却至室温时，开启压力釜，取出反应产物。

### 3. 反应条件的选择

（1）反应温度

反应温度低时，$SnCl_4$ 水解、脱水缩合和晶化作用慢。温度升高将促进 $SnCl_4$ 的水解和 $Sn(OH)_4$ 脱水缩合，同时重结晶作用增强，使产物晶体结构更完整，但也导致 $SnO_2$ 微晶长大。分别将反应温度控制为 120℃、140℃、160℃、180℃和 200℃进行实验，选择出反应的适宜温度范围。

（2）反应介质的酸度

当反应介质的酸度较高时，$SnCl_4$ 的水解受到抑制，中间物 $Sn(OH)_4$ 生成相对较少，脱水缩合后，形成的 $SnO_2$ 晶核数量较少，大量 $Sn^{4+}$ 离子残留在反应液中。这一方面有利于 $SnO_2$ 微晶的生长，同时也容易造成粒子间的聚结，导致产生硬团聚。

当反应介质的酸度较低时，$SnCl_4$ 水解完全，大量 $Sn(OH)_4$ 质点同时形成。在水热条件下，经脱水缩合和晶化，形成大量 $SnO_2$ 纳米微晶。此时，由于溶液中残留的 $Sn^{4+}$ 离子数量已很少，生成的 $SnO_2$ 微晶较难继续生长。因此产物具有较小的平均颗粒尺寸，粒子间的硬团现象也相应减少。本实验反应介质的酸度控制为 pH=1.45。

（3）反应物的浓度

反应物浓度愈高，产物 $SnO_2$ 的产率愈低。这主要是由于当 $SnCl_4$ 浓度增大时，溶液的酸度也增大，$Sn^{4+}$ 的水解受到抑制的缘故。

将温度控制在上述选出的范围内，酸度控制为 pH=1.45，分别将反应物浓度调整为 $0.1 \ mol \cdot L^{-1}$，$1.0 \ mol \cdot L^{-1}$ 和 $2.0 \ mol \cdot L^{-1}$ 试验出 $SnCl_4$ 的适宜浓度。

**4. 反应产物的后处理**

在选择出的适宜反应条件下，用实验步骤3.（2）中的方法得到反应产物。静置沉降，移去上层清液后减压过滤。过滤时应用致密的细孔滤纸，尽量减少穿滤。用大约 100 mL10％的乙酸加入 1g 乙酸铵的混合液洗涤沉淀物 4～5 次（防止沉淀物胶溶穿滤），洗去沉淀物中的 $Cl^-$ 和 $K^+$ 离子，最后用 95％的乙醇洗涤两次，于 80℃干燥，然后研细。

**5. 反应产物的表征**

（1）物相分析

用多晶 X 射线衍射法（XRD）确定产物的物相。在 JCPDS 卡片集中查出 $SnO_2$ 的多晶标准衍射卡片，将样品的 $d$ 值和相对强度与标准卡片上的数据相对照，确定产物是否为 $SnO_2$。

（2）粒子大小分析

用 Schererr 公式：

$$d_{hkl}=(K \cdot \lambda)/(\beta \cdot \cos\theta_{hkl})$$

计算样品在 hkl 方向上的平均晶粒尺寸。其中 $\beta$ 为扣除仪器因子后 hkl 衍射的半高宽（弧度）。$K$ 为常数，通常取 0.9。$\theta_{hkl}$ 为衍射峰的衍射角，$\lambda$ 为 X 射线的波长。

用投射电子显微镜（TEM）直接观察样品粒子的尺寸与形貌。

（3）比表面积测定

用 BET 法测定样品的比表面积，并计算样品的平均等效粒径。

（4）等电点测定

用显微电泳仪测定 $SnO_2$ 颗粒的等电点。

## 思考题

1. 比较同一样品由 XRD、TEM 和 BET 法测定的粒子大小，并对各自测定结果的物理含义作分析比较。

2. 水热法作为一种非常规无机合成方法具有哪些特点？

3. 用水热法制备纳米氧化物时，对物质本身有哪些基本要求？试从化学热力学和动力学角度进行定性分析。

4. 水热法制备纳米氧化物过程中，哪些因素影响产物的粒子大小及其分布？

5. 在洗涤纳米粒子沉淀物过程中，如何防止沉淀物的胶溶？

## 参考文献

［1］林鸿溢. 纳米材料与纳米技术. 材料导报，1993，（6）：42-46.

［2］纳米技术及其检测仪器专辑. 现代科学仪器，1998，1-2.

［3］李泉，曾广赋，席时权. 二氧化锡气敏材料的研究进展. 应用化学，1994，11（6）：1-6.

［4］程虎民，马季铭，赵振国，等. 纳米 $SnO_2$ 的水热合成. 高等学校化学学报，1996，17（6）：833-837.

［5］唐有祺. 当代化学前沿. 北京：中国致公出版社，1997，6-7.

# Exp 34   Hydro-thermal Preparation of Superfine $SnO_2$ Powder

## Objectives

1. To understand the principle of hydro-thermal preparation of nanoparticle.

2. To learn how to control the hydro-thermal condition which influences the property of microcrystal.

## Principle

Nanosized particles usually is superfine particles with the diameter of about 1~100 nm. When substance is in nanosized scale, many properties neither are different from atom and molecule nor from macroscopical substance, form a "new state" of substance—mesoscopic state. Nanometer particles in mesoscopic state, their electron movement is limited in nanometer scale because of the bondage of particle boundary, when particle size corresponds with the De Broglie wavelength of electron (or cavity) size, electron movement present distinct wave-particle duality, then many new characteristic and effects of materials in photo, electricity and magnetism would appear. For example, band energy would increase and light absorption edge shift to blueness because of quantum size effects. Magnetism material will transfer from multi-domain to single-domain, from ferromagnetism to super paramagnet in magnetic materials. In the view of chemistry, in nanosized material, the atom number on surface and interface is enough to rival with the atom number in particle, as a result, total special surface area can increase greatly, the chemical characters of particle surface and interface is especially lively, this character usually be called surface and interface effects. Besides, the macro quanta tunnel effects and dielectric confinement effects would appear. These new characters of nanosized particles point out a new study area for physics, electronics, chemistry and material science, it will lead a new technique revolution in beginning of 21 century. It is one of main wet chemical methods to prepare nanosized particles by chemical method. At the same time, hydrothermal method itself develops endlessly; provide a new possible route to prepare non-oxide nanosized materials by hydrothermal method using organic solvent as medium.

$SnO_2$ is a kind of semiconductor oxide; it is widely used in aspect of sensor, catalyst and transparent conductive film and so on. Nanosized $SnO_2$ has very large specific surface area, is a very proper material for gas and humidity-sensitivity . There are many kinds of methods to prepare superfine $SnO_2$ powder, such as Sol-Gel, chemical deposition, laser decomposition, hydro-thermal method etc. There are many merits in hydro-thermal method to prepare nanosized oxide powder, for example, product is crystal state directly, need not be baked and crystallized, so particle reunion which is difficult to avoid decrease in this method, and usually has uniform granularity and regular shape. So hydro-thermal method is one of good methods to prepare nanosized oxide powder.

Hydro-thermal method is a method to use the chemical reaction between substance to synthesize compound in aqueous solution (generally speaking, solvent not always water) in the condition of exceeding 100℃ and corresponding pressure (higher than ordinary pressure).

In the condition of hydro-thermal method (relatively high temperature and pressure), the reaction activity of water enhances, its steam pressure increases, ion product increases, but density, surface tension and viscosity decreases, the oxidation reduction potential of system change. In a word, the thermodynamics characters of substance in hydro-thermal condition are all different from normal state; so, in hydro-thermal condition, because of improved activity of reactant and solvent, it is propitious to form some special intermediates and phase, so it is possible to synthesize new compound which has some special structure; in hydro-thermal is propitious to grow some crystal, obtain crystal materials which has high purity, regular tropism, perfect shape and less non-balance state lacuna; By changing hydro-thermal reaction condition, it is possible to form the products which has different crystal structure and shape, also is propitious to obtain the compound which has low valence, middle valence and special valence. Hydro-thermal synthesis chemistry obtains recognition increasingly in the area of chemistry and material science.

In this experiment, nano microcrystal SnO$_2$ is prepared through hydro-thermdal reaction, its principle is as following:

SnCl$_4$ is hydrolyzed in gradually basic condition to form amorphous Sn(OH)$_4$, reaction equation is as following:

$$SnCl_4 + 4H_2O \longrightarrow Sn(OH)_4 \downarrow + 4HCl$$

The Sn(OH)$_4$ forms nanosized SnO$_2$ microcrystal after dehydration and crystallization.

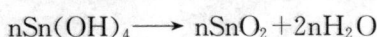

$$nSn(OH)_4 \longrightarrow nSnO_2 + 2nH_2O$$

## Instruments and Reagents

Instruments: 100 mL stainless steel pressure kettle (with polytetrafluoroethylene), electric furnace tube and temperature control device, electric stirrer, water pump, pH meter

Reagents: SnCl$_4$·5H$_2$O(A. R. ), KOH(A. R. ), acetic acid(A. R. ), ammonium acetate(A. R. ), $\omega = 95\%$ ethanol(A. R. )

## Experimental Procedures

### 1. Preparation of raw material

Prepare 1.0 mol·L$^{-1}$ SnCl$_4$ and 10 mol·L$^{-1}$ KOH solution with distilled water. Add 50 mL 1.0 mol·L$^{-1}$ SnCl$_4$ solution each time into 100 mL beaker, add 10 mol·L$^{-1}$ KOH solution dropwise with stirring, adjust the necessary pH value (i. e. 1.45) of reaction solution, lays aside the obtained raw material. Observe and record the change of pH value with the state of reaction solution.

### 2. Hydro-thermal reaction

To pour the prepared raw solution into stainless steel kettle with polytetrafluoroethylene lining, heat pressure kettle by electric furnace tube. Control the temperature in pressure kettle by temperature controller to keep hydro-thermal reaction in desired temperature for a period of time (~2h). Hydro-thermal reaction should be kept stirring in order to ensure u-niform reaction. Stop heating after finishing reaction. Open pressure kettle and take out products when the temperature in pressure kettle cool to room temperature.

### 3. Choice of reaction condition

(1) Reaction temperature

At low reaction temperature, the hydrolysis, dehydrate and crystallization $SnCl_4$ is slow. Increasing temperature would improve the hydrolysis of $SnCl_4$ and dehydrate of Sn $(OH)_4$ dehydrate, recrystallization increase at the same time, make the crystal structure of products more perfect, but also cause the growth of $SnO_2$ minicrystal. In this experiment the reaction temperature 120~160℃ is better.

(2) The acidity of reaction medium

In high acidity of reaction medium, the hydrolysis of $SnCl_4$ would be depressed, inter-mediate $Sn(OH)_4$ obtained is relative less, after dehydration, less number of $SnO_2$ crystal nucleus would be obtained, a great deal of $Sn^{4+}$ ion stays in reaction solution. On the one hand, it is useful for the growth of $SnO_2$ microcrystal, on the other hand it easily makes the particles gather together, lead hard reunion appear, it should be avoided in nanometer parti-cles preparation.

In lower acidity of reaction medium, $SnCl_4$ hydrolyzes completely, large quantity of small size $Sn(OH)_4$ particle form. In hydro-thermal condition, by dehydrating and conden-sing, crystallization, form amount of $SnO_2$ nanosized microcrystal. Here, because of the less amount of left $Sn^{4+}$ ion, it is difficult to keep the growth of formed $SnO_2$ microcrystal. So the products have smaller average particle size, the hard reunion between particles is re-duced. The acidity in the reaction medium of this experiment be controlled at pH=1.45.

(3) Concentration of reactant

When checking the effect of reactant concentration solely, more high the reactant con-centration, more less the rate of $SnO_2$ produced. This is mainly because when increasing the concentration of $SnCl_4$, the acidity of solution is also high, the hydrolysis of $Sn^{4+}$ ion is de-pressed.

Control the temperature in the range of step 3. (1), the pH=1.45, adjust the concen-tration of reactant at 0.1 mol·$L^{-1}$, 1.0 mol·$L^{-1}$, 2.0 mol·$L^{-1}$ respectively then determine the proper concentration.

### 4. Subsequent treatment of reaction products

Lay aside the products to let it sedimentation; filter the precipitate under reduced pressure after decantation upper clear solution. Use fine filter paper for filtration, to reduce penetration as far as possible. To wash precipitate four or five times by using the mixture of about 100 mL 10% acetic acid and 1g acetic ammonium (in case of precipitate peptization), wash out the $Cl^-$ and $K^+$ ion in precipitate, at last, wash it twice with $\omega=95\%$ ethanol, dry it at 80℃, then comminute it.

### 5. Characterization of reaction products

(1) Phase analysis

Determine the phase of products by poly-crystal X ray diffraction (XRD). To look up polycrystalline standard diffraction card of $SnO_2$ in JCPDS cards, contrast the data $d$ value of sample with that in standard card, make sure if product is $SnO_2$.

(2) Analysis on particle size

By using the semi full width at half maximum of polycrystalline X ray diffraction peak, schererr formula:

$$d_{hkl}=(K \cdot \lambda)/(\beta \cdot \cos\theta_{hkl})$$

To calculate the average particle size of sample in direction of hkl orientation. Where $\beta$ is full width at half maximum (degree of arc) of hkl diffraction taken out instrument facts. $K$ is constant, usually is 0.9. $\theta_{hkl}$ is diffraction angle of diffraction peak. $\lambda$ is wavelength of X ray.

To observe the size and shape of particle sample direct by transmission electron microscope (TEM).

(3) Determination of special surface

To determine the special surface of sample by BET method, calculate the average equivalent particle size of sample.

(4) Determination of isoelectric point

Determine the iso-electric point of $SnO_2$ particles by micro electrophoresis.

## Questions

1. To compare the particle size data of same sample obtained from determination by XRD, TEM and BET, then analyses and compare the physical significance of all experiment results.

2. As a normal method in inorganic synthesis, what characters does hydro-thermal method have?

3. When prepare nanometer oxide by hydro-thermal method, what basic demand is required for substance itself? Try to give a qualitative analysis in the view of chemical thermodynamics and dynamics.

4. In the course of preparing nanometer oxide by hydro-thermal method, which facts

affect the particle size and distributing of products?

5. How to prevent precipitate from peptization in the course of washing nanometer particle precipitate?

# References

[1] Lin Hongyi. Nanomaterial and nanotechnology [J]. Material Review, 1993, 6: 42 -45

[2] Album of nanotechnology and detecting instruments. Modern Science Instruments, 1998, 1-2.

[3] Li Quan, Zeng Jiming, Xi Shiquan. Tin dioxide gas sensing material [J]. Chin J. Appl Chem, 1994, 11 (6): 1-6

[4] Cheng Humin, Ma Jimin, Zhao Zhenguo, et al. Hydrothermal synthesis of nanosized SnO_2. Chem Res Chin Univ, 1996, 17 (6): 833-837.

[5] Tang Youqi. Contemporary frontier of chemistry. [M] Beijing: Chinese Zhigong Press, 1997, 6-7.

# 实验三十五　$C_{60}$衍生物的光化学合成和表征

## 实验目的

1. 了解富勒烯基本化学反应特性。
2. 了解光化学合成，液相柱色谱分离提纯方法。
3. 熟悉 NMR，UV−Vis，IR 等测试手段的运用。

## 实验原理

富勒烯（Fullerene）是全部由碳原子组成的一大类分子的总称。其中最具代表性的富勒烯分子是 1985 年首次报道的足球状 $C_{60}$。随后又陆续发现了橄榄状、管状、洋葱状同系物。富勒烯是继石墨、金刚石之后被发现的第三种碳的同素异构体。以 $C_{60}$ 为代表的富勒烯及其衍生物的制备、性质研究是富勒烯科学的一个重要分支。富勒烯类新材料的许多不寻常特性使它在现代科技和工业部门中得到很多的实际应用，包括在润滑剂、催化剂、高强度碳纤维、半导体、非线性光学材料、超导材料、光导体、高能电池、燃料、传感器、分子器件等许多领域都具有潜在应用价值。

$C_{60}$ 被认为是三维欧几里德空间可能存在的对称性最高、最圆的分子。$C_{60}$ 分子的表面由 12 个五边形和 20 个六边形组成，整个分子的外形为具有 60 个顶点的球形 32 面体，其分子属 $I_h$ 点群，所有 60 个碳原子全部等价，每个碳原子周围只有 3 个碳原子。这样的结构使 $C_{60}$ 分子非常坚固和稳定，它可以 $2.7 \times 10^4 \mathrm{km/h}$ 的速度与刚性物体相撞而不破裂；在常压、空气条件下，$C_{60}$ 固体加热到 450℃ 才开始燃烧。

$C_{60}$ 分子的成键特征比金刚石和石墨复杂。由于球状表面的弯曲效应和五元环的结构，引起分子杂化轨道的变化。与石墨相比，$\pi$ 电子轨道不再是纯的 p 原子轨道，而是含有部分 s 轨道的成分，因此 $C_{60}$ 分子中 C 原子的杂化轨道处于 $sp^2$（石墨晶体）和 $sp^3$（金刚石晶体）杂化之间。$C_{60}$ 分子中每个碳原于以 $sp^{2.28}$ 杂化形成 3 个 $\sigma$ 键，再以 $s^{0.09}$P 杂化形成离域 $\pi$ 键，$\sigma$ 键沿球面方向，而 $\pi$ 键分布在球的内外表面，从而形成具有芳香性的球状分子。与苯分子中所有化学键等长，所不同的是，$C_{60}$ 分子的化学健分为两类，长键（五元环与六元环间），键长为 146pm，短键（两个六圆环间），健长为 139 pm（与苯环中的碳碳键长相同）。$C_{60}$ 分子的这种结构使它比苯更易于发生加成反应，生成一系列的加成化合物[1]。

由于 $C_{60}$ 分子是一个非极性分子，只在一些芳香性溶剂中有一定溶解度，但在极性有机溶剂中溶解度很小，在水中的溶解度则几乎为零，这在很大程度上限制了它的应用。氨基酸有很强的亲水性，它与 $C_{60}$ 通过加成反应生成的衍生物能溶解于水。如 Wudl 等人报道用胺类化合物与富勒烯加成合成了一个水溶性 $C_{60}$ 衍生物，并具有生物效应[2]。

本实验将亚氨基二乙酸甲酯在光照条件下直接与 $C_{60}$ 反应，选择性地生成单加成衍生物。

亚氨基二乙酸甲酯与 $C_{60}$ 的光化学反应方程式如下：

本实验的产物中含有一个吡咯环，氮原子并不直接与 $C_{60}$ 球成键。一个可能的机理如下[3]：

该机理中第一步进攻 $C_{60}$ 球的是碳而不是氮，在氨基酸中由于既有氨基的推电子作用，又有羧基的拉电子作用，因此以碳为中心的自由基可以稳定存在。

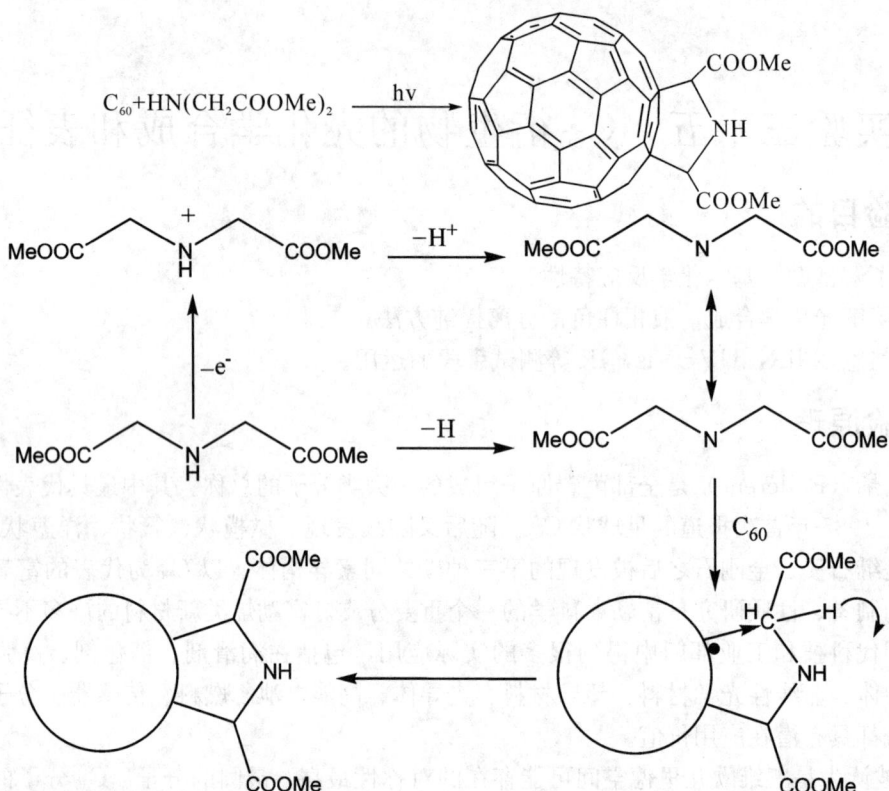

$$C_{60}+HN(CH_2COOMe)_2 \xrightarrow{h\nu}$$

## 仪器和试剂

仪器：红外光谱仪（VECYOR 22 FT-IR），紫外-可见分光光度计（CARY IE），旋转蒸发仪，电子天平（毫克级），超声波清洗器，电磁搅拌器，灯箱（箱内配 150W 自镇流荧光高压汞灯两个，为防止反应过程中灯箱过热，可在灯箱中安装 1 台小电风扇），色谱柱 1 个（带活口塞，Φ20 mm×200 mm，玻璃砂 100 目），回流冷凝管 1 个，磨口圆底烧瓶（25 mL ×2 个，50 mL× 2 个，100 mL×2 个，250 mL×1 个），烧杯（50 mL× 2 个），锥形瓶（100 mL×2 个），锥形瓶（250 mL×2 个），量筒（50 mL×1 个，10 mL×1 个），站架，烧瓶夹，不锈钢药勺（1 个），滴管（2 支），双连球（1 个）

试剂：$C_{60}$（纯度为 98%），亚氨基二乙酸（分析纯），甲苯（分析纯），无水甲醇（分析纯），氢氧化钠（分析纯），盐酸（分析纯），二氯亚砜 $SOCl_2$，氘代氯仿（分析纯），硅胶（柱层析用 200～300 目），pH 试纸（1-14）

## 实验步骤

### 1. 亚氨基二乙酸甲酯盐酸盐的合成

将 1.0 克亚氨基二乙酸和 10 mL 无水甲醇中加入 1 个 25 mL 磨口圆底烧瓶，搅拌下慢慢滴入 8 滴二氯亚砜 $SOCl_2$（约 0.4 mL），滴加过程中会产生大量氯化氢气。用水浴加热回流 2 h。于旋转蒸发仪上蒸干，所得固体即为亚氨基二乙酸甲酯盐酸盐，用于下步反应。

### 2. $C_{60}$ 衍生物的光化学合成

（1）在 50 mL 烧杯中将亚氨基二乙酸甲酯盐酸盐（2.0 mmol）与等物质的量的 NaOH

固体混合，加入约 10 滴水溶解，再加 15 mL 甲醇，超声致出现浑浊。将 pH 试纸用蒸馏水润湿后测定该溶液的 pH，应为 8.5 左右，若超出可用 NaOH 固体或稀盐酸调节。

（2）称取 70 mg $C_{60}$ 固体置于 250 mL 圆底烧瓶中。加入 100 mL 甲苯超声使 $C_{60}$ 全部溶解，加入上面亚氨基二乙酸甲酯甲醇溶液。摇动使其混合均匀，此时清液应为紫色。如为棕色可再加少量甲苯使其变回紫色。制备好的溶液为混浊状（沉淀是什么？）。

（3）将反应瓶光照至溶液由紫色完全消失变为红色（小于 60 min），反应过程中应适当摇动反应液。

（4）反应完毕加入 5 mL 蒸馏水于反应瓶中，摇动，用一个滴管分离除去水相，有机相用旋转蒸发仪蒸干。固体加 20 mL 甲苯，超声波处理，分出清液，若仍有固体未溶，可再用 20 mL 甲苯萃取。

### 3. $C_{60}$ 和衍生物的柱层析分离和收率的测定

（1）在色谱柱中加 30 mL 甲苯，将硅胶用甲苯浸润后再慢慢倒入色谱柱。装柱时应避免气泡留在硅胶上。装好后检查硅胶上有无明显气泡和缺陷，如有可用双连球加压使甲苯快速从活塞流出以赶出气泡，或取下硅胶柱适当摇动。打开旋塞，放出上层甲苯，当甲苯液面逐渐下降至硅胶柱上层平面后，关上旋塞。硅胶高度约 4~5 cm。用 1 个滴管从色谱柱柱壁慢慢加入前面反应的萃取液，待全部加完后，打开旋塞。当萃取液液面逐渐下降至硅胶柱上层平面时，再加甲苯淋洗，若淋洗速率太慢，可用双连球加压。按色带分别收集未反应 $C_{60}$ 和反应产物。

（2）用紫外-可见分光光度计分别测定未反应 $C_{60}$ 和反应产物甲苯淋洗液浓度并计算 $C_{60}$ 和产物质量。由于富勒烯及其衍生物吸光度都很大，测定时需要进行稀释（如何尽量少用溶剂避免浪费？）。标准曲线由实验课老师提供。

### 4. 产物的表征

（1）用旋转蒸发仪分别蒸干 $C_{60}$ 和产物的甲苯淋洗液。

（2）测定 $C_{60}$ 和产物的 IR 谱。

（3）产物的 [1]HNMR 谱。

## 实验结果和讨论

1. 计算 $C_{60}$ 转换率和产物产率。

2. 根据 $C_{60}$ 和产物的红外光谱图，分析、讨论氨基酸与 $C_{60}$ 加成反应的结果及产物结构。

3. 根据 $C_{60}$ 和产物的紫外可见光谱图，分析，讨论氨基酸与 $C_{60}$ 加成反应的结果及产物结构。

4. 分析产物的核磁共振谱图，讨论氨基酸与 $C_{60}$ 加成反应产物的结构。

## 实验说明

1. 甲醇对眼睛有害，富勒烯的毒性目前尚不十分清楚，应尽量避免直接接触皮肤。

2. 反应要在通风橱中进行，光照一段时间后反应液温度会迅渐升高至沸腾，此时应打开灯箱上的风扇适当降温；但温度太低也不利于反应进行。

## 思考题

1. 实验步骤2.(1)中，为什么需要弄潮 pH 试纸后再测 pH？溶液 pH 过高或过低各有什么不好？

2. 实验步骤2.(2)中，什么情况下会出现棕色，棕色物质是什么？

3. 预测产物的 $^{13}$CNMR 图谱。

4. 在所得产物上再加一个相同的取代基，会产生多少种异构体？

## 参考文献

[1] 韩汝珊. 一个新的足球新家族. 长沙：湖南教育出版社，1994.

[2] Wudl F, Hirsch A, khemani KC, *et al*. Fullerenes：Synthesis，properties and Chemistry of large carbon Clusters，ACS Sym. 1992，48（11）：161.

[3] Hirsch A. The chemistry of the fullerenes. New York：Thieme Medical Publishers Inc，1994.

# Exp 35    Photochemical Synthesis and Characterization of C$_{60}$ Derivative

## Objectives

1. To learn basic chemical reaction characteristics of fullerene.

2. To learn the method of photochemical synthesis, liquid chromatography separation and purification.

3. To know well of NMR, UV-Visible, IR etc application of testing means.

## Principles

Fullerene is a general designation composed solely of carbon atoms. Footballene C$_{60}$ is the most representational in the fullerene. After this olive shaped, tubular and onion-like homolog (ue) were found in succession. Fullerene is the third allotrope of carbon which following the founding of graphite and diamond.

Many unusual properties of new materials like fullerene almost can obtain effective application in many areas of modern science and technology and industry, including lubricant, catalyst, abrasive, high strength carbon fiber, semiconductor, non-linear optical materials, superconductive materials, light conductor, high efficient battery, and fuel, sensor, and molecule apparatus and have potential applicable value.

The band character of C$_{60}$ molecule is more complex than diamond and graphite. Because of curve effect and five-membered cyclic structure of spherical surface, lead the change of molecule hybrid orbit. Comparing with graphite, π electron orbit on longer is pure p atom orbit, but has partly component of s orbit, thus the hybrid orbit of carbon atom is between sp$^2$ (graphite crystal) and sp$^3$ (diamond crystal) hybrid in C$_{60}$ molecule. Every carbon atom in C$_{60}$ molecule form three σ bond by sp$^{2.28}$ hybrid, then form delocalized π bond by s$^{0.09}$P hybrid, σ bond in the direction of spherical surface, but π bond distribute on inside and outside surface of sphere, thereby form aromaticity spherical molecule. They have the length of chemical bond with benzene molecule, the difference is, the chemical bond of C$_{60}$ molecule can be divided into two parts, long bond (between five circle and six circle), length of bond is 146pm, short bond (between two six circles), length of bond is 139 pm (same with the C-C bond length in benzene circle). The structure of C$_{60}$ molecule made it easier to have addition reaction than benzene, and formed a series of addition compounds[1].

Because C$_{60}$ molecule is not a polar molecule, it just has a certain solubility in some aromatic solvent, but has small solubility in polar organic solvent, its solubility is almost zero in water, this limit its application to a great extend. Amino acid has very strong hydrophilic, the ramification formed by addition reaction of amino acid and C$_{60}$ can dissolve in water, this kind of compound is important in the area of life science. For example, Wudl[2] *et al* synthesized a aqueous $C_{60}$ ramification, found that the ramification has certain restrainable function on HIV protein enzyme. The reaction between amino acid and $C_{60}$ reported in literature[3], use either amino acid react firstly with a assistant agent to create active intermediate, then react with $C_{60}$. Such as 1, 3 dipole addition reported by Prato *et al*, or amino acid react with aldehyde to get intermediate, then react with $C_{60}$; Or use the functional group of ramification to react with amino acid farther, such as the first $C_{60}$ peptides ramification reported by Wudl *et al*, which has biological effect.

In this experiment, let iminodiacetic dimethyl ester react directly with $C_{60}$ in light condition, form selectivity single addition ramification.

$$C_{60} + HN(CH_2COOMe)_2 \xrightarrow{hv}$$

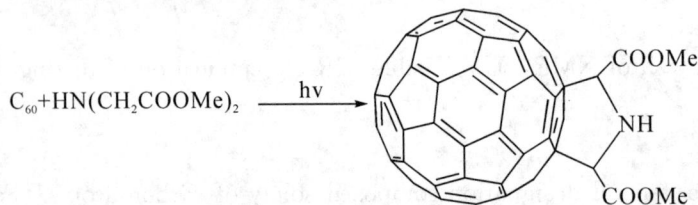

It's obvious that the above mechanism can't explain this experiment result. In this experiment, the products have a pyrrole cycle; N atom is not directly connected with $C_{60}$ spheres to form a bond. One possible mechanism is as follow:

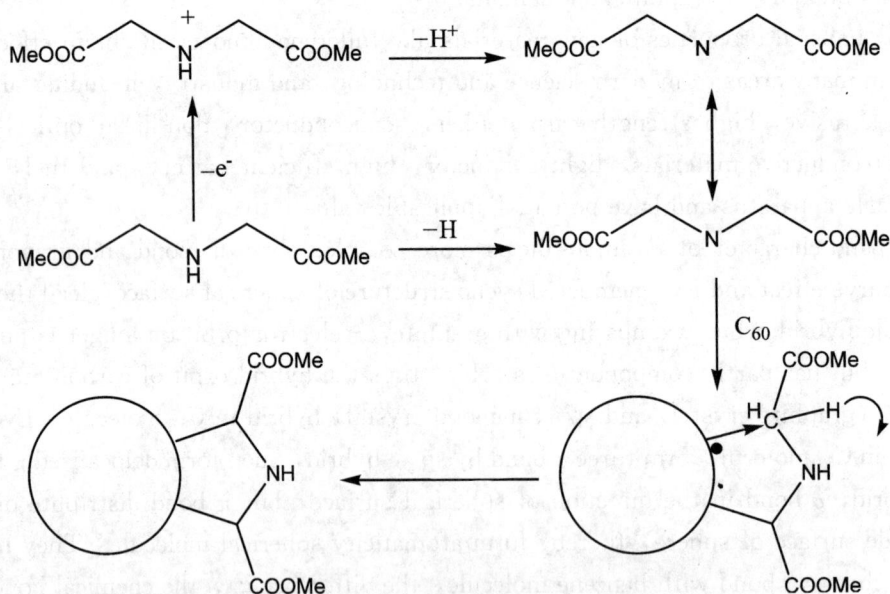

The former mechanism is that it is not nitrogen but carbon that first attack $C_{60}$ sphere. In amino acid, because not only the electron donating effect of amino group, but also the electron withdrawing effect of carboxyl group, the free radical centered by carbon can exist stable.

## Instruments and Reagents

Instruments: Fourier transformation infra-red spectrometer (VECYOR 22 FT-IR), ultraviolet-visible spectrophotometer (CARY -IE), rotary evaporator, electronic balance (milligram level), ultrasonic cleaner, electromagnetism electro-magnetic stirrer, light box (there are two 150W self-ballasted fluorescent high pressure mercury (vapor) lamp in box and a small electric fan is fixed to prevent light box from overheating in the course of reaction,), a chromatogram column (with stopcock, $\Phi 20$ mm $\times$ 200 mm, glass bead 100 mesh), reflux condensing pipe (reflux condenser), ground round bottom flask (25 mL$\times$2, 50 mL$\times$ 2, 100 mL$\times$2, 250 mL$\times$1), beaker (50 mL$\times$2), conical flask (100 mL$\times$2, 250 mL$\times$2), measuring cylinder (50 mL$\times$1, 10 mL$\times$1), stand, flask nip, a stainless steel scoop, two burettes, a bi-join ball

Reagents: $C_{60}$ (purity 98%), iminodiacetic acid (A. R.), toluene (A. R.), absolute alcohol (A. R.), sodium hydroxide (A. R.), hydrochloric acid (A. R.), thionyl chloride $SOCl_2$, chloroform (A. R.), silica gel (using for column chromatography 200~300 mesh), pH test paper (1-14)

## Experimental Procedures

### 1. Preparation of iminodiacetic acid hydrochloride

Add 1. 0 gram iminodiacetic acid and 10 mL absolute alcohol into a 25 mL ground-glass stoppered flask, add slowly 8 drops of thionyl chloride $SOCl_2$ (about 0. 4 mL) while stirring, some gaseous hydrochloric acid would arose in the course of adding. Heat up and reflux for 2h by water bath. To dry up by rotary evaporator, the solid obtained is iminodiacetic acid methyl ester hydrochloride, it can be directly used in following reaction.

### 2. Photochemical synthesis of $C_{60}$ derivative

(1) To mix dimethyl iminodiacetate hydrochloride (2. 0 mol) with equivalent molar amount of NaOH(s) in 50 mL beaker, add about 10 drops of water for dissolving, then add 15 mL methanol, use ultrasonic device to make it turbid. To moisten pH test paper by distilled water and determine the pH value of solution, it should be about 8. 5 or so, adjust it by NaOH(s) or dilute hydrochloric acid if the value exceed.

(2) Weigh 70 mg $C_{60}$ solid, put it in 250 mL round flask. Add 100 mL toluene and make $C_{60}$ dissolve completely with ultrasonic device, add above solution of dimethyl iminodiacetate, and mix them well by shaking. The clear solution should be violet. Add small amount of toluene to turn the color into violet if it is brown. The solution prepared well is in turbid state (what is precipitate?).

(3) Illuminate reaction bottle to make the color of solution turn from purple to fading away, into red (less than 60 min), the reaction solution should be shaken properly in the course of reaction.

(4) Add 5 mL distilled water in reaction bottle after reaction finished, shake; Remove aqueous phase by a burette, dry up organic phase by rotary evaporator. Add 20 mL toluene to solid, deal with ultrasonic device, separate out clear solution. Extract it by 20 mL toluene if there is solid undissolved.

### 3. Column chromatogram separation and yield determination of $C_{60}$ and its derivative

(1) Add 30 mL toluene into chromatogram column; pour silica gel soaked by toluene slowly into chromatogram column. Prevent air bubble from staying at silica gel column. Check silica gel column if there is any air bubble and lacuna or not after finishing loading, increase pressure by using bi-join ball to make toluene flow out fast from stopcock to extrude air bubble if there is any air bubble, or take down silica gel column and shake properly. Turn on faucet and let above layer toluene out, turn off faucet when the liquid face of toluene decline gradually to the above layer plane of silica gel column. The height of silica gel is about 4~5 cm. Add the above reaction solution slowly by using a burette from the column wall of chromatogram column, turn on faucet after adding completely. Add toluene once again to wash when the liquid face of toluene decline gradually to the above layer plane of silica gel column, increase pressure by using bi-join ball if washing speed is too slow. Collect the under reacted $C_{60}$ and reaction products according to color band.

(2) Determine the concentration of unreacted $C_{60}$ and reaction product in toluene elution by ultraviolet-visible spectrophotometer and calculate the mass of $C_{60}$ and products respectively. Because the absorbency of fullerene and its derivative are both strong, it is necessary to dilute (how to reduce the spending of solvent mostly?) when determination. Standard curve can be obtained from the teacher.

### 4. Characterization of products

(1) Dry up $C_{60}$ and the products in toluene washer respectively by rotary evaporator.
(2) Determination of IR spectra of $C_{60}$ and products.
(3) [1]HNMR spectra of products.

## Results and Discussion

1. Calculate the conversion rate of $C_{60}$ and the yield of products.
2. According to the infra-red spectrum of $C_{60}$ and products, analyze and discuss the results of the addition reaction between amino acid and $C_{60}$ and the structure of products.
3. According to the ultraviolet-visible spectrum of $C_{60}$ and products, analyze and discuss the results of the addition reaction between amino acid and $C_{60}$ and the structure of products.
4. To analyze the NMR spectra of products, discuss the structure of products of addition reaction between amino acid and $C_{60}$.

## Notes

1. Methanol is harmful to eyes, the toxicity of fullerene is not clear nowadays, avoid skin contacting with fullerene if possible.
2. To carry out the reaction in ventilator, the temperature of reaction solution will in-

crease rapidly to boiling after a period of illumination, here the fan in lamp box should be turned on to lower temperature; but too low temperature is not fit for reaction.

## Questions

1. In the procedure 2. (1), why the pH test paper need to be moistened before pH determination? What disadvantage would appear if pH value is too high or too low?

2. In the step 2. (2), when brown color will appear ? what is brown substance?

3. Forecast the $^{13}$CNMR spectra of products.

4. To add a same substituent in obtained products, how many isomers it will have?

## References

[1] Rushan Han. A new family of fullerenes. Changsha: Hunan education publisher, 1994.

[2] Wudl F, Hirsch A, khemani KC, *et al*. Fullerenes: Synthesis, properties and Chemistry of large carbon Clusters. ACS Sym. 1992, 48 (11): 161.

[3] Hirsch A. The Chemistry of the Fullerenes. New York: Thieme Medical Publishers Inc, 1994.

# 附录　常见离子的鉴定方法

## 一、常见阳离子的鉴定方法

### 1. $Na^+$

(1) $Na^+$ 与醋酸铀酰锌 $[Zn(Ac)_2 \cdot UO_2(Ac)_2]$ 在中性或醋酸介质中反应，生成淡黄色晶体状醋酸铀酰锌钠沉淀：

$$Na^+ + Zn^{2+} + UO_2^{2+} + Ac^- + HAc + H_2O \longrightarrow NaAc \cdot Zn(Ac)_2 \cdot 3UO_2(Ac)_2 \cdot 9H_2O\downarrow + H^+$$

在碱性介质中，$UO_2(Ac)_2$ 可生成 $(NH_4)_2U_2O_7$ 或 $K_2U_2O_7$ 沉淀，在强酸性介质中结晶状醋酸铀酰锌钠沉淀的溶解度增加。因此鉴定反应必须在中性或微酸性溶液中进行。

实验步骤：取 1 滴试液于试管中，加氨水（$6.0\ mol \cdot L^{-1}$）中和至碱性，再加 HAc 溶液（$6.0\ mol \cdot L^{-1}$）酸化，然后加 1 滴 EDTA 溶液（饱和）和 2～3 滴醋酸铀酰锌，充分摇荡，放置片刻，若有淡黄色晶体状沉淀生成，表示有 $Na^+$ 存在。

(2) $Na^+$ 在弱碱性溶液中与 $K[Sb(OH)_6]$ 饱和溶液生成白色晶状沉淀。

### 2. $K^+$

(1) $K^+$ 与 $Na_3[Co(NO_2)_6]$（俗称钴亚硝酸钠）在中性或稀醋酸介质中反应，生成亮黄色 $K_2Na[Co(NO_2)_6]$ 沉淀：

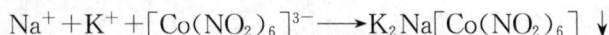

$$Na^+ + K^+ + [Co(NO_2)_6]^{3-} \longrightarrow K_2Na[Co(NO_2)_6]\downarrow$$

强酸或强碱均能使试剂分解，妨碍鉴定，因此，在鉴定时必须使溶液呈中性或微酸性。$NH_4^+$ 也能与试剂反应生成橙色 $(NH_4)_3[Co(NO_2)_6]$ 沉淀干扰鉴定。为此，可在水浴上加热 2 分钟，使之完全分解：

$$NO_2^- + NH_4^+ \longrightarrow N_2\uparrow + H_2O$$

以消除铵根离子的干扰。加热时，黄色的 $K_2Na[Co(NO_2)_6]$ 无变化。

(2) $K^+$ 离子与四苯硼钠（$Na[B(C_6H_5)_4]$）反应生成白色沉淀：

$$K^+ + [B(C_6H_5)_4]^- \longrightarrow K[B(C_6H_5)_4]\downarrow$$

反应须在碱性、中性或稀酸溶液中进行。

$NH_4^+$ 与试剂有类似反应，须事先转化为 $NH_4NO_3$ 再加热分解而除去。$Ag^+$、$Hg^{2+}$ 的影响可用 KCN 消除。当溶液 $pH \approx 5$，在有 EDTA 存在时，其他离子不干扰。

实验步骤：取 3～4 滴试液于试管中，加入 4～5 滴 $Na_2CO_3$ 溶液（$0.5\ mol \cdot L^{-1}$），加热，使有色离子变为碳酸盐沉淀。离心分离，在所得清液中加入 HAc 溶液（$6.0\ mol \cdot L^{-1}$），再加入 2 滴 $Na_3[Co(NO_2)_6]$ 溶液，最后将试管放入沸水浴中加热 2 分钟，若试管中有黄色沉淀，表示有 $K^+$ 存在。

### 3. $NH_4^+$

(1) $NH_4^+$ 与 Nessler 试剂（$K_2[HgI_4]$ + KOH）反应生成红棕色的沉淀：

$$NH_4^+ + [HgI_4]^{2-} + 4OH^- \longrightarrow HgO \cdot HgNH_2I\downarrow + I^- + H_2O$$

Nessler 试剂是 $K_2[HgI_4]$ 的碱性溶液，如果溶液中有 $Fe^{3+}$、$Cr^{3+}$、$Co^{2+}$ 和 $Ni^{2+}$ 等离子，能与 KOH 反应生成深色的氢氧化物沉淀，而干扰 $NH_4^+$ 离子的鉴定，可用下法：在原试液中加入 NaOH 溶液，并微热，用滴加 Nessler 试剂的滤纸条检验逸出的氨气，由于 $NH_3(g)$ 与 Nessler 试剂作用，使滤纸上出现红棕色斑点。

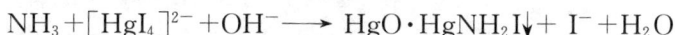

$$NH_3 + [HgI_4]^{2-} + OH^- \longrightarrow HgO \cdot HgNH_2I\downarrow + I^- + H_2O$$

（2）取试液，加 NaOH$(1.0\ mol \cdot L^{-1})$ 溶液碱化，微热，用红色的石蕊试纸检验逸出的气体，试纸呈蓝色，表示有 $NH_4^+$ 存在。

实验步骤：①取 2 滴试液于试管中，加入 NaOH 溶液 $(2.0\ mol \cdot L^{-1})$ 使呈碱性，微热，用滴加 Nessler 试剂的滤纸检验逸出的气体。如有红棕色斑点出现，表示有 $NH_4^+$ 存在。

②取 2 滴试液于试管中，加入 NaOH 溶液 $(2.0\ mol \cdot L^{-1})$ 碱化，微热，用润湿的红色石蕊试纸（或用 pH 试纸）检验逸出的气体，如试纸显蓝色，表示有 $NH_4^+$ 存在。

### 4. $Mg^{2+}$

（1）$Mg^{2+}$ 离子与镁试剂 I（对硝基偶氮间苯二酚）在碱性介质中反应，生成蓝色沉淀。有些能生成深色氢氧化物的金属离子对鉴定有干扰，可以用 EDTA 试剂来配合掩蔽。

（2）在氨性介质中与磷酸二氢钠 $(NaH_2PO_4)$ 作用，有白色的磷酸铵镁 $(MgNH_4PO_4)$ 沉淀生成，表明有 $Mg^{2+}$ 离子的存在。

实验步骤：取 1 滴试液于点滴板上，加 2 滴 EDTA 溶液（饱和），搅拌后，加 1 滴镁试剂 I、1 滴 NaOH 溶液 $(6.0\ mol \cdot L^{-1})$，如有蓝色沉淀生成，表示有 $Mg^{2+}$ 存在。

### 5. $Ca^{2+}$

（1）$Ca^{2+}$ 与乙二醛双缩 [二-羟基苯胺]（简称 GBHA）在 pH 为 12~12.6 时反应，生成红色螯合物沉淀。

（2）$Ca^{2+}$ 离子与可溶性草酸盐，在中性或碱性条件下作用，可生成白色的细微晶状草酸钙沉淀。

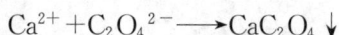

$$Ca^{2+} + C_2O_4^{2-} \longrightarrow CaC_2O_4\downarrow$$

所得沉淀不溶于醋酸，但可溶于稀 HCl 或稀 $HNO_3$：

$$CaC_2O_4 + H^+ \longrightarrow Ca^{2+} + HC_2O_4^-$$

虽 $SrC_2O_4$ 和 $BaC_2O_4$ 也是难溶化合物，但 $BaC_2O_4$ 能溶于醋酸，$SrC_2O_4$ 可微溶于醋酸。所以在醋酸溶液中 $Ba^{2+}$ 不干扰鉴定，其他能形成难溶性草酸盐的金属离子（如 $Pb^{2+}$、$Zn^{2+}$）须预先除去。

实验步骤：取 1 滴试液于试管中，加入 10 滴 $CHCl_3$，加入 4 滴 GBHA（0.2%）、2 滴 NaOH 溶液 $(6.0\ mol \cdot L^{-1})$、2 滴 $Na_2CO_3$ 溶液 $(1.5\ mol \cdot L^{-1})$，摇荡试管，如果 $CHCl_3$ 层显红色，表示有 $Ca^{2+}$ 存在。

### 6. $Sr^{2+}$

由于挥发性的锶盐如 $SrCl_2$ 置于酒精喷灯氧化焰中燃烧，能产生猩红色火焰，故利用焰色反应鉴定 $Sr^{2+}$。若试样是不易挥发的 $SrSO_4$，应把它转化为碳酸盐，后再转化为氯化锶，进行实验。

实验步骤：取 2 滴试样于试管中，加入 2 滴 $Na_2CO_3$ 溶液 $(0.5\ mol \cdot L^{-1})$，在水浴上加

热得 $SrCO_3$ 沉淀，离心分离。在沉淀中加 1 滴 HCl 溶液（$6.0\ mol \cdot L^{-1}$），使其溶解为 $SrCl_2$，然后用清洁的镍丝或铂丝蘸取 $SrCl_2$ 置于煤气灯的氧化焰中灼烧，如有猩红色火焰，表示有 $Sr^{2+}$ 存在。

### 7. $Ba^{2+}$

在弱酸性介质中，$Ba^{2+}$ 与 $K_2CrO_4$ 反应生成黄色 $BaCrO_4$ 沉淀：

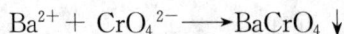

$$Ba^{2+} + CrO_4^{2-} \longrightarrow BaCrO_4 \downarrow$$

沉淀不溶于醋酸，但可溶于强酸。因此鉴定反应必须在弱酸中进行。

$Pb^{2+}$、$Hg^{2+}$ 和 $Ag^+$ 等离子也能与 $K_2CrO_4$ 反应生成不溶于醋酸的有色沉淀，为此，可预先用金属锌使 $Pb^{2+}$、$Hg^{2+}$ 和 $Ag^+$ 等还原成单质金属而除去。

实验步骤：取 1 滴试样于试管中，加 $NH_3 \cdot H_2O$（浓）使呈碱性，再加锌粉少许，在水浴中加热 1~2 分钟，并不断搅拌、离心分离。在溶液中加醋酸酸化，加 1~2 滴 $K_2CrO_4$ 溶液，摇荡，在沸水中加热，如有黄色沉淀，表示有 $Ba^{2+}$ 存在。

### 8. $Al^{3+}$

$Al^{3+}$ 与铝试剂（金黄色素三羧酸铵）在 pH 为 6~7 的介质中反应，生成红色絮状螯合物沉淀。$Cu^{2+}$、$Bi^{3+}$、$Fe^{3+}$、$Cr^{3+}$ 和 $Ca^{2+}$ 等离子干扰反应，$Bi^{3+}$ 和 $Fe^{3+}$ 可预先加入 NaOH 使它们生成 $Fe(OH)_3$ 和 $Bi(OH)_3$ 而除去。$Cr^{3+}$、$Cu^{2+}$ 与铝试剂的螯合物能被 $NH_3 \cdot H_2O$ 分解。$Ca^{2+}$ 与铝试剂的螯合物可被 $(NH_4)_2CO_3$ 转化为 $CaCO_3$。

实验步骤：取 2 滴试液于试管中，加 NaOH 溶液（$6.0\ mol \cdot L^{-1}$）碱化，并过量 2 滴，加 1 滴 $H_2O_2$（3%），加热 2 分钟，离心分离。用 HAc 溶液（$6.0\ mol \cdot L^{-1}$）将溶液酸化，调 pH 为 6~7，加 2 滴铝试剂，摇荡后，放置片刻，加 $NH_3 \cdot H_2O$（$6.0\ mol \cdot L^{-1}$）碱化，置于水浴上加热，如有橙红色（$CrO_4^{2-}$ 存在）物质生成，可离心分离。用去离子水洗涤沉淀，如沉淀为红色，表示有 $Al^{3+}$ 存在。

### 9. $Sn^{2+}$

（1）与 $HgCl_2$ 反应

$SnCl_2$ 溶液中 $Sn^{2+}$ 主要以 $SnCl_4^{2-}$ 的形式存在。$SnCl_4^{2-}$ 与适量 $HgCl_2$ 反应生成白色沉淀 $Hg_2Cl_2$：

$$[SnCl_4]^{2-} + HgCl_2 \longrightarrow [SnCl_6]^{2-} + Hg_2Cl_2 \downarrow$$

如果 $SnCl_4^{2-}$ 过量，则沉淀变为灰色，即 $Hg_2Cl_2$ 与 Hg 的混合物，最后变为黑色，即 Hg。

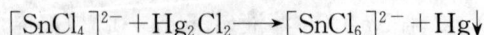

$$[SnCl_4]^{2-} + Hg_2Cl_2 \longrightarrow [SnCl_6]^{2-} + Hg \downarrow$$

加入铁粉，可使许多电极电势大的离子还原为金属，预先分离，以消除干扰。

实验步骤：取 2 滴试液于试管中，加 2 滴 HCl 溶液（$6.0\ mol \cdot L^{-1}$），加少许铁粉，在水浴上加热至作用完全，至气泡不再发生为止。吸取清液于另一干净试管中，加入 2 滴 $HgCl_2$，如有白色沉淀生成，表示有 $Sn^{2+}$ 存在。

（2）与甲基橙反应

$SnCl_4^{2-}$ 与甲基橙在浓 HCl 介质中加热下进行反应，甲基橙被还原为氢化甲基橙而褪色。

实验步骤：取 1 滴试液于试管中，加 1 滴 HCl（浓）及甲基橙（0.01%），加热，如甲基橙褪色，表示有 $Sn^{2+}$ 存在。

### 10. $Pb^{2+}$

$Pb^{2+}$ 与 $K_2CrO_4$ 在稀 HAc 溶液中，发生反应生成难溶的 $PbCrO_4$ 黄色沉淀：
$$Pb^{2+} + CrO_4{}^{2-} \longrightarrow PbCrO_4 \downarrow$$
沉淀溶于 NaOH 溶液及浓硝酸，难溶于稀 HAc，稀硝酸及 $NH_3 \cdot H_2O$。
$$PbCrO_4 + OH^- \longrightarrow [Pb(OH)_3]^- + CrO_4{}^{2-}$$
$$PbCrO_4 + H^+ \longrightarrow Pb^{2+} + HCrO_4{}^-$$

$Ba^{2+}$、$Bi^{3+}$、$Hg^{2+}$ 和 $Ag^+$ 等离子在 HAc 溶液中也能与 $CrO_4{}^{2-}$ 作用生成有色沉淀，所以这些离子的存在对 $Pb^{2+}$ 的鉴定有干扰。可先加入稀硫酸，再用过量的 NaOH 浓溶液，使 $PbSO_4$ 转化为 $[Pb(OH)_3]^-$，进一步转化为 $Pb(Ac)_2$，使 $Pb^{2+}$ 分离出来，再进行鉴定。

实验步骤：取 2 滴试液于试管中，加 1 滴 $H_2SO_4$ 溶液（$6.0\ mol \cdot L^{-1}$），加热几分钟，摇荡，使 $Pb^{2+}$ 沉淀完全，离心分离。在沉淀中加入过量 NaOH 溶液（$6.0\ mol \cdot L^{-1}$），并加热 1 min，使 $PbSO_4$ 转化为 $[Pb(OH)_3]^-$，离心分离。在清液中加 HAc 溶液（$6.0\ mol \cdot L^{-1}$），再加 1 滴 $K_2CrO_4$（$0.1\ mol \cdot L^{-1}$）溶液，如有黄色沉淀，表示有 $Pb^{2+}$ 存在。

### 11. $Bi^{3+}$

$Bi^{3+}$ 在碱性溶液中能被 $Sn^{2+}$ 还原为黑色 Bi：
$$Bi(OH)_3 + [Sn(OH)_4]^{2-} \longrightarrow Bi \downarrow + [Sn(OH)_6]^{2-}$$
实验步骤：取 2 滴试液于试管中，加入浓氨水，$Bi^{3+}$ 变为 $Bi(OH)_3$ 沉淀，离心分离。洗涤沉淀。在沉淀中加入少量新配制的 $Na_2[Sn(OH)_4]$ 溶液，如沉淀变黑，表示有 $Bi^{3+}$ 存在。

$Na_2[Sn(OH)_4]$ 溶液的配制方法：取几滴 $SnCl_2$ 溶液于试管中，加入 NaOH 溶液至生成的 $Sn(OH)_2$ 白色沉淀刚好溶解，便得到澄清的 $Na_2[Sn(OH)_4]$ 溶液。

### 12. $Sb^{3+}$

Sb（Ⅲ）在酸性溶液中能被金属锡还原为金属锑：
$$[SbCl_6]^{3-} + Sn \longrightarrow Sb \downarrow + [SnCl_4]^{2-}$$
当砷（Ⅲ，Ⅴ）存在时，也能在锡箔上形成黑色斑点（As），但 As 与 Sb 不同，当用水洗去锡箔上的酸后加新配制 NaBrO 的溶液则溶解。注意一定要将 HCl 洗净，否则在酸性条件下，NaBrO 也能使 Sb 的黑色斑点溶解。

$Hg^{2+}$、$Bi^{3+}$ 等离子也干扰 $Sb^{3+}$ 的鉴定，可用 $(NH_4)_2S$ 预先分离。

实验步骤：取 2 滴试液于试管中，加 $NH_3 \cdot H_2O$ 溶液（$6.0\ mol \cdot L^{-1}$）碱化，加 2 滴 $(NH_4)_2S$ 溶液（$0.5\ mol \cdot L^{-1}$），充分摇荡，于水浴上加热 5 分钟左右，离心分离。在溶液中加 HCl 溶液（$6.0\ mol \cdot L^{-1}$）酸化，使呈微酸性，并加热 3~5 min，离心分离。沉淀中加 1 滴 HCl（浓），再加热使 $Sb_2S_3$ 溶解。取此溶液在锡箔上，片刻锡箔上出现黑斑。用水洗去酸，再用 1 滴新配制的 NaBrO 溶液处理，黑斑不消失，表示有 $Sb^{3+}$ 存在。

### 13. As（Ⅲ，Ⅴ）

砷常以 $AsO_3{}^{3-}$ 和 $AsO_4{}^{3-}$ 的形式存在。

$AsO_3^{3-}$ 在碱性溶液中能被金属锌还原为 $AsH_3$ 气体：

$$AsO_3^{3-}+OH^-+Zn+H_2O \longrightarrow [Zn(OH)_4]^{2-}+AsH_3\uparrow$$

$AsH_3$ 气体能与 $AgNO_3$ 作用，生成的产物由黄色逐渐变为黑色：

$$AgNO_3 + AsH_3 \longrightarrow Ag_3As\cdot AgNO_3\downarrow（黄）+ HNO_3$$

$$Ag_3As\cdot 3AgNO_3 + H_2O \longrightarrow H_3AsO_3 + HNO_3 + Ag\downarrow$$

这是鉴定 $AsO_3^{3-}$ 的特效反应，若是 $AsO_4^{3-}$ 应预先用亚硫酸还原。

实验步骤：取 2 滴试液于试管中，加 NaOH 溶液（$6.0\ mol\cdot L^{-1}$）碱化，再加少许 Zn 粒，立刻用一小团脱脂棉塞在试管上部，再用 5% $AgNO_3$ 溶液浸过的滤纸盖在试管口上，置于水浴中加热，如滤纸上 $AgNO_3$ 斑点渐渐变黑，表示有 $AsO_3^{3-}$ 存在。

### 14. $TiO^{2+}$

$TiO^{2+}$ 在酸性介质中能与 $H_2O_2$ 生成橙红色的配合物：

$$TiO^{2+}+H_2O_2 \longrightarrow [TiO(H_2O_2)]^{2+}$$

$Fe^{3+}$，$CrO_4^{2-}$ 和 $MnO_4^-$ 等离子干扰鉴定，须预先分离。

### 15. $Cr^{3+}$

在碱性介质中 $Cr^{3+}$ 可被 $H_2O_2$ 氧化为 $CrO_4^{2-}$：

$$Cr^{3+} +H_2O_2+OH^- \longrightarrow CrO_4^{2-}+ H_2O$$

在酸性条件下（以重铬酸根形式存在），当戊醇存在时，加入 $H_2O_2$，振荡后戊醇呈现蓝色：

$$Cr_2O_7^{2-} + H_2O_2 \longrightarrow CrO(O_2)_2 + H_2O$$

蓝色的 $CrO(O_2)_2$ 在水溶液中不稳定，在戊醇中较稳定。溶液酸度应控制在 pH＝2～3，当酸度过大时（pH<1），则

$$CrO(O_2)_2+ H^+ \longrightarrow Cr^{3+}+O_2\uparrow+ H_2O$$

溶液变为蓝绿色（$Cr^{3+}$ 颜色）。

实验步骤：取 1 滴试液于试管中，加 NaOH 溶液（$2.0\ mol\cdot L^{-1}$）至生成沉淀又溶解，再多加 1 滴。加 $H_2O_2$ 溶液（3%），微热，溶液呈黄色。冷却后再加 2 滴 $H_2O_2$ 溶液（3%），加 5 滴戊醇（或乙醚），最后慢慢加 $HNO_3$ 溶液（$6.0\ mol\cdot L^{-1}$）。注意：每加 1 滴 $HNO_3$ 都必须充分摇荡，如戊醇层呈蓝色，表示有 $Cr^{3+}$ 存在。

### 16. $Mn^{2+}$

$Mn^{2+}$ 在稀 $HNO_3$ 或稀 $H_2SO_4$ 的介质中可被 $NaBiO_3$ 氧化为紫红色 $MnO_4^-$：

$$Mn^{2+} + NaBiO_3+H^+ \longrightarrow MnO_4^-+Bi^{3+}+ Na^++H_2O$$

过量的 $Mn^{2+}$ 会与生成的 $MnO_4^-$ 反应生成 $MnO(OH)_2(s)$。$Cl^-$ 及其他还原剂存在，对 $Mn^{2+}$ 的鉴定有干扰，因此不能在 HCl 的介质中鉴定 $Mn^{2+}$。

实验步骤：取 2 滴试液于试管中，加 $HNO_3$ 溶液（$6.0\ mol\cdot L^{-1}$）酸化，加少量 $NaBiO_3$ 固体，摇荡后，静置片刻，如溶液呈紫红色，表示有 $Mn^{2+}$ 存在。

### 17. $Fe^{2+}$、$Fe^{3+}$

（1）$Fe^{2+}$

$Fe^{2+}$ 与 $K_3[Fe(CN)_6]$ 溶液在 pH>7 溶液中反应,生成深蓝色沉淀:

$$Fe^{2+} + K_3[Fe(CN)_6] \longrightarrow [KFe(Ⅲ)(CN)_6 Fe(Ⅱ)]_x \downarrow$$

$[KFe(Ⅲ)(CN)_6 Fe(Ⅱ)]_x$ 沉淀能被强碱分离,产生红棕色的 $Fe(OH)_3$ 沉淀。

实验步骤:取 1 滴试液于点滴板上,加 1 滴 HCl 溶液($2.0\ mol \cdot L^{-1}$)酸化,加 1 滴 $K_3$ $[Fe(CN)_6]$溶液($0.1\ mol \cdot L^{-1}$),如出现蓝色沉淀,表示有 $Fe^{2+}$ 存在。

(2) $Fe^{3+}$

①与 $SCN^-$ 离子反应

$Fe^{3+}$ 与 $SCN^-$ 在酸性介质中反应,生成可溶性深红色 $[Fe(SCN)_n]^{3-n}$ 离子:

$$Fe^{3+} + nSCN^- \longrightarrow [Fe(SCN)_n]^{3-n} \quad (n=1\sim6)$$

$[Fe(SCN)_n]^{3-n}$ 能被碱分离,生成红棕色的 $Fe(OH)_3$ 沉淀。浓硫酸或浓硝酸能使试剂分解:

$$SCN^- + H_2SO_4 + H_2O \longrightarrow NH_4^+ + COS\uparrow + SO_4^{2-}$$

$$SCN^- + HNO_3 + H^+ \longrightarrow CO_2\uparrow + NO\uparrow + SO_4^{2-} + H_2O$$

②$Fe^{3+}$ 与 $[Fe(CN)_6]^{4-}$ 反应

$Fe^{3+}$ 与 $K_4[Fe(CN)_6]$ 反应生成蓝色溶液。

$$Fe^{3+} + K_4[Fe(CN)_6] \longrightarrow [KFe(Ⅲ)(CN)_6 Fe(Ⅱ)]_x \downarrow$$

沉淀不溶于稀酸,但能被 HCl 分解,也能被 NaOH 沉淀为 $Fe(OH)_3$。

实验步骤:取 1 滴试液于点滴板上,加 1 滴 HCl 溶液 ($2.0\ mol \cdot L^{-1}$) 及 1 滴 $K_4[Fe(CN)_6]$,如立即出现蓝色沉淀,表示有 $Fe^{3+}$ 存在。

## 18. $Co^{2+}$

$Co^{2+}$ 在中性或微酸性溶液中与 KSCN 反应生成蓝色的 $[Co(SCN)_4]^{2-}$ 离子:

$$Co^{2+} + 4SCN^- \longrightarrow [Co(SCN)_4]^{2-}$$

所生成的配离子在水溶液中不稳定,在丙酮溶液中较稳定。$Fe^{3+}$ 离子的存在对鉴定有干扰,可用 NaF 掩蔽来消除。大量 $Ni^{2+}$ 离子的存在,使溶液呈浅蓝色而干扰。

实验步骤:取 2 滴试液于试管中,加入数滴丙酮,再加入少量 KSCN 和 $NH_4SCN$ 晶体,充分摇荡,若溶液呈鲜艳的蓝色,表示有 $Co^{2+}$ 存在。

## 19. $Ni^{2+}$

$Ni^{2+}$ 与丁二酮肟在弱碱性溶液中反应,生成鲜红色的螯合物沉淀。

大量的 $Co^{2+}$、$Fe^{3+}$、$Fe^{2+}$ 和 $Cu^{2+}$ 等离子的存在,能与试剂(丁二酮肟)反应生成带色的沉淀,干扰离子的鉴定,须预先除去。

实验步骤:取 1 滴试液于试管中,加入 1 滴氨水 ($2.0\ mol \cdot L^{-1}$) 碱化,加丁二酮肟溶液 (1%),若出现鲜红色沉淀,表示有 $Ni^{2+}$ 存在。

## 20. $Cu^{2+}$

$Cu^{2+}$ 离子与 $K_4[Fe(CN)_6]$ 在中性或弱碱性介质中反应,生成红棕色 $Cu_2[Fe(CN)_6]$ 沉淀。

$$Cu^{2+} + [Fe(CN)_6]^{4-} \longrightarrow Cu_2[Fe(CN)_6] \downarrow$$

沉淀难溶解于 HCl、HAc 及稀 $NH_3 \cdot H_2O$，但易溶于浓 $NH_3 \cdot H_2O$：

$$Cu_2[Fe(CN)_6] + NH_3 \longrightarrow [Cu(NH_3)_4]^{2+} + [Fe(CN)_6]^{4-}$$

沉淀易被 NaOH 溶液转化为 $Cu(OH)_2$：

$$Cu_2[Fe(CN)_6] + OH^- \longrightarrow Cu(OH)_2\downarrow + [Fe(CN)_6]^{4-}$$

$Fe^{3+}$ 干扰 $Cu^{2+}$ 离子的鉴定，可用 NaF 掩蔽 $Fe^{3+}$ 离子，或在氨性溶液中使 $Fe^{3+}$ 转化为 $Fe(OH)_3$ 沉淀，分离除去。此时，$Cu^{2+}$ 离子以 $[Cu(NH_3)_4]^{2+}$ 的形式留在溶液中。用适量 HCl 酸化后，再用 $K_4[Fe(CN)_6]$ 鉴定 $Cu^{2+}$ 的存在。

实验步骤：取 1 滴试液于点滴板上，加 2 滴 $K_4[Fe(CN)_6]$ 溶液（$0.1 \text{ mol} \cdot L^{-1}$），若生成红棕色沉淀，表示有 $Cu^{2+}$。

### 21. $Zn^{2+}$

$Zn^{2+}$ 离子在强碱性溶液中与二苯硫腙反应生成粉红色螯合物，在水中难溶，显粉红色，在 $CCl_4$ 易溶，显棕色。

实验步骤：取 1 滴试液于试管中，加入 3 滴 NaOH 溶液（$6.0 \text{ mol} \cdot L^{-1}$），加 5 滴 $CCl_4$，加 1 滴二苯硫腙溶液，如水层显粉红色，$CCl_4$ 层由绿色变棕色，表示有 $Zn^{2+}$ 存在。

### 22. $Ag^+$

$Ag^+$ 与稀 HCl 反应生成沉淀 AgCl。沉淀能溶解于浓 HCl 形成 $[AgCl_2]^-$ 和 $[AgCl_3]^{2-}$ 等配离子。AgCl 沉淀还能溶于稀氨水形成 $[Ag(NH_3)_2]^+$ 配离子。可利用这两个反应与其他阳离子的难溶氯化物沉淀进行分离。在溶液中加入硝酸溶液，重新得到 AgCl 沉淀。或加入可溶性的碘化物，以形成更难溶解的黄色 AgI 沉淀。

实验步骤：取 5 滴试液于试管中，加入 5 滴 HCl 溶液（$2.0 \text{ mol} \cdot L^{-1}$），置于水浴上温热，使沉淀聚集，离心分离。沉淀用热的去离子水洗一次，然后加入过量氨水（$6.0 \text{ mol} \cdot L^{-1}$），摇荡，如有不溶沉淀物存在时，离心分离。取一部分溶液于试管中加 $HNO_3$ 溶液（$2.0 \text{ mol} \cdot L^{-1}$），如有白色沉淀生成，表示有 $Ag^+$ 存在。或取一部分溶液于试管中，加入 KI 溶液（$0.1 \text{ mol} \cdot L^{-1}$），如有黄色沉淀生成，表示有 $Ag^+$ 存在。

### 23. $Cd^{2+}$

$Cd^{2+}$ 与 $S^{2-}$ 反应生成黄色 CdS 沉淀。沉淀溶于 HCl 溶液（$2.0 \text{ mol} \cdot L^{-1}$）和稀 $HNO_3$，但不溶于 $Na_2S$，$(NH_4)_2S$，NaOH，KCN 和 HAc。

可用控制溶液酸度的方法与其他离子分离并进行鉴定。

实验步骤：取 1 滴试液于试管中，加入 2 滴 HCl 溶液（$2.0 \text{ mol} \cdot L^{-1}$），加 1 滴 $Na_2S$ 溶液（$0.1 \text{ mol} \cdot L^{-1}$），可使 $Cu^{2+}$ 沉淀，$Co^{2+}$、$Ni^{2+}$ 和 $Cd^{2+}$ 均无反应，离心分离。在清液中加 $(NH_4)_2Ac$ 溶液（30%），使酸度降低，如有黄色沉淀析出，表示有 $Cd^{2+}$ 存在。在该酸度下，$Co^{2+}$ 和 $Ni^{2+}$ 不会生成硫化物沉淀。

### 24. $Hg^{2+}$ 和 $Hg_2^{2+}$

（1）$Hg^{2+}$ 能被 $Sn^{2+}$ 逐步还原，最后还原为金属汞，沉淀由白色（$Hg_2Cl_2$）变为灰色或黑色（Hg）：

$$[SnCl_4]^{2-} + HgCl_2 \longrightarrow [SnCl_6]^{2-} + Hg_2Cl_2 \downarrow$$
$$[SnCl_4]^{2-} + Hg_2Cl_2 \longrightarrow [SnCl_6]^{2-} + Hg \downarrow$$

实验步骤：取 1 滴试液，加入 1~2 滴 $SnCl_2$ 溶液($0.1\ mol \cdot L^{-1}$)，如生成白色沉淀，并逐渐转变为灰色或黑色，表示有 $Hg^{2+}$ 存在。

(2) $Hg^{2+}$ 能与 KI、$CuSO_4$ 溶液反应生成橙红色 $Cu_2[HgI_4]$ 沉淀。

$$Hg^{2+} + I^- \longrightarrow [HgI_4]^{2-}$$
$$Cu^{2+} + I^- \longrightarrow CuI \downarrow + I_2 \downarrow$$
$$CuI + [HgI_4]^{2-} \longrightarrow Cu_2[HgI_4] \downarrow + I^-$$

实验步骤：取 1 滴试液加 1 滴 KI 溶液（1%）和 1 滴 $CuSO_4$ 溶液，加少量 $Na_2SO_3$ 固体（为了除去 $I_2$ 的黄色），如生成橙红色 $Cu_2[HgI_4]$ 沉淀，表示有 $Hg^{2+}$ 存在。

(3) $Hg_2^{2+}$

可将其氧化为 $Hg^{2+}$，再进行鉴定。

欲将 $Hg_2^{2+}$ 从混合离子中分离出来，常常采用下法：

加入稀 HCl，将其转化为 $Hg_2Cl_2$ 沉淀，若有 $Ag^+$、$Pb^{2+}$ 等离子存在，其氯化物亦难溶于水。由于 $PbCl_2$ 溶解度较大，并可溶于热水，可先分离。在 $Hg_2Cl_2$、AgCl 的混合沉淀中加入 $HNO_3$ 和稀 HCl 时，$Hg_2Cl_2$ 溶解，同时被氧化为 $HgCl_2$，而 AgCl 不溶，则可分离。

$$Hg_2Cl_2 + HNO_3 + HCl \longrightarrow HgCl_2 + NO \uparrow + H_2O$$

实验步骤：取 3 滴试液于试管中，加入 3 滴 HCl 溶液（$2.0\ mol \cdot L^{-1}$），充分摇荡，置水浴上加热 1 分钟，趁热分离。沉淀用热 HCl 水[1 mL 水加1滴 HCl 溶液（$2.0\ mol \cdot L^{-1}$）配成]洗两次。于沉淀中加 2 滴 $HNO_3$（浓）及 1 滴 HCl 溶液（$2.0\ mol \cdot L^{-1}$），摇荡，并加热 1 分钟，则 $Hg_2Cl_2$ 溶解，而 AgCl 沉淀不溶解，离心分离。于溶液中加 2 滴 KI 溶液（4%）、2 滴 $CuSO_4$ 溶液（2%）及少量 $Na_2SO_3$ 固体。如生成 $Cu_2[HgI_4]$ 橙红色沉淀，表示有 $Hg_2^{2+}$ 存在。

### 25. $Fe^{3+}$，$Al^{3+}$ 和 $Cr^{3+}$ 分离鉴定（见附录图1）

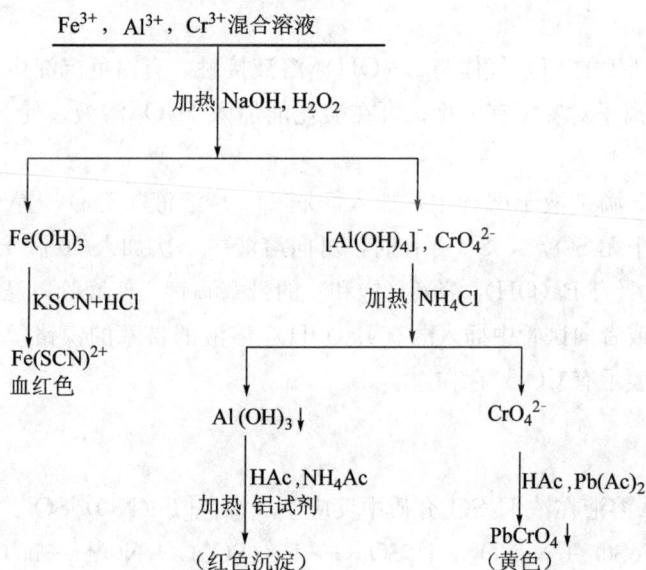

附录图 1

**26. $Ag^+$，$Hg_2^{2+}$，$Fe^{3+}$，$Co^{2+}$，$Ni^{2+}$ 分离鉴定（见附录图 2）**

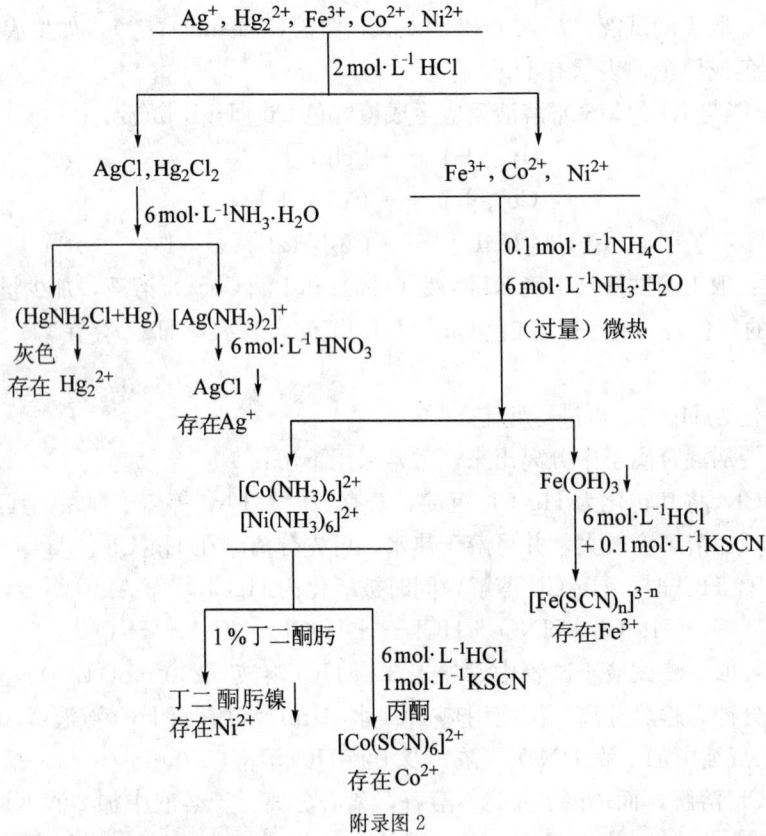

附录图 2

# 二、常见阴离子的鉴定方法

## 1. $CO_3^{2-}$

将试液酸化后产生的 $CO_2$ 气体与 $Ba(OH)_2$ 溶液接触，有白色沉淀生成表示有 $CO_3^{2-}$ 存在。$SO_3^{2-}$、$S^{2-}$ 等离子对鉴定有干扰，可在酸化前加入 $H_2O_2$ 溶液，使 $SO_3^{2-}$、$S^{2-}$ 氧化为 $SO_4^{2-}$，消除干扰。

实验步骤：取 5 滴试液于试管中，加入 5 滴 $H_2O_2$ 溶液（3％），置于水浴上加热 3 分钟，如果检验溶液中无 $SO_3^{2-}$、$S^{2-}$ 存在时，可向溶液中一次加入 20 滴 HCl 溶液（6.0 mol·$L^{-1}$），并立即插入吸有 $Ba(OH)_2$ 溶液（饱和）的带塞滴管，使滴管口悬挂 1 滴溶液，观察溶液是否变混浊。或者向试管中插入蘸有 $Ba(OH)_2$ 溶液的带塞的镍铬丝小圈，若镍铬丝小圈的液膜变混浊，表示有 $CO_3^{2-}$ 存在。

## 2. $NO_3^-$

$NO_3^-$ 与 $FeSO_4$ 溶液在浓 $H_2SO_4$ 介质中反应生成棕色 $[Fe(NO)]SO_4$：

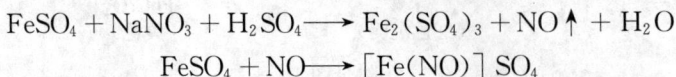

$$FeSO_4 + NaNO_3 + H_2SO_4 \longrightarrow Fe_2(SO_4)_3 + NO\uparrow + H_2O$$
$$FeSO_4 + NO \longrightarrow [Fe(NO)]SO_4$$

$[Fe(NO)]^{2+}$ 在浓硫酸与试液层界面处生成（硫酸密度大，在加入时不摇动，则沉在底部，与试液形成界面），呈棕色的形状，俗称"棕色环法"。$Br^-$，$I^-$ 及 $NO_2^-$ 等干扰 $NO_3^-$ 的鉴定，可加入稀 $H_2SO_4$ 及 $Ag_2SO_4$ 溶液，使 $Br^-$ 和 $I^-$ 生成沉淀分离除去。在溶液中加入尿素，并微热，可除去 $NO_2^-$ 的干扰：

$$2NO_2^- + CO(NH_2)_2 + 2H^+ \longrightarrow 2N_2\uparrow + CO_2\uparrow + 3H_2O$$

实验步骤：取 2 滴试液于试管中，加入 1 滴 $H_2SO_4$ 溶液（$2.0\ mol \cdot L^{-1}$），加入 4 滴 $Ag_2SO_4$ 溶液（$0.02\ mol \cdot L^{-1}$），离心分离。在清液中加入少量尿素固体，并微热。在溶液中加入少量 $FeSO_4$ 固体，摇荡溶解后，将试管斜持，慢慢沿试管壁滴入 1 mL 浓硫酸。若硫酸层与水溶液层的界面处有"棕色环"出现，表示有 $NO_3^-$ 存在。

### 3. $NO_2^-$

(1) $NO_2^-$ 与 $FeSO_4$ 在 HAc 介质中反应，生成棕色 $[Fe(NO)]SO_4$：

$$Fe^{2+} + NO_2^- + HAc \longrightarrow Fe^{3+} + NO\uparrow + H_2O + Ac^-$$

$$Fe^{2+} + NO \longrightarrow [Fe(NO)]^{2+}$$

实验步骤：取 2 滴试液于试管中，加入 4 滴 $Ag_2SO_4$ 溶液（$0.02\ mol \cdot L^{-1}$），若有沉淀生成，离心分离。在清液中加入少量 $FeSO_4$ 固体，摇荡溶解后，加入 4 滴 HAc 溶液（$2.0\ mol \cdot L^{-1}$），若溶液呈棕色，表示有 $NO_2^-$ 存在。

(2) $NO_2^-$ 与硫脲在稀 HAc 介质中反应，生成 $N_2$ 和 $NCS^-$

$$CS(NH_2)_2 + HNO_2 \longrightarrow N_2\uparrow + H^+ + NCS^- + H_2O$$

生成的 $NCS^-$ 在稀 HCl 介质中与 $FeCl_3$ 反应生成 $[Fe(NCS)_n]^{3-n}$。

$I^-$ 干扰 $NO_2^-$ 的鉴定，可使其生成 AgI 沉淀分离除去。

实验步骤：取 2 滴试液于试管中，加入 4 滴 $Ag_2SO_4$ 溶液（$0.02\ mol \cdot L^{-1}$），离心分离，加入 2 滴 HAc 溶液（$6.0\ mol \cdot L^{-1}$）和 4 滴硫脲溶液（8%）摇荡，再加 2 滴 HCl 溶液（$2.0\ mol \cdot L^{-1}$）及 1 滴 $FeCl_3$ 溶液（$2.0\ mol \cdot L^{-1}$），若溶液呈红色，表示有 $NO_2^-$ 存在。

### 4. $PO_4^{3-}$

$PO_4^{3-}$ 与 $(NH_4)_2MoO_4$ 溶液在酸性介质中反应，生成黄色的磷钼酸铵沉淀：

$$PO_4^{3-} + NH_4^+ + MoO_4^{2-} + H^+ \longrightarrow (NH_4)_3PO_4 \cdot 12MoO_3 \cdot H_2O\downarrow + H_2O$$

$S^{2-}$、$S_2O_3^{2-}$ 和 $SO_3^{2-}$ 等还原性离子存在时，能使 Mo（VI）还原成低氧化态化合物。因此，预先加 $HNO_3$，并于水浴上加热，以除去这些干扰离子。

$$
\begin{array}{llll}
S_2O_3^{2-} & & NO\uparrow & SO_4^{2-} \\
S^{2-} + & H^+ + NO_3^- \longrightarrow & NO\uparrow + H_2O + S\downarrow \\
SO_3^{2-} & & NO_2\uparrow & SO_4^{2-}
\end{array}
$$

实验步骤：取 1 滴试液于试管中，加入 2 滴 $HNO_3$（浓），并置于沸水浴中加热 1~2 分钟。稍冷后，加入 4 滴 $(NH_4)_2MoO_4$ 溶液，并在水浴上加热至 40~50℃，若有黄色沉淀产生，表示有 $PO_4^{3-}$ 存在。

### 5. $S^{2-}$

$S^{2-}$ 与 $Na_2[Fe(CN)_5NO]$ 在碱性介质中反应生成紫色的 $[Fe(CN)_5NOS]^{4-}$：

$$S^{2-} + [Fe(CN)_5NO]^{2-} \longrightarrow [Fe(CN)_5NOS]^{4-}$$

实验步骤：取 2 滴试液于点滴板上，加 1 滴 $Na_2[Fe(CN)_5NO]$ 溶液（5%）。若溶液呈紫色，表示有 $S^{2-}$ 存在。

### 6. $SO_3^{2-}$

在中性介质中，$SO_3^{2-}$ 与 $Na_2[Fe(CN)_5NO]$ 和 $ZnSO_4$ 和 $K_4[Fe(CN)_6]$ 三种溶液反应生成红色沉淀，其组成尚不清楚。在酸性介质中，红色沉淀消失，因此，溶液为酸性时必须用氨水中和。$S^{2-}$ 干扰鉴定，可加入 $PbCO_3(s)$ 使 $S^{2-}$ 形成 PbS 沉淀除去。

实验步骤：取 5 滴试液于试管中，加入少量 $PbCO_3(s)$，摇荡，若沉淀由白色变为黑色，则需要再加少量 $PbCO_3(s)$，直到沉淀呈灰色为止。离心分离，保留清液。

在点滴板上加 $ZnSO_4$ 溶液（饱和）、$K_4[Fe(CN)_6]$ 溶液（0.1 mol·L$^{-1}$）及 $Na_2[Fe(CN)_5NO]$ 溶液（1%）各 1 滴，加 1 滴 $NH_3·H_2O$ 溶液（2.0 mol·L$^{-1}$）将溶液调至中性，最后加 1 滴除去 $S^{2-}$ 的试液。若出现红色沉淀，表示有 $SO_3^{2-}$ 存在。

### 7. $S_2O_3^{2-}$

$S_2O_3^{2-}$ 与 $Ag^+$ 反应生成白色的 $Ag_2S_2O_3$ 沉淀，但沉淀能迅速分解为 $Ag_2S(s)$ 和 $H_2SO_4$，颜色由白色变为黄色、棕色，最后变为黑色。

$$Ag^+ + S_2O_3^{2-} \longrightarrow Ag_2S_2O_3 \downarrow$$
$$Ag_2S_2O_3 + H_2O \longrightarrow Ag_2S \downarrow (黑色) + H_2SO_4$$

$S^{2-}$ 离子的存在干扰 $S_2O_3^{2-}$ 离子的鉴定，须预先除去。

实验步骤：取 1 滴除去 $S^{2-}$ 的试液于点滴板上，加 2 滴 $AgNO_3$ 溶液（0.1 mol·L$^{-1}$），若见到白色沉淀生成，并很快变为黄色、棕色，最后变为黑色，表示有 $S_2O_3^{2-}$ 存在。

### 8. $SO_4^{2-}$

$SO_4^{2-}$ 离子与 $Ba^{2+}$ 离子反应生成白色沉淀。

$CO_3^{2-}$ 和 $SO_3^{2-}$ 等干扰鉴定，可先酸化，以除去这些离子，消除干扰。

实验步骤：取 2 滴试液于试管中，加 HCl 溶液（6.0 mol·L$^{-1}$）至无气泡产生时，再多加 1~2 滴，加入 1~2 滴 $BaCl_2$ 溶液（1.0 mol·L$^{-1}$），若生成白色沉淀，表示有 $SO_4^{2-}$ 存在。

### 9. $Cl^-$

$Cl^-$ 与 $Ag^+$ 反应生成白色 AgCl 沉淀。

$SCN^-$ 也能与 $Ag^+$ 生成白色沉淀 AgSCN，因此，当 $SCN^-$ 存在时干扰 $Cl^-$ 的鉴定。

但在 $NH_3·H_2O$ 溶液（2.0 mol·L$^{-1}$）中 AgSCN 难溶，AgCl 易溶，并生成 $[Ag(NH_3)_2]^+$，则可将分离 $SCN^-$ 除去。在清液中加入 $HNO_3$，提高酸度，使 AgCl 沉淀再次析出。

实验步骤：取 2 滴试液于试管中，加 1 滴 $HNO_3$ 溶液（6.0 mol·L$^{-1}$）和 3 滴 $AgNO_3$ 溶液（0.1 mol·L$^{-1}$），在水浴上加热 2 分钟。离心分离。沉淀用去离子水洗涤，使溶液 pH 接近中性。加入 2 滴 $(NH_4)_2CO_3$ 溶液（12%），并在水浴上加热 1 分钟，离心分离。在清液中加 1~2 滴 $HNO_3$ 溶液（2.0 mol·L$^{-1}$），若有白色沉淀生成，表示有 $Cl^-$ 存在。

### 10. Br$^-$ 和 I$^-$

Br$^-$ 与适量的 Cl$_2$ 水反应会被氧化，游离出单质溴使溶液呈橙红色。在有机相（如 CCl$_4$、CHCl$_3$）呈红棕色，而水相无色。在过量的 Cl$_2$ 水中，会因生成 BrCl 变为淡黄色。

$$Br_2 + Cl_2 \longrightarrow BrCl$$

I$^-$ 在酸性介质中能被 Cl$_2$ 水氧化为 I$_2$，在有机相（如 CCl$_4$、CHCl$_3$）呈紫红色，在过量的 Cl$_2$ 水中会被继续氧化为 IO$_3^-$ 使颜色消失。

$$I_2 + Cl_2 + H_2O \longrightarrow HIO_3 + HCl$$

若向含有一定浓度的 Br$^-$ 和 I$^-$ 混合溶液中逐滴加入氯水，由于 I$^-$ 的还原能力较强，则优先被氧化为 I$_2$，使有机相呈紫红色。如果继续加入氯水 Br$^-$ 被氧化为 Br$_2$，I$_2$ 进一步被氧化为 IO$_3^-$。使得有机相的紫红色消失，而呈现红棕色。如氯水过量时，颜色变为淡黄色（由于生成了 BrCl）。

实验步骤：取 5 滴试液于试管中，加 1 滴 H$_2$SO$_4$ 溶液（2.0 mol·L$^{-1}$）酸化，加 1 mL CCl$_4$，加 1 滴氯水，充分摇荡，CCl$_4$ 层呈紫红色，表示有 I$^-$ 存在。继续加入 Cl$_2$ 水，并摇荡，若 CCl$_4$ 层紫红色褪去，又呈现出棕黄色或黄色，表示有 Br$^-$ 存在。

# 参考文献

［1］李于善，贺艳，涂志英. 无机化学实验（英汉双语教材）. 北京：中国水利水电出版社，2007.

［2］曹凤歧. 无机化学实验与指导.（第二版）. 北京：中国医药科技出版社，2006.

［3］北京师范大学无机化学教研室. 无机化学实验.（第三版）. 北京：高等教育出版社，2001.

［4］胡世代，陈建林，罗凤秀. 无机化学微型实验. 成都：电子科技大学出版社，2000.

［5］中山大学无机化学教研室. 无机化学实验.（第三版）. 北京：高等教育出版社，1992.